Fish Processing and Preservation

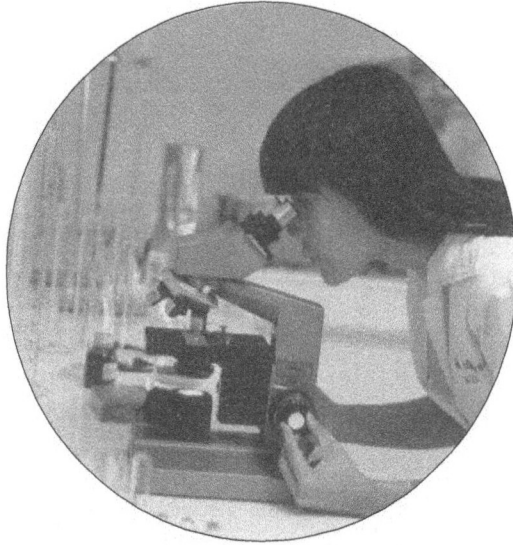

The Author

Dr. K.P. Biswas, M.Sc, Ph.D., D.F.Sc., (Bombay), E.F. (West Germany), F.Z.S., F.A.B.S (Kolkata), Former Joint Director Fisheries (L-I), Government of Orissa, Director Fisheries, A&N Islands, Government of India, Principal, Fisheries Training Institute, Fishery Technologist (ICAR), and at present (Faculty Member of Marine Science Department, University of Calcutta and West Bengal University of Animal and Fishery Sciences, is the author of 16 books in English and 4 books in Bengali and about 180 research and review papers in the field of fisheries, environment and ocean sciences.

His latest books, "*Advances in Fishing Technology*" and "*Advancement in Fish, Fisheries and Technology*" have been published in 2012.

Fish Processing and Preservation

K.P. Biswas

2014

Daya Publishing House®

A Division of

Astral International Pvt. Ltd.

New Delhi – 110 002

Publisher's note:

Every possible effort has been made to ensure that the information contained in this book is accurate at the time of going to press, and the publisher and author cannot accept responsibility for any errors or omissions, however caused. No responsibility for loss or damage occasioned to any person acting, or refraining from action, as a result of the material in this publication can be accepted by the editor, the publisher or the author. The Publisher is not associated with any product or vendor mentioned in the book. The contents of this work are intended to further general scientific research, understanding and discussion only. Readers should consult with a specialist where appropriate.

Every effort has been made to trace the owners of copyright material used in this book, if any. The author and the publisher will be grateful for any omission brought to their notice for acknowledgement in the future editions of the book.

Cataloging in Publication Data—DK
  Courtesy: D.K. Agencies (P) Ltd. <docinfo@dkagencies.com>

Biswas, K. P. (Kamakhya Pada), 1936-
  Fish processing and preservation / K.P. Biswas.
    p.   cm.
  Includes bibliographical references (p.   ) and index.

  ISBN 9789351301097 ((International Edition)

  1. Fishery processing. 2. Fishery products—Preservation.   I. Title.

DDC 664.94   23

Published by   :   **Daya Publishing House®**
                                     A Division of
                                     Astral International Pvt. Ltd.
                                     – ISO 9001:2008 Certified Company –
                                     81, Darya Ganj, Near Hindi Park,
                                     Delhi Medical Association Road,
                                     New Delhi - 110 002
                                     Phone: 011-4354 9197, 2327 8134
                                     Fax: +91-11-2324 3060
                                     E-mail: info@astralint.com
                                     Website: www.astralint.com

Laser Typesetting   :   Classic Computer Services, Delhi - 110 035
Printed at   :   Salasar Imaging Systems, Delhi - 110 035

PRINTED IN INDIA

Preface

Fish receives special attention as a food because of its content of first class animal protein with high nutritive value and richness in various essential amino acids. Its contribution of a fair quantity of water-soluble vitamins and many of the minerals and micro-nutrients is also well recognized. New knowledge on the role of Omega-3 fatty acids in human physiology and their high contents in fish has added a new dimension to their importance in health and nutrition.

Immediately after fish died, the autolysis will take place and later digestive juices will invade the flesh and start putrefaction. Then fat is attacked by oxygen and gives rancidity, especially in smoked and dried fish, Sweety flavour in fish, due to inorganic acid, when broken down during autolysis, produce bitter flavour and hypoxanthine formation. Later micro-organisms in the gut and at the surface of skin will multiply rapidly and will pick up more during handling, transportation and processing, accelerating the chemical changes in texture, taste, appearances and qualities.

Fish must be preserved by storing it in time of abundance for use in time of shortage. Therefore all methods or any technical method may be resorted to keep the products; (a) free of pathogenic and spoilage micro-organisms and their toxins, (b) free of chemical compounds causing problems, (c) nutritional quality is retained and (d) extending the shelf-life of fish and fishery products by icing and chilling, drying, smoking, salting, boiling and steaming, freezing, using chemical reagent and additives, packaging and radiation.

Processing is a function besides, handling, transport and storage involved in the products flow from the production sectors to the consumption sectors. Fish products, as commodities can be divided into primary or basic fish products (in the round form or dressed with minimum processing), low-cost minimum storage products and convenience products.

In the fish processing industry, the methodology of production can be divided as, (a) traditional methods (drying, salting, smoking, fermentation); (b) non-traditional methods (canning, refrigeration, fish meal and oil) and (c) modern methods (freeze-drying, irradiation).

The appropriate fish-processing technology must be developed for each set of circumstances in order that the best utilization of the catch or cultured fish raw materials can be obtained. Fish processing technological development, therefore, should have the aims of;

To increase the national income from the industrial scale fisheries products by providing technological services and quality control guidelines for the factories.

For fish-processing extension under the small-scale fisheries development, to concentrate mainly on the fisheries producers, who have been at the bottom of the socio-economic scale and for general household fish preservation in order to generate the production profit for majority of the target group.

To develop the production of suitable fish products as main sources of protein for low-income people, which is an integral part of the national food security and nutrition program.

To suitably utilize the under-utilized species (trash fish and some pelagic fish), which are the by-catch of various fisheries for human food and live-stock feed.

The book, "Fish processing and preservation" in its seventeen chapters have attempted to cover up these technological advances for better utilization of remaining 79 per cent of the catch (21 per cent of the fish catch is marketed as fresh or chilled) and form staple food along the coastal and inland landing centers of our country.

K.P. Biswas

Contents

Preface   ***vii***

Introduction   ***xv***

1. Nutritive Value of Fish   **1**
*Determination of Protein–Principle–Estimation of Nitrogen (Procedure)–
Estimation of Fat Content in Fish Flesh–Procedure–Determination of Ash
Content–Procedure–Fish in Human Diet.*

2. Changes in Fish after Death   **8**
*Changes after Death–Pre-rigor Condition–Rigor mortis Condition–Post-rigor
Conditions–Lipid Oxidation and Hydrolysis–Mechanism of Autolysis–Enzymes
Involved in Autolysis–Effects of Autolysis–Positive Effects of Autolysis–
Measurement of Autolysis–Alpha Amino Nitrogen (AAN)–Non-Protein
Nitrogen (NPN)–Free Fatty Acids–Control of Autolysis in Fish.*

3. Bacterial Flora in Fish   **20**
*Micro-Flora Present on Fish Skin–Micro-Flora in Fish Respiratory System–
Micro-Flora Present in Fish Digestive Tract–Bacterial flora of Indian Mackerel–
Bacterial Flora of Fresh Oil Sardines.*

4. Post-harvest Handling of Fish   **24**
Proper Handling of Fish–Chilling Procedure.

5. Measurement of Spoilage   **28**
*Sensory Scores (Basis)–Organoleptic Quality Testing of Sashimi Grade Tuna–
Handling and Selection of Sashimi Tuna (Sensory Evaluation)–Handling of
Tuna Onboard and Production of Quality Sashimi–Spoilage Index by Means of
Systematic Organoleptic Examination–Physical Method of Quality Assessment–*

On the Use of Electricity in Testing the Quality of Fresh Fish–Objective Assessment of Raw Fish Quality–Objective Chemical Methods–Methods–Evaluation of the Condition of Fish–Total Viable Bacterial Count–Eye-Fluids for Objective Tests for Fish Quality–Methods–Refractive Index–Optical Density–Biuret Colourimetric Measurement–Pyridine Colourimetric Measurement–Vacuum Distillation Procedure for Estimating the Quality of Fish–Methods–Preparation of Sample–Procedure.

6. Methods of Retarding Spoilage   **47**

7. Refrigeration in Retarding Fish Spoilage   **49**
Processing Procedures–Ice-Freshwater Ice–Salt Water Ice–Operational Environment–Advantages of Sea Water Ice–Ice-Making Machine–Freezing - Air Blast System–Spiral Design–Boxa Freezing–Fluidized Belt Freezing–Contact Plate Freezing–Vertical Plate Freezer–Heat Exchangers–Water and Glycol Chilling–Refrigeration in Fishing Vessels–Ice–Preservation of Ice–Insulation–Refrigerated Sea Water–Storage Temperatures–Methods of Heat Extraction–Contact Freezing–Immersion Freezing–Blast Freezer–Frozen Fish Storage–Choice of Refrigerants–Ammonia–Halogenated Hydrocarbons–Selection of Refrigerating Machinery–The Compression Machine–Changes during Freezing of Fish–Temperature Changes during Freezing–Texture Changes Resulting from Freezing–Reuter's Theory–Cell-Puncture Theories–Recent Theories–Effect of Ice Crystal Size–Quick Freezing–Advantages of Quick Freezing–Rate of Chilling–Refrigeration Capacity–Freezing Under Pressure–Choice of Freezing Equipments–Thawing and Refreezing–Salt Absorption–Other Changes Caused by Freezing–Cryogenic Freezing–Methods of Application and Freezing Equipments–Freezing Tunnel and its Operation–Liquid Nitrogen Freezing System Layout.

8. Preservation of Fish in Cold Temperature   **82**
Fresh Fish Preservation–Preservation and Transport of Oil Sardine–Brine Cooling of Fish Aboard a Fishing Vessel–Brine Tanks–Evaporators–Refrigeration Machines–Changes taking Place in Cold Storage of Fish–Changes in Texture of Stored Fish–Desiccation-Adverse Effects of Moisture Loss–Minimizing Desiccation–Temperature Differential–Humidity–Air Circulation–Fluctuation of Storage Temperature–Drip in Frozen Fish–Factors Controlling Drip–Cook Drip–Disadvantages of Drip–Retarding Drip Formation–Alteration in Texture–Changes in Colour and Flavour of Stored Fish–Colour Changes in Stored Fish–Flavour Changes in Fresh and Frozen Fish–Loss of Flavour–Development of Abnormal Flavour–Mechanism of Oxidation of Fish Oils in Frozen Fish–Types of Fish Most Susceptible to Colour and Flavour Changes–Minimizing Colour and Flavour Changes–Keeping Fish Temperature Low–Avoidance of Pro-Oxidants–Use of Antioxidants–Bacteriology of Frozen Fish–Minimizing Changes Occurring during Storage–Freezing Characteristic of Tropical Fishes, Tilapia–Changes Occurring during Preservation of Fish by Freezing and Prolonged Frozen Storage–Effect of Freezing, Glazing and Storage on Vitamins and Minerals in the Fish Flesh.

9. **Preservation by Reduction of Moisture**   **109**
    *Drying–Salting–Commercial Methods of Salting Fish–Dry-Salting–Brine-
    Salting–Spoilage of Salted Fish–Pink Spoilage–Dun Spoilage–Sun Drying of
    Bombay Duck–Reduction of moisture–Heat treatment–Smoking of Fish–Smoke
    Quality–Smoking of Eel Fillets–Salting–Pre-Drying Period–Smoke Kiln–Period
    of Smoking–Smoke Cured Fillet from Oil Sardine.*

10. **Preservation by Heat Treatment–Canning of Fish**   **122**
    *Historical Outline–Principles of Canning Seafoods–Micro-organisms in Relation
    to Canning–Sterilization–Condition of Raw Material–Freshness–Maturity–Fill
    of Container–Heat Penetration–Vacuum–"Cut-out" weight–Corrosion–Strain
    on Containers–Consistency of Product–Starch Content–Size of Particles–
    Composition of Container–Conduction–Thickness of Container Walls–Size of
    Container–Initial Temperature–'Coming-up" Time–Retort Temperature–Mode
    of Heat Transfer–Conduction–Convection–Time and Temperature
    Relationships–Water in the Retort–Air in the Retort–Temperature Pressure
    Relationship–Thermal Death Time–Vacuum–Vacuum in Canned Fishery
    Products–Method of Exhaust–Headspace or "Fill" of Can–Delay between
    Exhausting and Seaming–Canning of Oil Sardine–Canning of Oil Sardine-
    Natural Pack–Canning of Tuna in Oil–Canning of Lactarius–Canning of Smoked
    Eel–Canning of Smoked Sciaenids–Blackening of Canned Prawns–Types of
    Blackening–Source of Contaminants–Micro-Flora Involved in Spoilage of Canned
    Prawns–Bacteriological Investigations of Prawn Canneries–Maintenance of
    Bacteriological Quality in Canned Prawns–Bacteriological Quality of Cooked
    Frozen Prawns–Factory sanitation–Sources of Bacterial Contamination–
    Adequate Cleaning Program–Cleaning Schedule for Fish Processing Plants–
    Cleaning Schedule After Each Shift of Work–Cleaning of Boat Decks, Fish Holds,
    Wooden Boxes etc.–Deodourisation of Fish Containers, Fish Carrier Vans and
    Refrigerated Wagons.*

11. **Curing of Fish**   **157**
    *Traditional Curing of Fish in India–Bombay-Sind area–Madras Presidency–
    Curing with Salt in the East Coast–Curing with Salt in the West Coast–
    "Ratnagiri" Curing–Pickling–Fish Paste–Prawn Curing–Smoked Prawns–
    Semi-dried Prawns–Bengal Presidency–Sun Drying–Curing with Salt–Curing
    in Oil–Fish Pastes–Prawn Curing–Sun Drying–Dry Curing–Mona Cure–Wet
    Curing–Pit Cure–Colombo Cure–Smoke Cure–Between Catching and Curing
    Mackerel–Curing of Mackerel–Marinating Fish–Spoilage.*

12. **Radiation in Fish Preservation**   **171**
    *Radiation of Fish for Long Shelf Life–Increased Storage Life of Fish by Gamma
    Irradiation–Radiation Pasteurization of White Pomfret (Pampus argenteus)–
    Yellow Discolouration in Irradiated White Pomfret–Control of Radiation Induced
    Oxidative Changes.*

13. **Antibiotics in Fish Preservation**   **175**
    Antibiotics in the Preservation of Prawns.

14. Storage of Fish and Fish Products   **177**

Protective Covers for Frozen Fish–Glazes: Definition and Properties–Ice Glaze– Whole and Dressed Fish–Fish Frozen in Blocks–Fish Steaks–Shell Fish–Re- glazing–Additives to Ice Glazes–Glazing Fish for Locker Storage–Other Types of Glazes–Pectinate Films–Gelatin-Base Coatings–Combination of Chemical and Irish Moss Extractive–Waxes–Edible Oils–Films and Wraps–Cellophane– Polyethylene–Aluminum Foil–Vinylidene Chloride (Saran, Cryovac)–Rubber Hydrochloride Film (Pliofilm)–Coated Papers–Waxed Paper–Vinyl-and Saran- Coated Papers–Polyethylene-Coated Papers–Polyethylene-Coated Cellophane– Antioxidants in Papers for Food Wrappers–Modified Atmosphere Packing (MAP) of Seafood–Insulated Container for Long Distance Transport of Fish– Container–Transport of Iced Fish–Transport of Frozen Fish–Ice Box–Design of Ice Box–Stowage on Quality of Fish–Storage of Tuna Onboard after Catching– Storage of Cured Fish Products–Quality and Shelf Life of Dried Sharks Produced in India–Freeze Drying.

15. Fish Processing and Fish Products   **194**

The Alternative Uses of Fish–General Principle for Production of Good Quality Products–Quality Control in Food Service–Controlling Microorganisms in Raw Materials–Sanitation–Physical Parameters Involved in Fish Processing Technology–Physical Properties of Fish–Biophysics in Freezing–Temperature and Humidity–Drying and Salting–Processing of Indian Mackerel–Curing– Processing of Prawns and Shrimps–Freezing of Prawns–Canning of Prawns– Drying of Prawns–Dried Prawns–Semi-Dried Prawns–Other Prawn Products– Prawn By-products–Chitosan from Prawn Shell–Process–Handling and Processing Tuna–Fresh and Chilled Sashimi Grade Tuna–Handling of Tuna on Board–Tuna for Export–Standards for Fresh, Chilled Tuna Handling Centers– Product Specifications of Sashimi Grade Tuna–Processing of Tuna–Masmeen– Processing of Cephalopods–Processing of Major Carps–Chilled Products– Freezing–Thermal Process–Fish Mince–Fish Finger–Fish Balls–Fish Cutlet– Processing of Tilapia for Value Addition–Post-harvest Processing–Degutting– Filleting–De-boning of Skeleton after Filleting–Laminated Bombay Duck– Process–Processed Products from Sharks–Main Shark Commodities–Shark Fin Rays–Process for Shark Fins–Extraction of Fin Rays (Both fresh and dried fins can be used)–Shark Liver Oil–Flow Sheet–Extraction of Sardine Oil–Minced Fish–Meat Picking–Preparation of Mince–Storage Life–Advantage–Minced Meat from Fatty Fish–Sturgeons and Caviar–Caviar Substitutes and other Fish Roes–Prawn Wafers–Composition–Soup Powder from Trash Fish–Process– Composition of Ingredients–Preparation for Consumer–Beche De Mer–Fish Ensilage–Fermented Fish Products–Choice of Proteins for Protein Hydrolysates– Autolysis–Fermented Fish Products–Bacto-peptones–Proteolysed Liver Preparation–Fish Albumin–Alkali Solubilisation of FPC–Enzymatic Hydrolysis–Fish Protein Concentrates–By Biological Procedures–Fish Protein Concentrate from Oil Sardine–Fish Meal.

16. Plants and Machineries for Fish Processing Industry   248

Fish Protein Concentrate Industry–Fish Protein Concentrate (FPC) Plant–Equipments and their Function–Fish Meal Manufacture in India–Traditional Method of Fish Meal Production–Modern Methods–Working of a Modern Fish Meal Plant–Small Scale Fish Meal Drier–Determination of Moisture Content of Fish Meal–Component of the Instrument–Operational Procedure–Fish Ensilage Plant–Equipment and their Functions–Requirement of Space and Equipment–Mechanized Peeling Table for Prawn Processing Factories–Mechanized Fish Processing–Processing of White Fish–Filleting Machines–Flat Fish Filleting Machines–Mechanized Herring Processing–Processing of Herring for Food–Modern Fish Processing Plant.

17. Quality Control of Fish Products   272

Fish Food Quality–Seafood Borne Human Bacterial Infections–General Principles to Produce Good Quality Fish Products–Quality Control in Fish Food Service–Raw Materials – Controlling Microorganisms in Raw Materials–Sanitation–Hygienic Conditions with Respect to Bacteriological Characteristics of Cooked Frozen Prawn–Chemical Quality of Water Used in Fish Processing Industry–Seafood Borne Bacterial Pathogens–Method of Examination for Shrimp Spoilage–Classifications Used to Judge each Shrimp (or portion) Examined–Examination for Salmonella in Shrimp–Examination of Frozen Peeled and Deveined Shrimp for Insect Filth–Filtration Technique–Microscopic Examination of Filter Papers–Filth–Testing of Fish Ham and Sausage–Drained Weight of Fishery Products for Quality Control Inspection–Quality Control Laboratory for Seafood Processing Plants.

References   287

Index   293

Introduction

The processor, nutritionist, cook and the consumer all have a direct interest in the composition of fish. The processor should know the nature of the material, before he can correctly apply the techniques of preservation. The nutritionist wanted to know what portion of fish can make to the diet and to health. The cook must know whether the fish is normally lean or fatty in order to prepare it for the table. The consumer is interested not only in taste and flavour of fish, but also to some extent its nutritive value.

Fish is one of the most valuable sources of high grade protein available to man in this hungry world and a knowledge of its composition is essential to make the fullest use to be made of it.

Food elements in fish are protein, fat, mineral matter, and vitamins. The water content varies with the individual fish and the species, but it averages from 70 to 80 per cent. Protein constitute from 16 to 18 per cent of fish flesh on the average. The fat content of fish is dependent upon the individual fish, the species, the time of the year, the sex, the size, the maturity and other things and it varies from 1 to 20 per cent. Mineral constitutes approximately one per cent of the edible part of the fish and these minerals are comparable to that in other meats except that fish contains more iodine. For vitamin, no standard value can be assigned fish as to their vitamins content, it is a variable factor. The fatty fish and fish liver oils contain various and high amounts of fat soluble vitamins A and D. In general, the content of the various B vitamins in the flesh of fish approximates to that in various meat.

The proteins in fish are more readily digested than beef proteins, although no more completely digested, fish protein is from 85 to 95 per cent digestible. Fish proteins are complete, that is, they contain ten essential amino-acids in proportions that can be economically utilized by the body.

Fat gives good flavour to cooked fish and is a source of energy. The fat of fish is much more difficult to digest than the fat of beef. Unsaturated fatty acids are particularly plentiful in fish fat.

The bones of fish (soft in canned fish) may be eaten as a source of calcium and phosphorus. Shell fish are a good source of calcium and phosphorus, as well as, iron, copper and iodine. The saying that fish is a good "brain food" stems from the belief that it is a superior source of phosphorus. Fish eggs and sperm (roe and milt) are a good source of phosphorus and lecithin.

Fish is one of the most perishable and easily damaged of all fresh foods. As soon as fish dies, spoilage begins. Retarding of spoilage in fish is complicated by the fact that they are cold blooded animals. Enzymes and bacteria normally found in fish are adjusted to functioning at lower temperatures than those found in warm-blooded animals. The rate of spoilage can vary greatly between species.

Spoilage is the result of the whole series of complicated changes brought about in the dead fish tissue by its own enzymes, by bacteria and by chemical action. Changes are brought about by the enzymes of the living fish which remain active after its death. Million bacteria and other micro-organisms, many of them potential spoilers, are present in the surface slime, on the gills and in the intestines of living fish. They do not harm, because the natural resistance of a healthy fish keeps them at bay. Soon after the fish dies, however, bacteria begin to invade the tissues. It is believed that they enter through the gills and kidney, along veins and arteries, and directly through the skin and peritoneum, that is the lining of the belly cavity. In addition to enzymatic and bacterial changes, chemical changes involving oxygen from the air and the fat in the flesh produce rancid odours and flavours.

Spoilage of fish is influenced by catch methods, temperatures, cleanliness and subsequent handling throughout the distribution chain. Therefore, care must be taken to avoid these factors. Fish should be washed free of slime, soon after catching before packing into fish boxes, since washing can reduce the surface bacterial load up to 90 per cent. Fish also spoils more rapidly at higher ambient temperatures, or when exposed to direct sun light. The spoilage rate almost doubles (and hence the shelf life is halved) for each 5 degree Celsius rise in temperature above 0 degree Celsius. It is therefore desirable to cool fish quickly and as soon as possible after being caught or when cooling is impracticable, to prevent or slow down the absorption of heat.

Freshness in fish is usually judged in the trade by appearance, odour and texture of raw fish. The most important things to look for assessing freshness of fish are the general appearance of the fish including that of the eyes, gills, surface slime and scales and the firmness or softness of the flesh, the odours of the gills and the belly cavity, the appearance, and particularly the presence and absence of discolouration along the under side of the back bone, the presence or absence of *rigor mortis* or death stiffening and the appearance of the belly wall.

Drying is the oldest method of food preservation than any others. The requirement is to reduce the free water content by heating and evaporating, which is able to depress the growth of micro-organisms. There are three ways to propagate heat energy; convection, conduction and radiation. Convection heating means bodily transfer of heated substances. Conduction heating means heat transferred by molecular activity through one substance to another. Radiation heating is a transfer of heat energy in the same manner as light, and with the same velocity.

The second law of thermodynamics states that heat energy flows only in one direction, from hot to cold bodies. The mechanism of heat transfer during low (icing, chilling, freezing) and high (drying, smoking, boiling, canning, radiation) thermal processing have been taken advantage for preservation of fish to extend shelf-life of fish and fishery products

Fish processing technology has witnessed remarkable developments in the later half of 20th century. The advent of transportation facility, modern packaging materials and various fish processing techniques has all contributed to the remarkable development of fish processing industry. Modern consumers are highly quality conscious. It does not seem to have been realized in this country that the real value of fish does not consist in quantity, but on the condition in which it is made available to the consumers. The global demand for processed fish products is changing fast. The importers are now insisting on new kinds of value addition and ready-to-eat products. They want stringent quality standards. Diversification and value addition have become the key words of fishery products in ready-to-eat market.

The present exploitation (2009) in fisheries sector of the country accounted for 3.2 million tones (82 per cent of the potential), of which about 26 per cent is marketed fresh or chilled and forms staple food along the coastal and inland landing centers. About 10 per cent of the catch goes for drying and curing. Frozen fish production accounts for 25 per cent, while 26 per cent goes for reduction to fish meal and 1.6 per cent for other miscellaneous purposes. The utilization in fish canning industry is only 11 per cent.

There are about 620 registered exporters in the country. The post-harvest infrastructure includes 251 ice plants, 481 shrimp peeling plants, 371 freezing plants, 495 cold storage units, 7 canning plants, 16 fish meal plants, 11 surimi plants and one agar-agar production unit. 95 per cent of the seafood processing units are concentrated in 20 major clusters in 9 states.

Value added products of different kinds are slowly becoming popular as "convenience food". Though basically aimed for export market, there also have a promising potential in the domestic markets. The value added products include extruded products, battered and breaded products, surimi and derivatives, pickle and cured products in retortable pouches.

Chapter 1

Nutritive Value of Fish

The main constituents of fish flesh is water, which usually accounts for about 80 per cent of the weight of fresh fish, while the average content of most fatty fish is about 70 per cent. The water content of the muscle is slightly higher at the tail than at the head, this slight but consistent increase from head to tail is balanced by a slight reduction in protein content.

The amount of protein in fish muscle is usually some where between 15 to 20 per cent, but values lower than 15 per cent or as high as 28 per cent, is also generally encountered. All proteins including those from fish are chains of chemical units linked together to make one large molecule. These units of which there are about 20 types, are called amino acids and certain of them are essential in the human diet. If a diet is to be fully and economically utilized, amino acids must not only be present, but also occur in correct proportions. Two essential amino acids, lysine and methionine are high in fish compared to cereal foods.

The fat content of fish can vary very much more widely than the water, protein and mineral content. The ratio of highest to the lowest value of protein or water contnt encountered is not more than three to one, the ratio between highest and lowest fat values is more than three hundred to one. Investigations carried out have clearly established the beneficial effect of marine fish and its oil in lowering serum cholesterol level both in experimental animals and man. C 20:6, C 20:6 hexanoic acids present in fish lipids for lowering the serum cholesterol level and hence, canned fish in fish oil was advocated for formulating a suitable diet for people suffering from hypercholesterolemia. Human brain normally contains considerable quantity of W-3 (22:6) fatty acids and subjects suffering from multiple sclerosis contain less than half the amount of this fatty acid. Fish lipid is a rich source of W-3 fatty acids and also 22:6, W-3 acid. Thus where fish constitutes, one of the components of the diet, multiple sclerosis is almost absent.

The amount of carbohydrate present in most fishes is generally less than one per cent. However, in mollusks, it contain up to 5 per cent of carbohydrates, glycogen.

The ash, consists largely of a number of minerals and the total rarely exceeds 1 to 2 per cent of the edible portion, fish muscle. The main mineral constituents are phosphorus, sulphur, chlorides, iodine, fluorine, potassium, sodium, calcium, magnesium, iron, copper, manganese, zinc and cobalt.

Vitamins can be divided into two groups, that are soluble in fat, such as, vitamins A, D, E, and K and that those are soluble in water, such as, vitamins B and C. All the vitamins necessary for good health in human and domestic animals are present to some extent in fish, but the amounts varies widely from species to species and throughout the year.

The vitamin content of individual fish of the same species and even of different parts of the same fish can also vary considerably. Often the parts, such as, the liver and the gut contain much of the oil soluble vitamins, the livers of cod, sharks and halibut for example contain almost all of the vitamins A and D present in those species. In contrast, the same two vitamins in eels for example are present mainly in the flesh. The minerals and vitamins are not markedly affected by careful processing or by preservation, provided storage is not very prolonged.

The main flavouring compounds present in the fish are glycine, histidine, serine, betaines, inosine monophosphate, ammonic trimethylamine and other volatile bases. These are also found to vary in relation to different species.

Inspite of its relatively low caloric value and its high amount of waste in dressing, fish is an economical food.

Determination of Protein

Protein can be estimated quantitatively by estimating the total nitrogen in the sample by Kjeldhal's method. In this procedure, the total nitrogen present in the tissue, *i.e.* protein bound, as well as non-protein nitrogen can be analyzed. Total nitrogen percentage when multiplied by 6.25 gives the protein content.

Principle

The principle involved in this method is the conversion of the nitrogenous compound into ammonium sulphate, by boiling the tissue with concentrated sulphuric acid. The ammonium sulphate solution obtained can be treated with excess of sodium hydroxide and the liberated ammonia can be absorbed in the standard acid and the excess acid is titrated back with standard alkali (Macro Kjeldhal's method). In the modification of the above method, if the ammonia liberated is small, it can be absorbed in 2 per cent boric acid solution. The ammonia can then be titrated directly with a standard acid. Amount of nitrogen can be calculated from the ammonia determined. The apparatus used for this determination is the micro-Kjeldhal's unit specially meant for small quantity of sample. It consists of a sample distillation tube, a steam generator, steam trap and a receiver.

Estimation of Nitrogen (Procedure)

Digestion should be done inside a hood with good draught. About 2 gram of fish flesh are accurately weighed and transferred to a Kjeldhal's digester flask to which 20 ml of concentrated sulphuric acid and 1.6 gram of digestion mixture (10 parts of potassium sulphate and 1 part of copper sulphate ground together) were added and the mixture was heated gently till the effervescence ceased. The solution is then boiled till it became clear or light green in colour. It was then cooled and the volume was made up to 250 ml in a volumetric flask. Allow the water in the steam generator to boil gently and open the stop cock at the bottom of the steam trap so that the steam can escape. One ml of the sample solution was transferred to the distillation tube through the small funnel, immediately followed by washing of the funnel with two small portion of distilled water and then followed by the addition of 1-2 drops of phenolpthelein indicator plus 2 ml of 40 per cent NaOH solution and again followed by second washing of the funnel. The whole operation should be done as quickly as possible to avoid back suction of the reaction moisture into the steam trap. The receiving flask containing 10 ml of 2 per cent boric acid with 1-2 drops of Tashiro's indicator (pink in colour) was taken and placed under the condenser so that the distillate nozzle was dipping into the acid.

Close the stop cock of the small funnel and on the steam trap, thus compelling the steam to pass through distillation tube. The steam from the generator was passed through the Micro Kjeldhal's distillation tube for about 7-10 minutes when the pink boric acid solution in the receiving flask turn green.

The receiving flask was then lowered, so that the tip of the condenser was 1 cm above the surface of the distillate. The distillation was continued for one more minute, rinse the tip of the condenser tube with a few drops of water. The resultant green solution was then titrated against N/50 sulphuric acid, till the end point was reached from green to pink colour. A blank was run simultaneously. Clean the distillation tube by removing the flame under the steam generator. This created a vacuum in the steam trap and the material in the distillation tube was sucked into the trap. Open the steam trap to reject this moisture. Replace the flame, add 10 ml of water to the distillation tube and repeat the cleaning process.

Given that 1 ml of 0.02 Ammonium sulphate = 0.28 mg of Nitrogen

Total nitrogen per cent 0.02 × 6.25 = 12.91

1 ml of N/50 Sulphuric acid = 0.28 mg of Nitrogen

Volume of digested material was made up to 250 ml

Therefore, 2 g = 250 ml

Titrate value = 0.59

1 ml of made up solution is used for distillation for 4 minutes

$$\text{Total Nitrogen per cent} = \frac{0.59 \times 250 \times 0.28 \times 100}{1 \times 2.0000 \times 1000} = 2.065$$

2 g = 20000 mg Protein per cent = 2.065 X 6.25 = 12.91 per cent

Estimation of Fat Content in Fish Flesh

Apparatus and reagents required

1. Soxhlet continuous extraction apparatus
2. Solvent ether (Boiling point 30-40°C)
3. Fish muscle

The Soxhlet continuous extraction apparatus is designed so that a fresh portion of the solvent comes in contact with the material to be extracted over relatively long period of time.

The apparatus consists of a flask containing a volatile solvent, resting on some type of heating device, preferably a multi-regulated electric hot-plate. The flask is connected by means of ground joints with an extraction tube having a siphon arrangement and a side arm. The extraction tube is connected by means of ground joints with a condenser

Procedure

The flask of the Soxhlet apparatus was cleaned, dried and weighed accurately. About 5 grams of fish flesh was dried by keeping in a dessicator over concentrated sulphuric acid, under partial vacuum over night. The dried material was weighed into a thimble made of some suitable porous material, such as cotton, some special type of porous porcelain or filter paper. The thimble was then placed in the extraction tube and the tube is then connected with the weighed flesh and also with the condenser

The apparatus is then set up (with sufficient ether in the flask) and heated with the help of an electric bulb (40-60 watts) up to a temperature about 40°C. The heat vaporizes the volatile solvent, which passes up the side arm and is condensed in the condenser. The condensed solvent falls drop by drop into the thimble. When sufficient solvent has been transferred to the extraction tube to fill the siphon arm, it siphons back ether in the weighed flask.

The top of the thimble should be above the level of the siphon to prevent floating particles from passing in to the flask. It may be necessary to prevent splashing of the solid due to the fall of the solvent, by covering the solid with a porous sieve plate disc or filter paper.

This process was continued until the extraction was complete for about 8 to 10 hours, varying at times from about 8 to 24 hours. The flask was removed and the volatile solvent was evaporated. The residue is the extracted fatty material. The flask with the fatty material was then dried in a hot air oven till the weight remained constant.

Determination of Ash Content

Ash is the residue remaining after a food stuff is ignited until it is carbon free, usually at a temperature not exceeding dull red heat.

Procedure

Weighed into a porcelain crucible a quantity of substance representing about 2 grams of the dry material. Dried as usual at 100°C, burnt at a low red heat, ash in a muffle furnace at a dull red heat until free from carbon. Remove from the furnace, allowed to cool for some time, placed in a dessicator until cooled and weighed. The process of heating and cooling and weighing was repeated till the weight is constant.

Recorded the results and calculated the percentage of ash in the material.

Fish in Human Diet

Fish are a rich source of protein, fatty acids, essential vitamins and minerals, such as, vitamin A, calcium, zinc and iodine. The vitamin A, calcium and iron found in small fish species are particularly bio-available, that is easily absorbed by the body. According to the FAO, rising incomes and high consumer preference for fish, especially in Asia, have caused global fish consumption to double in the past 30 years to 15 kg per person per year. This trend is mainly attributable to demand from growing urbal populations in China and other Asian countries. Studies in rural Bangladesh and Cambodia have shown that small fish make up between 50 to 80 per cent of all fish eaten in the production season. Although small fish are consumed in low quantities, they are eaten whole. This helps as they are particularly rich in micro-nutrients. Their bones are an excellent source of calcium. In some species, accumulation of vitamin A has been found in their eyes and intestines. A study of poor rural households in Bangladesh in 1997 revealed that small fish intake provided about 40 per cent of the vitamin A and 32 per cent of the calcium requirements of an average house hold in the peak fish production season. The long chained Omega-3 poly-unsaturated fatty acids (PUFA) found in marine fish have a range of health benefits.

Proteins are important for growth and development of the body, maintenance and repairing of worn out tissues and for production of enzymes and hormones required for many body processes. The importance of fish in providing easily digestible protein of high biological value is well documented. In the past this has served as a justification for promoting fisheries and aquaculture activities in several countries. On a fresh weight basis, fish contains good quality protein (about 18-20 per cent) and contains all the eight essential amino acids including sulphur containing methionine and cystine.

The fat content of fish varies depending on the species, as well as the season, but in general, fish have less fat than red meats. The fat content ranges from 0.2 per cent to 25 per cent. However, fats from fatty fish species contain the poly-unsaturated fatty acids (PUFAs) namely, EPA (eicosapentaenoic acid) and DHA (docosanexaenoic acid, fatty acids), which are essential for proper growth of children and which prevent occurrence of cardio vascular diseases, such as, coronary heart disease. The fat also contributes to energy supplies and assists in the proper absorption of fat soluble vitamins, namely, A, D, E, and K.

Fish is a rich source of vitamins, particularly vitamins A and D from fatty species, besides thiamin, riboflavin and niacin (vitamins B1, B2, and B3). Vitamin A from fish is more readily available to the body than from plant foods. Vitamin A is required for normal vision and for bone growth. Fatty fish contains more vitamin A than lean fish species. As sun drying destroys most of the available vitamin A in fish, better processing methods are required to preserve this vitamin.

Vitamin D present in fish liver and oil is crucial for bone growth, since it is essential for the absorption and metabolism of calcium. Thiamin, Niacin and Riboflavin are important for energy metabolism. Fish also contains little vitamin C, which is important for proper healing of wounds, normal health of body tissues and aid in the absorption of iron in the human body. However, they have to be eaten fresh to absorb vitamins

The minerals present in the fish include, iron, calcium, zinc, iodine (from marine fish), phosphorus, selenium and fluorine. These minerals are highly bio-available. Iron is important in the synthesis of haemoglobin in red blood cells required for transporting oxygen to different parts of the body. Iron deficiency is associated with anemia, impaired brain function and in infants it is associated with poor learning ability and poor behavior. Due to its role in the immune system, its deficiency may also be associated with increased risk of infection.

Calcium is required for strong bones (formation and mineralization) and for the normal functioning of the muscle and the nervous system. It is also important in the blood clotting process. Vitamin D is required for its proper absorption. The intake of calcium, phosphorus and fluorine is higher, when small fish are eaten with their bones, rather than when the fish bone is discarded. Deficiency of calcium may be associated with with rickets in young children and osteomalacia (softening of bones) in adults and older people. Fluorine is also important for strong bones and teeth.

Zinc is required for most of the body processes, as it occurs together with proteins in essential enzymes required for metabolism. Zinc plays an important role in growth and development as well as the proper functioning of the immune system and for a healthy skin. Zinc deficiency is also associated with poor growth, skin problems, loss of hairs among other problems.

Iodine present in seafood is important for hormones that regulate body metabolism and in children it is required for their growth and normal mental development. Deficiency of iodine may lead to goiter (enlarged thyroid gland) and mental retardation in children.

Intake of anti-oxidants, including vitamin C was considered to lower the risk of cancers. While the vitamin K was found to help bones retain calcium to control the onset of osteoporosis. Recognition of these beneficial effects led to the dissemination of dietary guidelines aimed at reducing the frequency of chronic nutrition related diseases, such as obesity, cardio-vascular disease, hypertension, type II diabetes, osteoporosis and several forms of cancer.

It is evident that fish contributes more to the people's diets than just the high quality protein. Fish should therefore, be an integral component of the diet, preventing malnutrition by making these macro- and micro-nutrients readily available to the body.

The benefits of omega-3 fatty acids and the risks of sudden death and heart attacks is reduced by increased consumption of fin fish. It improves symptoms of rheumatoid arthritis, decreases the risk of bowel cancer and reduces insulin resistance in skeletal muscles. DHA supplements promote brain cell and synapse growth and improve disposition. In pregnant women, the presence of PUFAs in their diets has been associated with proper brain development among unborn babies. Omega-3 fatty acids have also been associated with reduced risk of premature delivery and low birth weight. Consuming two or more serving of fish with high omega-3 fatty acids may lower the risk of age-related muscular degeneration, which may cause blindness or vision impairment.

Eating oily fish regularly could help in fighting obesity. Lower prevalence of impaired glucose tolerance and type II diabetes in populations consuming large amounts of omega-3 PUFA. Eating fish not only improves the overall health, but also enhances the efficiency of antiretroviral drugs. Fish can contribute significantly to the nutritional regime of those living with HIV, particularly in terms of the high quality protein and micro-nutrients that fish provide in a readily accessible form.

Chapter 2

Changes in Fish after Death

Fish being a cold-blooded animal perish too fast towards spoilage. This is mainly due to the deterioration followed by changes its constituents, like protein, fat, non-protein nitrogen compounds, moisture and minerals.

The bio-chemical composition of fish muscle includes protein (15-24 per cent), fat (0.1-22 per cent),carbohydrate (1-3 per cent), inorganic substances (0.8-2 per cent) and water (66-84 per cent). The quantity of fat in fish meat varies according to species, in relation to their age, body parts, pre or post spawning season and the availability of food.

Changes after Death

The biochemical changes responsible for anabolism stop after the death of the fish, but catabolism will continue for a period of time. The irreversible changes mainly consists of i) slime secretion, ii)*rigor mortis*, iii) autolysis, iv) bacterial invasion, v) oxidation and hydrolysis leading to spoilage and loss of quality of fish.

The duration of these changes that take place will vary depending on the storage conditions, mostly in respect of temperature.

Slime is formed in certain cells of fish skin and the process becomes very active just after the death of fish. Slime contains large amounts of nitrogenous compounds, which provide very good nourishment for micro-organisms originating from the environment. As such, slime spoils the fish quickly, by giving unpleasant smell to the fish and opening the way for bacterial invasion to the fish.

Changes that occur in fish muscle after death and prior to processing is called death stiffening or *rigor mortis*. Rigor is a physiological reaction to death. Immediately after death, the muscle is totally relaxed and the limp elastic texture usually persists for some hours. Afterwards muscle will contract, when it becomes hard and stiff, the whole body becomes inflexible and this stage of the fish is referred to as *rigor mortis*.

It is defined as the phenomenon of stiffening of the muscle that take place shortly after death. The muscle becomes hard, non-elastic and inflexible, unable to be stretched out. Fish exhibits a shorter *rigor mortis* period commencing, 1 to 7 hours after death. The sequences of *rigor mortis* is; limp/elastic muscle in pre-rigor followed by stiff and hard muscle in rigor and again become limp/elastic muscle in post-rigor. The on-set, duration and resolution of *rigor mortis* is affected by; species, size, temperature and handling of fish; physical condition of the fish and; catching and killing methods.

The mechanism of stiffening phenomenon are; anaerobic pathway (glycolysis); 1 glucose > 2ATP and the aerobic pathway 1 glucose > 36ATP.

Pre-rigor Condition

On death of the fish aerobic oxidation of carbohydrates stops and anaerobic oxidation of glucose will lead to the formation of lactic acid due to stoppage of blood circulation (glycolysis). Thr pH of the muscle decreases. Glycolysis continues until the glycogen (stored carbohydrate) containing glucose is completely used up. The synthesis of ATP stops and hydrolysis begins as per the reaction ATP –ATPase→ ADP +Pi.

The level of phosphoric acid increases, although certain amount of ATP is synthesized by the hydrolysis of creatine phosphate pathway by the enzyme creatine kinase. Crp + ADP ———→ Cr + ATP, but the creatine reserves in the tissues are not infinite and they soon get exhausted and ATP concentration begins to fall near to zero. Normally 39 ATP molecules are produced under aerobic conditions and 2 ATP are produced under anaerobic conditions. The glycogen reserve of the muscle is directly related to the struggle the fish has to undergo during capture and death. The greater the struggle the fish has to undergo, the lesser are its glycogen reserves and the quicker is the onset of *rigor mortis* Therefore, *rigor mortis* appears quickly in very active fish and slowly in less active fish. At this stage, thick and thin filaments of myofibriller proteins are free to slide on each other, which indicates that muscles are extensible and can contract on stimulation.

Rigor mortis Condition

At the final phase of ATP breakdown, myosin cross bridges, interaction is established firmly between the thin and thick filaments, thereby making it non-extensible and hence non-contractile. Major proteins, actin and myosin, combine in the presence of calcium ions to form actomyosin. ATP then supplies the energy for contraction, and later also the energy for the removal of the calcium ions via calcium pump. This breaks the actomyosin complex, leaving the muscle ready for further contraction.

On death the circulatory system stops and the ATP level drops. Calcium ions leak, forming actomyosin. However, there is insufficient ATP for the calcium to operate, and so the actomyosin complex remains unbroken. The muscle is now in a continual state of rigidness, known as *rigor mortis*. Fish is edible in this condition. *Rigor mortis* has many technological consequences, such as, if the bones are removed prior to *rigor mortis*, the length of the fillet would shorten by 30 per cent and at the same time the fillet becomes wider, thicker because its volume does not change. This tightness very often causes the connective tissue of the individual myomeres to break. This process is termed as gaping and it results in muscle separation, which is considered as a quality defect. Gaping depends on temperature, the higher the temperature of the fish at the beginning of the *rigor mortis* process, the greater the gaping of the muscle. Therefore, during *rigor mortis* fish temperature should be as low as possible.

The onset of *rigor mortis* and its duration depends on the fish species, on the catching methods and on fish temperature. It was also found that for fast swimmer fishes *rigor mortis* is faster, but for a shorter duration than slow swimmers.

Before rigor, the fish flesh is soft but flexible. After rigor, the fish flesh is soft and much less flexible than that of the fish before rigor. This characteristic can be used to determine if the fish is pre-rigor or post-rigor.

Post-rigor Conditions

Fish spoilage begins due to autolysis and bacterial growth. Autolytic changes make catabolites available for microbial growth. Autolysis is also known as self digestion, controlled by enzymes.

In fish, there are different enzyme systems. When the fish is alive, these enzyme systems are necessary for building up the tissue and organs and for metabolisms (to digest food). After the fish is dead, these enzyme systems still operate and take part in the degradation process of fish, especially the digestive enzymes and enzymes in muscle, causing soft muscle in fish reducing the quality. Degradation products of the enzyme are the nutritional source of micro-organisms and the activation of intracellular enzyme systems will enhance spoilage of fish.

The reduction of trimethylamine oxide (TMAO), an osmoregulatory compound in many marine teleost fish, to trimethylamine (TMA) is usually due to bacterial action, but in some species an enzyme is present in the muscle tissue which is able to breakdown TAMO into dimethylamine (DMA) and formaldehyde (FA).

It is important to note, that the amount of formaldehyde produced is equivalent to the dimethylamine formed, but is of greater commercial significance. Formaldehyde induces cross-linking of the muscle proteins making the muscle tough and readily lose its water holding capacity. The enzyme responsible for formaldehyde-induced toughness is called TAMO-ase or TAMO demethylase and is most commonly found in gadoid fishes (cod family). Most of the TMAO demethylase enzymes reported to date were membrane-bound and become most active when the tissue membranes are ruptured by freezing or artificially by detergent solubilization. Dark (red) muscle has a higher rate of activity than white muscle, whereas other tissues like, kidney, spleen and gall bladder are extremely rich in the enzyme. Thus it is important that minced fish is completely free from organ tissue, such as, kidney from gadoid species if toughening in frozen storage is to be avoided. It is often difficult to ensure that the kidney is removed prior to mechanical deboning, since this particular organ runs the full length of the backbone and is adherent to it.

Enzymes in the flesh and gut previously involved in metabolism, now catalyze autolytic reactions, in which various compounds decompose. Enzymes in the flesh breakdown desirable compounds into tasteless or bitter ones, whilst gut enzymes attack the internal organs, turning them into soupy mess and allowing the bacteria to enter the flesh.

The chemical changes takes place during autolysis are;

(a) Protein denaturation, involves the destruction of its secondary, tertiary and quaternary structure, reducing the protein to a simple polypeptide chain. A number of factors, including slow freezing and variability of storage conditions cause this denaturation. A denatured protein has not only lost its ability to function as enzyme, but also its water-holding ability. As such, in denatured fish flesh, excessive dripping takes place, when thawed and appear white, dull and spongy and upon chewing become fibrous and tasteless.

(b) Decreasing flesh pH; a living fish has flesh pH of 7.0. However, after death, residual glycogen is broken down via glycolosis to pyruvic acid and then lactic acid. As this happens, the flesh becomes more acidic. If the pH remains above 6.6, the texture is reasonably soft, but below this level, the flesh becomes firm and eventually unacceptably tough.

(c) Total volatile base (TVB) is a measure of the total amount of a variety of nitrogen containing substances, which are produced during storage. The volatile base present in the flesh is a trimethylamine (TMA), which is formed from the reduction of trimethylamine oxide. Marine fish contain a small amount of trimethylamine oxide, the function of which is unknown. This odourless and tasteless compound is reduced by invading bacteria to TMA, which is characterized by its "fishy smell". TMA is useful as a quality index during the middle and late stages of spoilage after the bacteria have invaded the fish. TMAO is converted in the muscle tissue into dimethylamine (DMA) and formaldehyde by enzyme action during frozen storage. This formaldehyde is able to cross link with protein, denaturing the muscle structure. This fish loses water when it is thawed, and tough and fibrous texture, when cooked.

(d) Nucleotide breakdown involves the enzymatic breakdown of the energy carrier ATP.

(e) Bacterial flora on freshly caught fish depends on the environment in which it is caught, rather than on the fish species. Fish caught in very cold waters carry lower counts, whereas fish caught in warm waters have slightly higher counts. Bacteria varies in relation to raw material contamination during processing.

Raw material involves, skin, gills, gut, contaminated by environment, air, soil, water and during processing by equipments, staff and pests

Micro-organisms are found on all the outer surfaces (skin and gills) and in the intestines of live and freshly caught fish. The total number of organisms vary enormously. A normal range of 100 – 10000000 cfu (colony forming units)/sq.cm on the skin surface. The gills and the intestine both contain between 1000 and 1000000000 cfu/g. Spoilage bacteria are *Pseudomonas, Shewanella putrefaciens, Photobacterium phosphoreum, Enterobacteriaceae.* Pathogens found in natural environment of fish are; *Clostridium botulinum, Listeria monocytogenes, Vibrio parahaemolyticus, V. vulnificus, Aeromonas hydrophila* and *Plesiomonas shigelloides.* Other pathogens that can contaminate from the environment are; *Staphylococcus aureus, Salmonella* spp. and *Escherichia coli.*

Lipid Oxidation and Hydrolysis

Two distinct reactions in fish lipids of importance for quality deterioration are; (i) oxidation; enzymatic oxidation (lipoxygenase) cause (a) generation of characteristic fresh fish odour, (b) autoxidation, (c) oxygen reacts with double bonds of unsaturated fatty acids and (d) affects nutritional value, taste, odour, colour and texture; (ii) The enzymatic hydrolysis of lipids (fats) to produce free fatty acids and glycerol. Formation of free fatty acids. Normally this does not cause problems in fish but causes an off-flavour in oils (soapy).

They result in production of a range of substances among which some have unpleasant taste and smell. Some may also contribute to texture changes by building covalently to fish muscle proteins. The various reactions are either enzymatic or non-enzymatic. The relative significance of these reactions mainly depend on fish species and storage temperature. The highly unsaturated lipids of fish easily become oxidized, resulting in alterations in smell, taste, texture, colour and nutritional value. The process starts immediately after catch, but becomes particularly important for shelf-life only at temperature below 0°C (Harris and Tall, 1989), when oxidation rather than microbial activity becomes a major spoilage factor. The initiation of lipid oxidation arises from various early post-mortem changes in the tissue.

Mechanism of Autolysis

Among enzymatic, microbial and chemical spoilages of fish, autolysis is the one which usually sets in first, immediately after the termination of *rigor mortis,* (death stiffening).

Autolysis, the process of breakdown of fish tissue after death is caused by its own body enzymes mainly through softening the tissue. The five most important tissue components broken down during autolysis are proteins, lipids, nucleic acids, carbohydrates and nucleotides.

The breakdown of the carbohydrate (glycogen) and the nucleotide (ATP) starts immediately after death. But the gradual softening of the muscle starts only after *rigor mortis* is terminated. The low pH attained from breakdown of glycogen to lactic acid after death enhances the activity of endogenous tissue enzymes, which can break the remaining three tissue components, such as, proteins, lipids and nucleic acids into their simpler building units.

Hydrolysis of proteins is of the greatest significance, among the three components of the tissues. Proteins are the main tissue components of the fish constituting about

80 per cent on dry weight basis. The proteins, such as, actin, myosin and collagen are the main structural components of fish tissue. In live condition, these protein remain in their natural state supporting the original tissue structure. The low pH obtained during *rigor mortis* enhances the activity of proteolytic enzymes, which degrade the proteinaceous structural tissue components of fish through the process, called "proteolysis", resulting in softening of the tissue.

Enzymes Involved in Autolysis

Endogenous enzymes are the agents of autolysis. Live fish contains numerous enzymes required for the metabolic processes going on inside the fish body. These enzymes are distributed throughout the body. Out of the several functions carried out by these enzymes in different ways, one function is to hydrolyse complex substances into simpler forms. The group of enzymes involved in hydrolysis process is called hydrolases. It is only this group of enzymes that is responsible for the autolytic breakdown of tissue components. When a fish is alive, these enzymes have a controlling activity. They are put to use as and when required and prevented from their action in rest of the time.

The hydrolases involved in autolysis are of two types, such as, (a) digestive or gut enzymes and (b) muscle and internal organ enzymes.

The digestive or gut enzymes are secreted into the gut to digest the food taken by fish. These enzymes are synthesized within the secretary cells of the digestive system and are stored within these cells as inactive precursors called "proenzymes" or "zymogens". When food enters into the gut, the neuro-endocrine system of fish sends information to these cells to secrete the zymogens to outside. Outside the cell, they are converted into active enzymes by lowering of pH or by action of other enzymes. These active enzymes now act on the complex food components and digest them into simpler forms, which are absorbed into the wall of the intestine for further assimilation. Thus in living fish, the hydrolases do not act on their own tissue due to their presence as inactive zymogens. Their presence as inactive zymogens or active enzymes is controlled by the neuro-endocrine system. The zymogens of the digestive proteolytic enzymes, pepsin, trypsin and chymotrypsin are pepsinogen, trypsinogen and chymotrypsinogen respectively.

After death of fish, some enzymes which would have already been secreted into the gut remains inside it as residual enzymes. Besides, the neuro-endocrine system,which controls the activity of enzymes ceases to function. Due to this, more of zymogens are secreted and converted into active enzymes. These freshly secreted enzymes along with the residual enzymes, accumulate inside the gut. These gut enzymes, which are mostly proteolytic in nature, hydrolyse the proteins of the gut walls and enter into the muscle. There they hydrolyse the muscle proteins. As a result, the muscle adjacent to gut becomes very soft and flabby. The enzymes move gradually from the gut towards the body surface through the muscle, due to which muscle degradation and softening proceeds from the gut towards the skin.

The hydrolytic enzymes of the muscle and the internal organs (liver, kidney, spleen, gonads) are intracellular ones (enzymes) and are present inside the cells, confined within minute membrane-bound organelles, called lysosomes present in the cell cytoplasm. In the muscle cells, they are present in the peripheral sarcoplasm. The activity of lysosomal enzymes is controlled by the lysosomal membrane, which encloses them and thereby keeps them separated from the cytoplasm and cytoplasmic organelles. The enzymes, otherwise would digest them. When required the membrane ruptures and releases the enzymes to digest the given material intracellularly or extracellularly. There are 50 types of hydrolases present in the lysosomes, which can digest most of tissue components, such as, proteins, fat, nucleic acids etc. However, they can not hydrolyze the lysosomal membrane. When alive, this membrane does not allow the hydrolases to hydrolyze own cells. The important lysosomal enzymes are, cathepsins, lipases, carbohydrases, RNase and DNase. In live tissue, these enzymes have three functions.

1. Endocytosis–The cell engulfs a foreign particle into the cytoplasm. The lysosomes release their enzyme secretion on the particle and digest it intracellularly.

2. Exocytosis–The lysosomes move to the cell membrane and rupture upon touching it. Thus the enzyme is released to outside the cell where it digests any particle extracellularly.

3. Autophagy–The lysosomes secrete their enzymes on intracellular organelles like mitochondria or golgi complex present within the same cell and digest it. For this reason, lysosomes is called the "suicide bag of the cell".

After death of fish, when its chemical condition changes, the membranes of the lysosomes get damaged, liberating the hydrolytic enzymes into the cytoplasm. The lysosomal membrane breaks easily, when oxygen supply to the cells stop after death. Liberation of lysosomal enzymes leads to breakdown of the cell. The low pH produced in the fish muscle after death favours most of the hydrolytic enzymes. Now these enzymes hydrolyze the cellular components, such as, protein, carbohydrates, fats and nucleic acids.

Cathepsin is the most important enzyme responsible for the degradation of proteins of the muscle and internal organs. It is a group of acid proteases. In the muscle cell, the cathepsin breakdown the myofibrillar proteins (actin and myosin) and sarcoplasmic proteins (myogen and globulin) to amino acids. Different types of cathepsins have been identified (cathepsin A, B, C, D, E, H, L etc.), out of which cathepsin A, B, C, D and L are known to be present in fish muscle. Fish muscle contains more cathepsins than muscle of other animals. This is one of the reasons why fish spoils faster than meat of other animals. The concentration of cathepsins in the internal organs is much more than in the muscle. Therefore, the internal organs are removed from the fish body along with the gut immediately after harvest so as to prevent autolysis.

A second group of intracellular proteases, called calpains or calcium activated factor (CAF) also cause autolytic hydrolysis primarily causing post-mortem tenderization of fish through hydrolysis of Z line proteins, myosin and troponin.

The third group of intracellular protease is collagenase, which hydrolyses the protein collagen. As the connective tissue, which holds together the myotomes is made of collagen, its hydrolysis by collagenase results in undesirable separation of the myotomes into flakes.

The autolytic breakdown of the important tissue components takes place in the following five steps, saccharolysis and nucleotidolysis takes place immediately after death. The process of true autolysis, which follows *rigor mortis* and which manifests itself in softening of muscle include proteolysis, lipolysis and nucleolysis.

In proteolysis, the tissue proteins of fish are broken down sequentially to proteoses, peptones, polypeptides and finally to amino acids by a group of endogenous proteolytic (proteo=proteins; lytic=breaking) enzymes collectively called "proteases". The tissue proteins are hydrolysed both by the gut enzymes and the enzymes of the muscle and internal organs. The gut enzymes, such as, pepsin, trypsin and chymotrypsin hydrolyze the proteins of the gut and muscle adjacent to the gut. On the other hand, cathepsins, the most important lysosomal enzyme of the muscle and the internal organs, hydrolyze the proteins of the muscle and internal organs.

In lypolysis, the tissue lipids are broken down by a group of lipolytic (lipo=lipid; lytic=breaking) enzymes, collectively called "lipases". The triglycerides are successively degraded to diglycerides, monoglycerides and finally to free fatty acids and glycerol. The gut lipases can hydrolyze the linkage at all the three positions (1st, 2nd,3rd) unlike many other lipases, which often do not hydrolyze the linkage at the 2nd position. Another important group of lipids, called phospholipids are broken down by the enzymes phospholipases to glycerol, free fatty acids, phosphoric acid and an alcoholic group (choline, ethanolamine, serine or inositol). Both lipases and phospholipases are found in gut, muscle and internal organs of fish.

In nucleolysis, the nucleic acids of fish tissue are hydrolyzed to nucleotides by a group of endogenous enzymes collectively called "nucleases". DNA is hydrolyzed by the enzyme DNase and RNA by the enzyme RNase. Nucleases are present in the gut, muscle and internal organs of fish.

Effects of Autolysis

Autolysis has both negative and positive effects. In fresh fish, autolysis has more pronounced negative effect on the quality. On the other hand, there are some processing methods, where autolysis has more pronounced positive effect. However, the most important way of fish utilization is by direct consumption of fresh fish. Autolysis adversely affects freshness. Thus the negative effects of autolysis are of greater importance than the positive effects. By understanding the effects of autolysis, the rate of autolysis can be enhanced or retarded as desired.

The negative effects of autolysis includes; softening of tissue (loss of firmness). The texture of fresh fish is firm (consistent) and elastic. It does not retain finger indentations. But autolysis breaks down the proteinous structural components (actin, myosin, collagen) which is manifested by softening of the muscle. The firmness and elasticity disappears. Now the fish retains finger dentations. Texture of fish is most important in organoleptic (sensory) evaluation of quality of fish, in determining the cutting efforts required and in determining the ease of handling fish. As the texture of fish becomes soft, it not only gets low sensory scores, but distracts the consumer. It also requires more cutting efforts in the process of handling and processing.

In live fish, microbes are present on the surfaces exposed to outside. Skin, gills and lumen of the gut contain high microbial load. The lumen of gut, though present inside the fish body, is exposed to outside through mouth and cloacal aperture. The muscle and internal organs of fresh fish, being unexposed to outside, are sterile. The natural microbial flora of the gut lumen can not enter through the gut walls into the muscle in live fish. However, after death, when the digestive enzymes of the gut hydrolyze the gut wall during autolysis, the microbial flora penetrates through the gut walls into the hitherto sterile muscle and internal organs. They act on the muscle and internal organs and putrefy them. Similarly, when the components of the skin are hydrolyzed by autolysis, it becomes soft and permeable to microbes. Bacteria of skin can now enter through the skin into the muscle and putrefy it.

During autolysis, the complex tissue components are hydrolyzed to simpler substances, which become a nutrient-rich medium for microbial growth. They utilize these readily available simpler substances for their growth and metabolize them to off-odour, off-flavour substances, presence of which renders the fish spoiled. If autolysis can be prevented (or even delayed), spoilage by microbes could be delayed, because in absence of autolysis, microbes would have to first secrete their own enzymes to the surrounding space to hydrolyze the tissue components to simpler substances after which they can utilize them. This process is difficult and time taking due to which, the spoilage could be delayed.

The non-protein nitrogenous substances (NPN), such as TMAO, urea and free amino acids are the most favoured substances for microbial growth. Small quantities of NPN are present in live fish, but during autolysis, due to hydrolysis of proteins, substantial quantity of free amino acids are produced. This increases NPN in fish tissue making it nutrient rich medium for microbial growth.

The development of rancidity is caused by the hydrolysis of fish lipids in the formation of free fatty acids (FFA) as one of the end products. Its accumulation in fish tissue results in the development of a typical rancidity called, "hydrolytic rancidity".

Some species of fish, in the post-spawning period, indulge in high food intake. When food enters into the gut, the neuroendocrine system induces secretary cells of the organ of digestive system to secrete a large amount of enzymes in the inactive zymogen form. The zymogens become active enzymes outside the secretary cells and act on the food components. If fish is caught during the period of high food intake, the large quantity of enzymes secreted into the gut remain active for quite a long time after death. These enzymes can degrade the gut wall completely within a few hours and enter into the muscle. They hydrolyze the muscle adjacent to the gut. The end products of the hydrolysis of gut and muscle are the simpler substances like, amino acids, which become a nutrient-rich medium for the activity of the natural microbial flora of the gut. Besides, there are a large number of microbes in the gut of fish, which has eaten a lit. All these microbes grow on the end products of hydrolysis with a resulting production of gases like, ammonia, carbon dioxide and hydrogen. The accumulation of these gases inside the belly cavity leads to the swelling of the belly. The belly on the other hand, becomes soft due to the hydrolysis of its tissue by autolysis. Now ultimately the belly bursts after a short period of storage of food. This phenomenon is called belly bursting. It is a common phenomenon in some pelagic fishes like, oil sardine, herring, capelin and sprat. Such fishes degrade quickly and therefore spoil easily soon after they are caught.

The problem of belly bursting can be overcome by gutting (removing gut) if possible or by keeping the fish alive after harvesting so as to allow the ingested food to pass out from the gut. However, the most effective method of preventing belly bursting is quick chilling of fish immediately after harvesting.

The skeletal muscle forming both the sides of fish body is made of several transverse muscle segments called myotomes, which are separated from each other by connective tissue sheets called myocommata. Within each myotome remains a large number of elongated muscle cells, which run parallel to each other longitudinally along the length of the fish body, each extending between two myocommata. The myocommata is made of collagen fibers. Thus the whole muscle is supported by the collagenous connective tissue, myocommata. During autolysis, the proteolytic enzyme collagenase breaks down the protein collagen resulting in the disintegration of the myocommata. This results in undesirable separation of myotomes into flakes. This phenomenon is called gaping. Gaping results in its poor appearance courting consumer's dislike.

The end products of autolysis are free amino acids, glycerol, free fatty acids etc, which do not impart much of objectionable off-odour and off-flavour to fish. A fish which has passed through autolysis, is still acceptable to the consumer. However, the end products of autolysis become a nutrient-rich medium for the microbes, which utilize them and convert them to off-odour and off-flavour substances responsible for the spoilage. Therefore, autolysis though not spoilage in itself, leads to easy spoilage by microbes.

Positive Effects of Autolysis

Taste and flavour development – the intermediate products of nucleic acid (DNA and RNA) break down during autolysis and nucleotide (ATP) breakdown immediately after death are adenosine mono-phosphate (AMP), inosine mono-phosphate (IMP) and guanosine mono-phosphate (GMP). The products of protein hydrolysis during autolysis are amino acids, like histidine, glutamic acid, glycine, valine, aniline, arginine, methionine etc.All these breakdown products are flavour potentiators. They impart pleasant flavour and taste to fish and thus are responsible for taste and flavour development in them.

Fish silage, a byproduct prepared from uneconomic fish and fish offal by allowing the fish flesh to liquefy completely. The endogenous body enzymes hydrolyze the tissue components, which results in complete liquefaction of fish body. The hydrolysis is enhanced by adding acids, which lower the pH of fish tissue to the point favourable for the activity of these endogenous enzymes.

In temperate countries, salted fish is stored at cooler temperature for months together (some times up to nine months) to undergo the process of maturation or ripening. It is specially important for salting of fish like anchovies, which have high fat content. During this period, the endogenous tissue enzymes hydrolyze the tissue components to simpler compounds, which impart the typical pleasant mature taste greatly relished by the people of these countries. Fish loose their raw flavour and odour, their flesh become tender, which separate readily from bones. Such fish is eaten without cooking. Cathepsin is the enzyme responsible for the textural changes (softening) during the fermentation of salted fish.

Autolytic hydrolysis of fish proteins results in its tenderization. Tenderized meat is soft and pliable for which it is organoleptically better accepted than tough pre-tenderized meat.

Measurement of Autolysis

The extent of autolysis undergone by a particular time after death is measured by estimating the concentration of end products of autolysis, such as, free amino acids and free fatty acids. Their concentration is measured in terms of the following parameters.

Alpha Amino Nitrogen (AAN)

It is an estimate of the concentration of free amino acids in fish. Each amino acid has an amino group at its alpha position. In proteins, the amino acids, are bounded to each other in a chain by peptide bonds. These bonds involve the alpha amino groups of the amino acids. Autolytic hydrolysis of proteins breaks the peptide bonds and makes the amino acids free. Due to this, the alpha amino groups of each free amino acid become free. Each alpha amino group contains a nitrogen atom. Thus the concentration of alpha amino nitrogen in fish indicates the extent of autolytic hydrolysis.

Non-Protein Nitrogen (NPN)

When proteins are hydrolyzed to amino acids, the free amino acids produced no longer remain part of the proteins and therefore count towards non-protein component. Fish tissue also contains other non-protein nitrogenous substances, such as, TMAO, creatine, urea etc. Thus the quantity of non-protein nitrogenous substances in fish, expressed in term of non-protein nitrogen (NPN) indicates the extent of autolytic hydrolysis of proteins.

Free Fatty Acids

During autolysis, fat is hydrolyzed to free fatty acids and glycerol. The amount of free fatty acids (FFA) in fish tissue indicates the extent of hydrolysis of fats and therefore indicates the extent autolysis.

Control of Autolysis in Fish

The rate of autolytic degradation of tissue components, mainly depends on some factors, such as, temperature, presence of gut and internal organs, moisture content of tissue, salt content and pH.

If it is required to retard autolysis, as in case of fresh fish, which should not be allowed to soften by autolysis so as to retain its freshness, then the following steps are recommended.

1. Evisceration – It involves removal of gut and internal organs. This prevents the activity of gut enzymes and internal organ enzymes on the muscle tissue, thereby retarding autolysis.

2. Low temperature – Temperature of fish should be lowered quickly by chilling or freezing which retards the enzyme activity.

3. High temperature – Fish should be exposed to very high temperature by cooking, canning or frying, which denatures the tissue enzymes beyond a limiting temperature.

4. Drying – It retards autolysis by reducing the moisture content of tissue to such a low level where enzyme activity is almost negligible. Enzymes remain active only in their native three dimensional conformation, which needs water to be present in their structure.

5. Salting – It also retards autolysis, because the high concentration of salt used in salting preservation of fish denatures most of the enzymes.

On the other hand, autolysis can be enhanced if required, as in case of preparation of silage, by the following ways; (a) By lowering pH autolysis is enhanced through addition of acids, to the point, which favours maximum activity of the autolytic tissue enzymes. (b) As enzyme activity increases with increasing temperature up to the limiting temperature of denaturation, autolysis can be accelerated by exposing fish to high temperatures not reaching the limiting temperature.

A thorough understanding of the process of autolysis and its effect on quality of fresh fish and its processing is necessary as it has a tremendous influence on subsequent microbial spoilage, which renders the fish unacceptable to the consumer. Therefore, if autolysis can be controlled, the freshness of the fish can be retained for a prolonged period of time delaying the onset of spoilage. The role of post-mortem autolysis should never be undermined.

Chapter 3
Bacterial Flora in Fish

The skin, respiratory tract and digestive tract of fishes are open systems, constantly in contact with surrounding water and environment.

Micro-Flora Present on Fish Skin

Aeromonas hydrophila, Pseudomonas and *Vibrio* are the types of bacteria present on the skin of fishes. This bacterial population is generally influnced by the marine ecosystem. This population along with the slime on the scales of the fish body provides an efficient barrier against the entry of virulent micro-organisms through skin of fishes. Bacterial population associated with fish skin can be estimated by acridine orange epifluorescence microscopy, and by plate counts on several media.

Micro-Flora in Fish Respiratory System

Fishes breathe through their gills. In fish exterior is a long bony cover (operculum) over the gills that is used for pushing out water after the process of absorbing dissolved oxygen there from, taken through the mouth. In the swimming process of fishes, water flows into the mouth and is expelled out through the gills, where oxygen in the water is absorbed for circulation through blood vascular system. Freshwater fishes use a type of counter current flow to maximize the intake of oxygen that diffuses through the gills. Counter current flow occurs when deoxygenated blood moves through the gills in one direction, while oxygenated water moves through the gills in the opposite direction. This mechanism maintains the concentration gradient, thus increasing the efficiency of the respiration process as well. So the only type of microflora evidenced in fish respiratory system are symbionts which generally occur in their surrounding aquatic environment. Till date, no pathogenic microflora (or micro-flora of aquatic interest) in the fish respiratory system have been reported.

Micro-Flora Present in Fish Digestive Tract

It is known that the structure of the digestive tract in different fish species differs, although the differences are seen in the early stages of their development. So the first factor influencing the formation of gastro-intestinal bacteria communities is the structure of the digestive tract. The formation of regular microflora in the digestive tract of fish larvae and fry is a complex process. This depends on the zone of movement of fish, spawn, its food, nature of available micro-flora of the surrounding water. After studying the formation of micro-flora of the digestive tract of carps from the larval to the adult stage, it has been found that in the digestive tract of fish, the bacteria flora is formed gradually and becomes stable approximately on 67th day after hatching with *Aeromona, Pseudomonas, Clostridium* and *Bacteroides* bacteria predominating. In the digestive tract of fish, bacteria of the genus *Bacteroides* appear as late as on the 44th day after hatching. Later, they become predominant in the intestines of the adult fish. Result of investigation suggest that bacteria of the genera *Aeromonas, Pseudomonas* and *Flavobacterium cytophaga* prevail in the bacteriocenoses of the digestive tract of freshwater fish. The impact of the nature of food and its feeding intensity on the qualitative and quantitative composition of intestinal bacteriocenoses was studied by a number of scientists (Ringo and Oslen, 1999).The structure of intestinal bacteriocenoses of fish is influenced by farming conditions of fish too. *Aeromonas* and *Lactobacillus* bacteria prevail in the intestinal bacteriocenoses of fish inhabiting in natural water bodies; whereas, those belonging to *Enterobacteriaceae* which may make up up to 50 per cent of all bacteria are prevalent in the bacteriocenoses of fish raised in farms and fed on artificial food. Micro-flora of the digestive tract of fish was investigated intensively and in different aspects, but data about the impact of xenobiotics on the intestinal micro-flora of hydrobionts are few. If bacteria, non-typical of the living environment of hydrobionts, are abundant in the surrounding water, they make a negative influence on the immune system of fish by restraining it, thus imparting the animal's general physiological state.

Bacterial flora of Indian Mackerel

The total bacterial count of skin with muscle of mackerel showed variations with the season. Such seasonal variation was also exhibited in the fish during storage in chlortetracycline (CTC) incorporated ice. Almost all months of the year, about 90 per cent of the native flora was found to be sensitive to 5 ppm CTC, the maximum susceptible flora being obtained during the months of September and December, during which periods, the qualitative analysis of the flora showed that *Vibrios* constituted a lesser proportion of the total flora. But in February, the proportion of *Vibrios* in the total flora was maximum.

The native flora of skin with muscle of mackerel was mainly constituted by three genera, namely, *Vibrios, Pseudomonas* and *Achromobacter* which together comprised 75 per cent of the flora of the fresh fish. The *Bacillus* spp from the native flora of mackerel constituted only 5 per cent of the total flora. In the case of sardine, caught off Cochin, *Acromobactor, Vibrios* and *Pseudomonas* together accounted for 73 per cent of the native flora and *Bacillus* constituted only 1 per cent of the total flora.

Succession of bacterial genera during ice storage of fish is of significance from the stand point of fish spoilage. The native flora of fish undergoes considerable changes during the storage of the fish at low temperatures. When mackerel is stored in ice, *Vibrios*, which constituted 60 per cent of the original flora, underwent a drastic reduction in proportion and by the 7th day, it comprise only 5 per cent of the total flora and by the end of 21 days, *Vibrios* practically disappeared. Whereas, *Pseudomonas* and *Achromobacter* which respectively accounted for 12 per cent and 21 per cent of the initial flora increased to 24 per cent and 49 per cent respectively by 7th day. There after *Pseudomonas* rapidly increased and by the 21st day, it estimated itself as a dominant flora comprising 74 per cent of the total. *Achromobacter* on the other hand, showed only a slower rate of increase and by the 21st day, only 15 per cent of the flora was constituted by *Achromobacter*. However, at the time of spoilage of fish in ice *Pseudomonas* and *Achromobacter* accounted for nearly 90 per cent of the total flora.

Vibrios do not presumably have an active role in spoilage of fish. The organisms, which can grow at low temperature and exhibit proteolytic activity can be assumed to be responsible for spoilage. But putrefiers need no confine themselves to one major group alone, but they are to be found in all groups. Apparently only a small portion of bacteria can cause spoilage, the remainders probably exist as free riders or probably are involved in some synergism with weak spoilers

Bacterial Flora of Fresh Oil Sardines

Qualitatively, the bacterial flora of marine environments in different parts of the world show some differences. Season has a definite role in determining not only the population of bacteria present on marine fish, but also on the preponderance of the different genera of bacteria and their phosphorescent and bio-chemical characters.

Peak values of total bacterial counts were obtained in fresh oil sardine (*Sardinella longiceps*) during July –October for skin with muscle, March-April and September-November for gills and October for intestines. The higher bacterial counts during the July-October season may be attributed to the effect of monsoon. Peak values of total bacterial counts at 37°C were obtained during March and July-August for skin with muscle, March-May and August-November for gills and March-April and October for intestines. The common peak values for skin with muscle, gills and intestines during March show the presence of greater numbers of mesophiles. This may be due to the effect of summer.

The absence of phosphorescent bacteria on skin with muscle and gills from March to June may be due to the fact that phosphorescent character of bacteria is lost by high temperature of the season. The presence of phosphorescent bacteria in intestines from March to June may be due to the fact that complete destruction of phosphorescent bacteria is not effected because of the high initial load.

Higher percentage of *Pseudomonas* are obtained during May for skin and muscle, during December for gills and during March and September for intestines. *Vibrios* predominate in June in the case of skin and muscle, in March and June in the case of gills and in June and October in the case of intestines. Peak values of *Achromobacters* are obtained during August for skin with muscle, gills and intestines. Generally

Vibrios and *Pseudomonas* predominate almost during the same seasons of the year, namely, the summer and the end of monsoon, whereas, *Achromobacters* are present in great numbers only during August, the end of monsoon.

Generally monsoon season favours the presence of greater numbers of bacteria. Though mesophiles predominate in the warmer months, the numbers of phosphorescent bacteria are less. During winter, bio-chemically less active groups of bacteria are predominating in comparison with the other seasons.

Chapter 4

Post-harvest Handling of Fish

Proper Handling of Fish

The number and type of bacteria found at the time of capture of the fish vary with the season, locality, species, water temperature, and method of capture. The heaviest bacterial population in the ocean is usually found in the bottom mud. Bottom-feeding species will therefore often have a large bacterial load. However, the effect of the gear enters into the problem as well. A large part of bottom-feeding fish are taken by otter trawls. When the trawl is hoisted on deck, the fish are squeezed by the weight of mass in the net, and the contents of the intestines are expressed, adding to the contamination already contributed by the bacteria on the skin and gill surface. Hand line caught bottom fish usually have a lesser bacterial load, since they are landed singly and treated relatively carefully.

In line fisheries, sorting out is continuous as the fish come over the rail. The fish are generally handled quickly, and the delay on deck is of only short duration. In the trawl fishery, however, sorting is often a problem, since the fish are caught in large quantities and are dumped all at one time on the deck. Included with the desired fish are often items such as, rocks, mud, sea weed, star fish, scrap fish, shells etc. In addition to the sorting of the market fish from the unwanted material, different species and sizes are separated for icing. Small fish are more difficult to keep in good condition than are large ones, hence the small fish are commonly iced in separate pens. Some species of trawl fish, such as, hake and Pollock, are soft fleshed, and spoil more readily than do other species of comparable size. Thus they should also be iced separately. If the fish have considerable mud or debris on them, they are usually washed with clean water, preferably drawn by pump while the vessel is running.Harbor water should never be used, since it is often contaminated with oil, sewage, or garbage. During the sorting, and washing operations, many of the fish will still be alive, and care should be taken not to bruise them. The fish should not be stepped upon or thrown bodily. After death, the fish are not as susceptible to bruising, but they still must be treated with care to minimize crushing or tearing of the flesh. Pews, forks and fish hooks must be used only in the head, since a hole in the flesh allows the introduction of slime and bacterial contamination. If practical, it is better to sort the market fish by hand.

If a delay between the time the fish are landed and the time they are iced is unavoidable, care should be taken to see that the fish are protected from spoilage. If the sun is shining and the deck is hot, for instance, the fish should either be covered with a tarpaulin or kept cool and wet with clean sea water. When the fish are iced, small fish and the more perishable species should be handled first.

While most fish are iced after catching, not all are drawn (eviscerated). In determining whether or not the fish should be drawn, several factors are of importance;

(a) Time between catching and landing. Fish held for more than a couple of days are usually drawn. Some fish that are caught close inshore need not be eviscerated nor iced. However, fish that spoil rapidly should be gutted as soon as possible after catching.

(b) Size; Large fish, in comparison with small ones, require less time to eviscerate per unit of weight and are usually drawn. Large fish cool slowly in ice, and evisceration exposes more area to the cooling effect. Drawing is therefore, a matter of cooling efficiency as well as one of eliminating bacteria and enzymes from the intestines. Small fish are often iced without being drawn, as this is impractical and uneconomic due to great number in the catch.

(c) Requirements by the industry. Many plants prefer fish in undressed (round) condition, since automatic machinery used in the processing requires ungutted fish to operate properly.

(d) Economics – The economic return to the fisherman may not justify the extra labor of evisceration. If the demand for the fish is greater than the supply, the buyers may not insist on extra measures in handling.

However, the most important factor deciding how the fish are to be handled after capture is custom and tradition. Each area has its own practices and preferences which have been passed down through generations.

After the fish are dressed and before being iced, they should be washed free of blood and slime with clean sea water. The slime harbors spoilage bacteria, and is not as was once believed, a good barrier against spoilage of the flesh itself. The bathing action of melting ice during stowage in the hold will not remove all the blood and slime. Thorough cleaning and washing of the gut cavity are necessary to reduce to contamination and they materially help to improve the quality of the landed fish. The cleaned and washed fish should promptly be stowed out of the sun and with the belly cavities filled with ice to prompt rapid cooling.

Fish have different keeping qualities depending on species, season, and method of catching. For example fish with food in the stomach and intestinal tract will spoil rapidly. The same is the case if the fish struggled for a long time before it died, as when caught in a gill net.

Icing of fish is not a cure-all for quality preservation, but it does offer a considerable measure of protection from bacterial action. The importance of prompt and proper icing cannot be overstressed. Since a temperature drop of 5 degrees (from 37 to 32°F) can reduce the rate of spoilage by 50 per cent, it is very important to bring fish after capture rapidly to a suitable chill temperature of 32 to 34°F. Another reason for chilling fish rapidly is that the growth of bacteria goes through a lag phase. The length of this lag phase is increased as the temperature is lowered. The storage life of fish held at high temperatures for even a short time before icing is greatly reduced. Ice when properly used in adequate amounts, aids in preservation in two ways; (1) the temperature of the fish is lowered to approximately 32 degree to 36°F, which slows the bacterial and enzymatic changes, and (2) the melting of the ice bathes the fish in clean cold water, which with proper storage, washes away considerable amounts of slime, blood and bacteria. When ice melts, it absorbs heat from the surroundings. On a fishing boat, the surroundings will include the hold, the air, and the fish. It is desirable that as much heat as possible is taken from the fish, and the heat transfer from the hold etc., be kept to a minimum. This can be achieved by insulating the hold, providing as much fish surface as possible in contact with the ice. It will be understood that the ice should melt (due to latent heat of fusion) in order to provide proper chilling. Sub-cooled flake ice is probably the best means of icing. This ice will last long, and will flow freely. Salt water ice cools better than ice from freshwater, but also melts faster. Large amounts must therefore be used.

Experiments have shown that antibiotics are very effective in retarding the bacterial spoilage They should be used as early as possible after catching the fish, because they are bacteriostatic, and their effect is therefore best when the number of bacteria, on the fish is low. The antibiotics may be applied by dip, spray, or by mixing them in the ice. The last method will probably be most commonly used. Any bactericidal or bacteriostatic ice for use on fish will be of greatest value only if its use is combined with the best handling practices. The opportunity offered by antibiotics could be entirely lost if the potential of these products is used to extend operations and mask poor and unsanitary conditions.

Chilling Procedure

To correctly ice the fish in the hold and containers (icing procedure), five things should be accomplished;

1. The fish should be placed with ice around them to cool them as promptly as possible and to maintain their temperature as close to the melting point of ice (32°F) as is practical for the duration of the trips;

2. The ice and fish should be arranged, layer by layer, to allow water, blood and slime to drain through the mass into the bilge;

3. Top and bottom layer should have sufficient ice in order to protect heat transfer. Two or more inches of ice between layer to layer is required;

4. Minimize the thickness of layer of fish in order to take less time to cool the fish;

5. The fish should not be stacked too high (great pressure), otherwise, the physical damage as well as the shrinkage or loss of weight be excessive.

For long trips, mechanical refrigeration of the fish hold for keeping ice on the outbound trip without loss by melting has been quite successful. Only enough refrigeration need be supplied to absorb the heat entering the hold through the sides of the vessel and through the deck head. The actual freezing of the fish when ice is not intended, the ice should be allowed to melt and supply the refrigeration needed to cool the fish to 32 to 35°F.

The cooling of fish in circulating chilled sea water at 32°F is more efficient than is cooling in crushed ice. Crushed ice will often not completely surround the fish, as the water will do. Chilling with sea water is accomplished by pumping the chilled water into the "wells" in the hold and slowly circulating it through the mass of fish. In addition to the good chilling quality of refrigerated sea water, the use of such a method reduces weight losses in the fish.

So far, the importance of chilling of the fish as soon as possible after capture has been stressed. Proper chilling does not help very much, however, if the "housekeeping" aboard the vessel is poor. The fisherman who keeps his fish hold, gear and deck clean is also apt to be quality-conscious when he comes to icing his fish. The problem of sanitation and housekeeping aboard a boat depends greatly on its design and construction. Sharp corners and cracks should be eliminated and wood should have a smooth cover of paint. Bilges should be cleaned frequently during the fishing season, using a good detergent or bilge-cleaning compound to remove accumulated dirt, oil and slime. Pen boards should be scrubbed and allowed to dry after each trip. Deck areas used for sorting and cleaning fish should be scrubbed frequently. The washing-down should be done as soon as possible before the slime, scales, blood etc. get a chance to dry and stick to the surfaces. The ice should be protected from contamination. Shovels and scoops should be cleaned at frequent intervals, an gloves should be washed often and rinsed in chlorinated water if used over long periods. Of course, personal cleanliness of the fisherman is essential. As a primary handler of food products, he bears great responsibility.

Chapter 5

Measurement of Spoilage

Freshness is an important quality criterion. An experienced person can judge the freshness of a batch of fish by appearance and odour, particularly in relation to the intended use or market for it, without quantifying the degree of freshness. However, there are many situations where freshness must be expressed in some degree and defined way such as number on a scale or by a grade. Numeric scales are extensively used in scientific work as they enable the effects of various processing conditions to be quantitatively compared. They are also used in quality control in industry, *e.g.* for defining limits for acceptance or rejection of products or for monitoring the performance of processing operations.

It is very difficult to standardize sensory methods since it is not possible to define or measure freshness in an absolute sense. Thus while sensory grades and scores can be used in a meaningful way within a limited environment, say within factory or company they are not suitable for laying down standards to trade between companies and in particular, between countries. A number of non-sensory methods which can be standardized in a laboratory and based on changes in chemical and physical properties of fish at it spoils have been proposed to overcome this problem. It must be emphasized that these tests do not measure freshness directly as a human assessor would. They provide measures of some of the basic changes which accompany spoilage and thus can estimate the score or grade a sensory judge would give. This prediction cannot be exact and any freshness score derived from a non-sensory measurement has an inherent error. This error can be allowed for by suitable sampling procedures and this type of test is quite suitable for giving the average quality of a batch of fish.

Sensory Scores (Basis)

	Score Marks

General appearance (5 Marks)

Eyes perfectly fresh, convex black pupil, translucent cornea, bright red gills, No bacterial slime, outer slime water white or transparent, bright opalescent skin, no bleaching — 5

Eyes slightly sunken, grey pupil, slight opalescent cornea, some discolouration of gills and some mucus, outer slime opaque and some what milky, loss of bright opalescence and some bleaching — 3

Eyes sunken, milky white pupil, opaque cornea, thick knotted outer slime with some bacterial discolouration — 2

Eyes with completely sunken pupil, shrunken head covered with thick yellow bacterial slime, gills showing bleaching or dark brown discolouration and covered with thick bacterial mucus, outer slime thick yellow brown, bloom completely gone, marked bleaching and shrinkage. — 0

Flesh including belly flaps (5 marks)

Bluish translucent flesh, no reddening along the backbone and no discolouration of the belly flaps, kidney bright red — 5

Waxy appearance, no reddening along backbone, loss in original brilliance of kidney blood, some discolouration of belly flaps — 3

Some opacity, some reddening along backbone, brownish kidney blood and some discolouration of the flaps — 2

Opaque flesh, marked red or brown discolouration along the backbone, very brown to earthy brown kidney blood, and marked discolouration of the flaps — 0

Odours (10 marks)

Fresh seaweedy odours — 10

Loss of fresh seaweediness, shellfish odours — 9

No odours, neutral odours — 8

Slight musty, mousy, garlic peppery, milky or caprylic and like odours — 7

Bready, malty, beery, yeasty odours — 6

Lactic acid, sour milk or oily odours — 5

Some lower fatty acid odours (acetic or butyric acids, grassy, old boots, slightly sweet, fruity or chloroform-like odours) — 4

Stale cabbage water, turnipy, sour sink, wet matches, phosphene like odours — 3

Ammoniacal, with strong byre like (o-toluidine) odours	2
Hydrogen sulphide and other sulphide odours, strong ammoniacal odours	1
Indole, ammonia, faecal, nauseating, putrid odours	0

Texture (5 marks)

Firm, elastic to the finger touch	5
Softening of the flesh, some grittiness	3
Softer flesh, definite grittiness and scales easily rubbed of the skin	2
Very soft and flabby, retains the finger indentations, grittiness quite marked and flesh easily torn from the backbone	1

Cooked fishes, odour (10 marks)

Strong sea weedy odours	10
Some loss of sea- weediness	9
Lack of odour or neutral odours	8
Slight strengthening of the odour but no sour or stale odour, wood shavings, woodsap, vanillin or terpene-like odours	7
Condensed milk, caramel or toffee-like odours	6
Milk jug odours, or boiled potato or boiled clothes-like odours	5
Lactic acid and sour milk, or byre-like odours	4
Lower fatty acids (acetic or butyric acids), some grassiness or soapiness, turnipy or tallowy odours	3
Ammoniacal (trimethylamine and lower amines) odours	2
Strong ammoniacal (trimethylamine) and some sulphide odours	1
Strong ammonia and faecal, indole and putrid odours	0

Texture (5 marks)

Firm thick white curd, bluish white in appearance, no discolouration	5
Firm but wooly, lost its bluish whiteness, some yellowing	3
Softer, cheesy-like, marked discolouration	2
Sloppy, soapy, very marked browning along the backbone	1

Flavour (10 marks)

Fresh, sweet flavours characteristic of the species	10
Some loss of sweetness	9
Slight sweetness and loss of flavour characteristic of the species	8
Neutral flavour, definite loss of flavour, but no off flavours	7
Absolutely no flavour, as if chewing cotton wool	6

Trace of off flavours, some sourness but no bitterness	5
Some off flavours and some bitterness	4
Strong bitter flavours, rubber-like flavour, slight sulphide-like flavours	3
Strong bitterness but not nauseating	1
Strong off flavours of sulphides, putrid, tasted with difficulty	0

Organoleptic Quality Testing of Sashimi Grade Tuna

Generally the quality of sashimi produced from bluefin tuna is the highest followed by the quality of sashimi produced from bigeye tuna and yellowfin tuna. However, sashimi produced from top quality bigeye tuna is considered to be of better quality than sashimi produced from average quality of bluefin tuna.

The quality of tuna for the sashimi market is evaluated on the basis of objective criteria, such as species, the period and region of capture, the method of conservation (fresh/refrigerated or frozen) and the fishing device used. It is then evaluated on the basis of organoleptic criteria, such as the presence of fat, the appearance of the skin, protuberant, clear and moist eyes, intact stomach and fresh smell. The best sashimi is produced by large tuna individuals caught during the season preceding the reproductive season. Usually tuna is graded for sashimi as follows;

Grade 1+ – Tuna whose flesh is bright red, compact, clear and fat. This sashimi of outstanding quality is produced by tuna caught with hand line or long line, refrigerated on board.

Grade 1 – Tuna whose muscle tissue is red, with compact, clear and fat flesh, caught by long lines, refrigerated on board.

Grade 2 – Tuna whose muscle tissue is red, with compact, fairly clear but lean flesh, which can be used for production of steaks as well as for lower quality of sashimi. It can be either refrigerated or frozen at sea.

Grade 3 – Tuna whose muscle tissue can be both red or brown and whose flesh is compact, but opaque and lean. It is frozen at sea and used for the production of steaks.

Grade 4 – Tuna whose muscle tissue is grayish brown and whose flesh is soft and opaque. This tuna is sent to the processing industry.

The difference in quality between tuna for sashimi and tuna for steaks is not necessarily very marked. In fact, the decision to process tuna flesh into sashimi or steak would depend on the price level obtained in the market.

Different sashimi cuts from different species have various market values depending on fat contents; the higher the fat content, the lighter the colour and the more valued the sashimi will be. The best sashimi comes from Toro, the peripheral layer of ligher coloured tuna meat with a fat content of 25 per cent. The Toro is further divided into Otoro Pink, which gives the prime sashimi, and Chutoro, darker pink. Within Otoro, there is a smaller part called Sunazari, whose texture is marked with thin lines of fat content (around 15 per cent) is called Akami and fetches lower prices than Toro.

Feeding hopper.

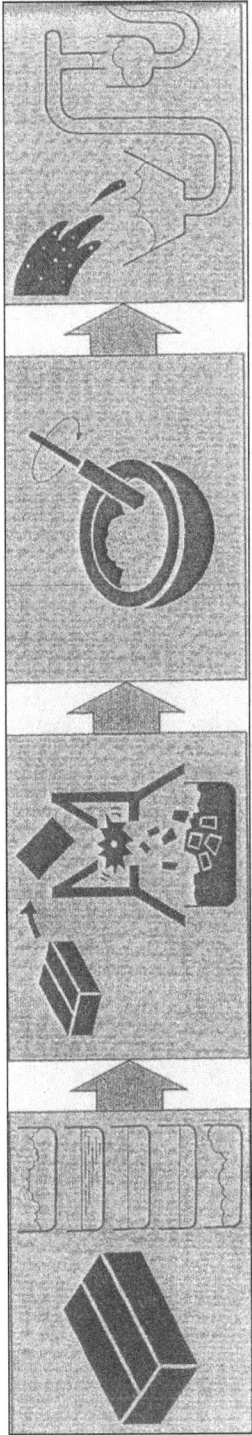

Fish meat, animal meat and all other additives are ground in a large mortar and thoroughly mixed.

Cutting in pieces.

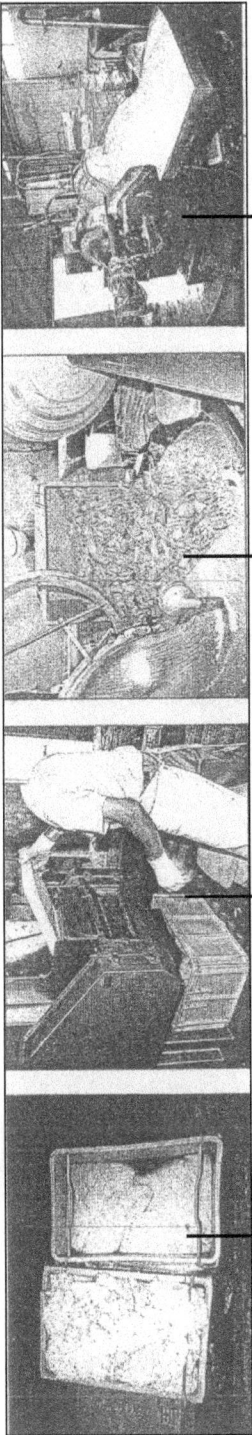

Preparation of various raw materials.

Processing Line of 'Surimi' Production

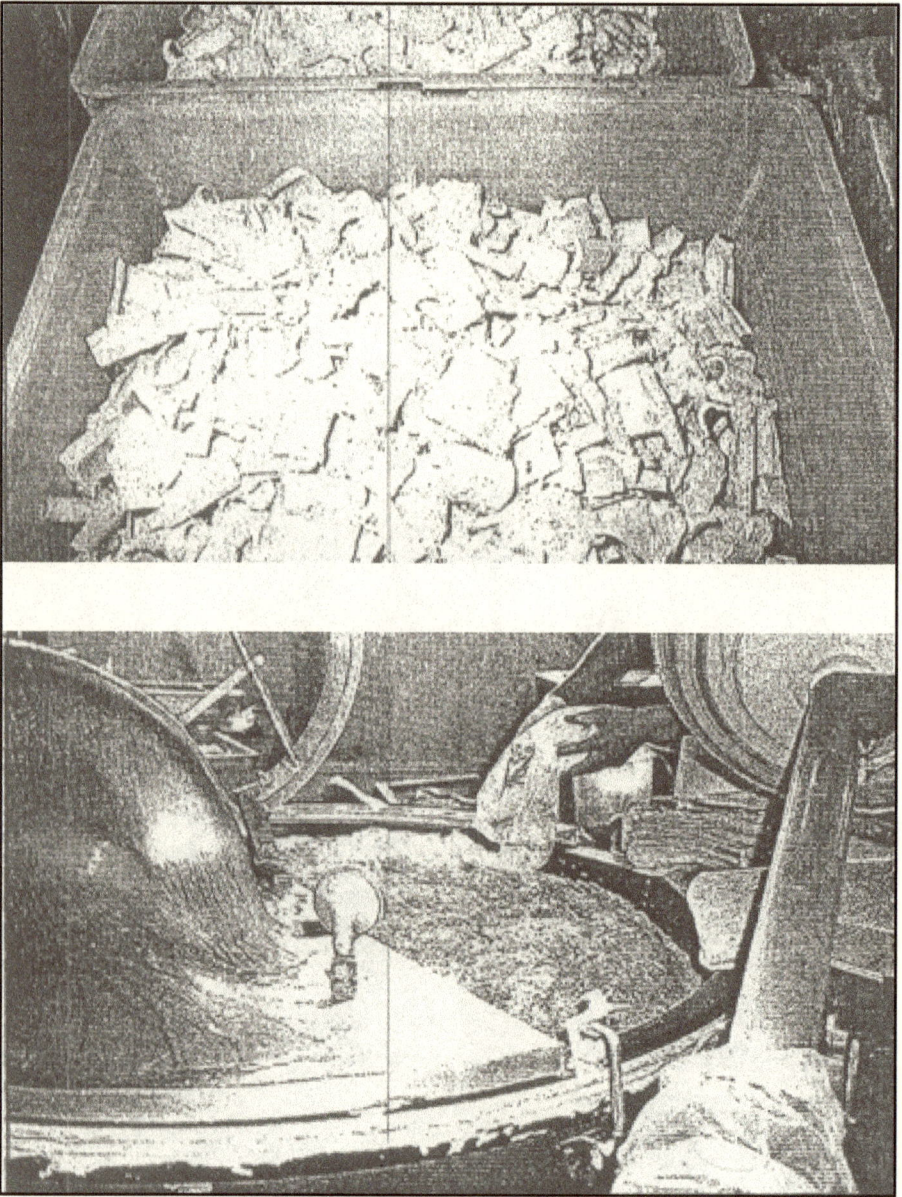

Processing Line of 'Surimi' Production

Frozen/chilled sashimi tuna which is for eating raw, must gurantee a bacterial number (survival number) of less than 100000/g tested material, test negative for coliform group bacteria, and have MPN (most potable number) of *Vibrio parahaemolyticus* of less than 100. The product must be labeled to indicate "for eating raw".

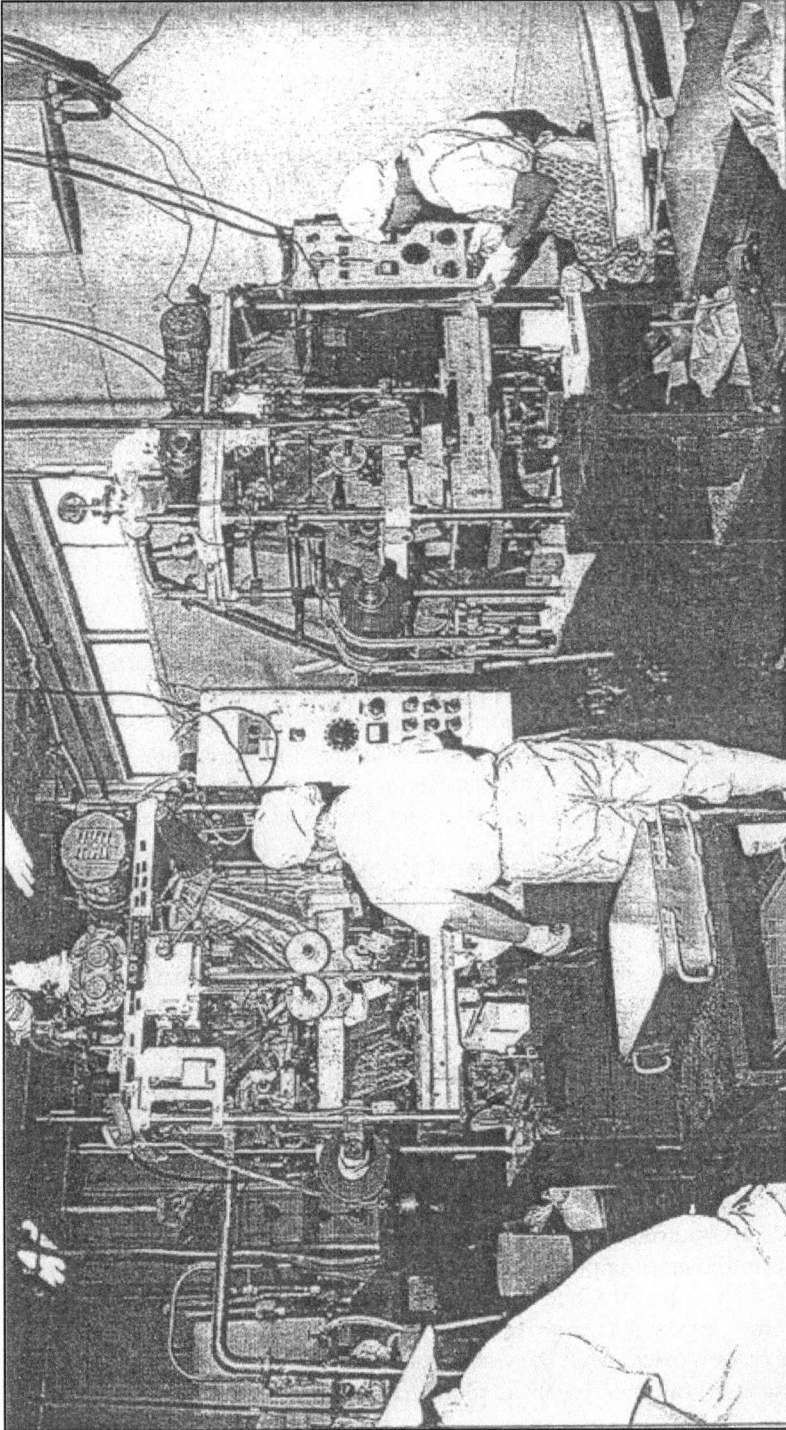

Seasoned 'Surimi' is put in Casing Tubes

Handling and Selection of Sashimi Tuna (Sensory Evaluation)

The fishermen catching the tuna generally do not follow the norms of post-harvesting handling of their catch for quality control measures and get the tunas to the shore as such. The local fish merchants buy the fish from the fishermen and supply the same to the processors, It is the processors, who arrange some kind of quality testing and take up immediate steps to prevent further deterioration of the tunas brought ashore.

All the tunas brought ashore are not accepted by the processors, as all of them do not meet the basic standards fixed by them. The selection of fish for procurement by the processor is done by a simple visual quality testing method. The processor, who is also an exporter, engage personnel, specialized in visually testing the quality of yellowfin tuna meat and grade them accordingly. The equipment used by these quality testers consist of a simple steel corer with a piston attached to it. It is locally called as "meat browser". The corer has length of 50 cm and an inner core diameter of 2.5 mm. Fishes are graded purely on the visual appearance of the meat drawn up by the corer. The corer is plunged rapidly into the body of the yellowfin tuna at the base of the first dorsal fin and a meat strip of about 15 cm long is drawn. The quality tester then places the meat strip on his palm and based on the overall visual appearance, like colour, firmness and smoothness of the meat strip grades the fish as, 'a', 'b', and 'c'. tuna meat graded as 'a' are supposed to be of the most superior quality and the meat can be used for "sashimi". The meat strip here is smooth, firm, unbroken and has fresh pale pinkish white colour. Tuna meat graded as, 'b' and 'c' too are good quality but does not meet the standards for consumption of 'sashimi'.The sampled meat strips here too are unbroken, but has a little bit of discolouration tinged with blood at times. The colour too is a darker pink shade compared to grade 'a'. If sampled meat is broken, not firm and bloody, the fish is rejected by the quality tester.

Handling of Tuna Onboard and Production of Quality Sashimi

Immediately after the harvest and taking fish onboard, a stinger should be pushed inside them behind the head in a way that it would get into the vertebral column and paralyze the entire nervous system. Immediately there after, before transfer to the fish hold (as per the Japanese system), the gills and the viscera should be removed. There after the fish should be hung for ten minutes for draining of blood. After this, each of the tuna should be cleaned with water for 1.5 minutes. The cleaned fish should then be put in an ice box or chilled refrigerated sea water tanks. Before this transfer, it should be ensured that (a) the fish under transfer would not struggle, (b) there would be no dropping of blood and (c) the 'Double Shine" (the process of totally incapacitating the brain and spinal chord) was perfectly done.

In several countries, the practice followed was to transfer the harvested fish from the vessels to the landing point at the fishing harbor concerned using a crane. The landed fishes are then washed for 30 seconds in cold water at the infrastructure facility for the purpose set up as part of the processing plant, a few meters distant, but within the crane's reach. After the washing is done, the cleaned fishes would be sent for grading at a point nearby at the plant. The grading would be done taking into account of skin colour and quality of flesh. The skin would be scanned all over with hand as a part of the process so as to check any damage. The fishes are there after conveyed by crane into the processing plant proper.

The checking of the quality as mentioned above is done by using a gadget, known as "Sashi bou". This is needed to be inserted into each of the fishes and later pulled out. When this is done some meat would also come out sticking all around "Sashi bou". This is then tested by a process that would also include testing by tasting. The "Sashi bou" test would indicate, how long the fish be kept under preservation. After "Sashi bou" testing the fishes would be packed in dry ice or jelly ice. The testing system with "Sashi bou" for assessing quality is followed in Philippines and Indonesia. Two sizes of "Sashi bou" are used, one for large fish and another for smaller ones.

Spoilage Index by Means of Systematic Organoleptic Examination

A characteristic index for the freshness of fish, called "spoilage index" which was determined by means of a systematic organoleptic examination expressed in numbers. Thirteen different characteristics are given scores from 0 to 5, including the pigmentation of the skin, the appearance of the slime and the clearness of the eyes, the colour and odour of the gills, the firmness of the body flesh and of the belly walls, the conditions of peritoneum, the colour of the flesh along the backbone and the adherence of this to the flesh and finally the odour and flavour of the cooked flesh. The average of 13 scores forms the spoilage index. A statistical study of tests made on six species of fish by 5 groups of judges in different localities (in all about 400 lots of fishes divided into 24 series) shows that the spoilage index can be applied equally to all species and that it is almost independent of the individual judge. The correlation between the spoilage index and the amount of volatile nitrogenous bases is very good.

Physical Method of Quality Assessment

On the Use of Electricity in Testing the Quality of Fresh Fish

The method is based on the relationship between freshness of fish and the electrical conductivity of the fish muscle, a fact which has been known since 1930. The electrical conductivity of fish muscle at first depends on its temperature. It increases by about 2.5 per cent, when the temperature rises by 1°C. Therefore the influence of temperature during the measurements had to be eliminated and better results were then obtained. Another difficulty was the polarization of the electrodes put into the fish muscle, and therefore, measurements were made with a current of 5000-8000 cycles per second and about 0.1 milliamperes. Furthermore, the conductivity of fish muscle was depended on the spot at which the electrodes are inserted. The electrical resistance is greater in the parts of the fish muscle near the head and towards the tail. Very high values were found, when measuring over the line where the dorsal and the ventral muscles of the fish come together. The myosepts, consisting of active tissues of fish muscle, are believed to contribute to the electrical resistance. The best arrangement of electrodes, which are constructed as small needles has been investigated. For measuring the resistance, an apparatus consisting in the name of a Wheatstone bridge, which allows instant reading of the value of the electrical conductivity or resistance has been developed. Measurements which have been carried out, during a fishing trip, brought results, which agree with the values for volatile bases. The investigations are made only on white fish- cod, haddock and Pollock, because the fat content of other fishes has an important influence on the electrical conductivity and will complicate the measurement of freshness in these cases.

Objective Assessment of Raw Fish Quality

Ever since the time when chemical analysis began to be applied to foods, attempts have been made to find and use an analytical procedure for a particular chemical substance or group of them, which would be accurate, reliable and specific and by which wholesome, fresh, acceptable or passable food stuffs could be readily and easily distinguished from those considered to be unfit for use as food. Among the earliest of these was Eber's use of the presence of ammonia in meat as an index of spoilage. Subsequently, a variety of products have been suggested and tried as spoilage indicators, including indole, histamine, succinic acid, volatile fatty acids TMA and others. The ideal sought was and still is, a method or procedure for a substance which would not be present in the fresh material, or only present in minimal amounts and which would increase to a significant level when spoilage became evident or detectable by the senses. Also this substance and the procedure for measuring it should be applicable to as many different kinds of food stuffs, as are in general use, and to all the processed variations which may occur. In other words, the ideal procedure should be able to distinguish between fresh fish and that in the first stages of spoilage, regardless of the species, processing treatment, storage conditions or any other factors which may be encountered. In addition, the method should involve relatively simple apparatus, which will be easily assembled and dismantled for cleaning and it should be quickly and readily carried out. If some of the methods proposed as indicators of spoilage are examined as to their compliance with these criteria, the following facts are evident from the data herein presented and from those in the literature. Histamine does not increase significantly during spoilage of crustacean, and its determination involves a rather specialized apparatus and technique. Indole depends upon the presence and activity of those bacteria capable of producing it under a particular set of conditions. Hence it has been proven to be unreliable as a spoilage index. The determination of basic volatile nitrogen compounds and of TMA by the micro-diffusion technique, for example, is relatively easy and simple. Unfortunately, under certain circumstances it fails to accurately indicate the condition or to do so soon enough, at the time when the spoilage first becomes noticeable. The measurement of the acidity of fish oil to evaluate the condition of fish has also proven to be unreliable. The presence of steam volatile acids in fish often parallels the organoleptic judgement, but at times, it also fails, particularly being dependent upon storage conditiond and bacterial flora. Its usefulness is further restricted in the cases where acids are added to the product. In addition, a further drawback is the rather lengthy and time-consumung determination, particularly, when an identification of individual fatty acids is attempted. Hydrogen sulphide is likewise even more variable and dependent upon bacterial growth during storage.

It has been said in many occasions, that no one universally applicable method is available to determine spoilage under all circumstances and no one ever will be found. For individual substances or groups of similar ones, this is probably the case. Taking into account the diversity of bacterial flora and the variety of fish products, it is to be expected that any method dependent upon a particular type of bacterial activity will meet the failure a certain portion of the time. However, in the case of VRS (Volatile Reducing Substances) method, the situation is definitely different. Here a procedure has been designed to determine all types of volatile constituents which may be present originally or produced in the fish through the spoilage process. It is generally agreed and accepted that when fish spoil, their odours increase in intensity. The VRS method measures this odour intensity, but does not differentiate it. Therefore, the usually close correlation between odour intensity and VRS content is understandable.

The facts remain that the VRS procedure has been found applicable to all the types of fish and the kinds of fish products examined, though it is not meant to imply that the VRS procedure is considered perfect or that it is infallible. It has been used to judge the quality of fish bought by the State of California for its institutions and it is regularly used to evaluate raw fish for canning purposes as well as the canned products. The method furthermore meets the other criteria of suitability for general and widespread use previously mentioned. The necessary apparatus is easily assembled and dismantled. Even though spherical ball and standard taper joints are used, the apparatus can be fabricated by a reasonably proficient glass blower. A determination can be carried out in a hour or less, depending on the aeration time adopted.

Taking all the facts into consideration VRS method offers the best means, at present available to evaluate and assess the freshness of fish and fish products. If desired supplementary tests for volatile nitrogen compound or for other substances may be carried out for comparison purposes. But final judgement of the quality of any batch is made on the basis of VRS content.

As regards autolysis and the changes resulting from this process in relation to spoilage it has been verified, that none of the marked changes usually associated with spoilage are produced as a result of autolysis. The fish may lose their firm texture and become soft, but no marked odours are produced. The deteriorative changes so noticeably evident in spoiled fish are evidently the result of bacterial activity in the fish. The presence of autolytic products enhances and speeds up the bacterial processes, but the final spoiled state is ordinarily the result of bacterial growth and its products. Therefore, when spoilage is discussed, the result of bacterial action is usually implied. Hence the search for methods to measure spoilage has mainly been directed towards the determination of particular bacterial products. No increase in VRS content has been observed in fish undergoing autolysis, unless some bacterial action was also present.

Fish spoilage is a many-sided process, involving number of factors which markedly affect the deteriorative pattern and hence the applicable methodology. Among the factors which have to be considered in discussing the means to objectively assess the freshness of fish and shell fish are (1) the types of fish, such as, bottom and pelagic fish, teleosts, elasmobranches, mollusks and crustacean, (2) variations in structure and composition, such as, white and red meat, lean and fatty fish, (3) the processing treatment of the fish, as whole, in the round or dressed, filleted, frozen, salted, cured, smoked and canned, (4) the variation in bacterial flora found in the environment of fish during spoilage and (5) the storage temperature. The effects of these factors are often inter-related and greatly influence the spoilage picture. In view of this complex situation and from practical considerations, any objective method will be of real use for the evaluation of the quality of raw fish must meet certain basic criteria for its consideration as a generally applicable procedure. These include the above five factors, and in addition, reliability and reproducibility, simplicity of the necessary apparatus, ease and speed of execution. For example, of two comparably applicable procedures, one that requires an hour or less for a result to be obtained would be considered more useful than the other which requires many hours or even possibly days before a final result is forthcoming.

Among the above factors influencing the spoilage picture and methodology of its detection, the diversity in the bacterial flora associated with spoilage is of prime importance. Any method used to evaluate the freshness or quality of fish (tuna for example) which depended, for example, upon the presence of hydrogen sulphide, volatile acids or indole would be found reliable a variable percentage of the time, depending among other things, upon the presence of the particular bacteria responsible for the production of the chemical indicator. Instances of this interdependence between bacterial flora present and usefulness of a chemical method is found in the case of trimethylamine (TMA). In addition to the fact that all fish do not contain TMA oxide, it has also been shown that there is a variation in the content of TMA oxidase among bacteria. Therefore, even if the oxide is present, only if the organisms reducing it to the TMA are also present, is a positive test for the TMA likely to be obtained. Therefore, any method which is solely dependent upon the presence of a specific chemical substance or group of substances is a priori doomed to failure a certain proportion of the time.

Attempts to device chemical methods for detecting and measuring the extent of spoilage have largely resolved around the development and use of analytical procedures for either a specific end-product, such as, indole or hydrogen sulphide or for a group of similar related end-products, such as, volatile nitrogenous compounds, volatile fatty acids or carbonyl compounds. Because of the reasons outlined above, the application of these type of methods has met with variable success in the hands of investigators and has not proven to be sufficiently general for wide spread use. The reason for the diverse opinions regarding their usefulness is also evident, taking into account the existing range of conditions and kinds of fish and shell fish. The range of usefulness of the so-called objective methods is based on the organoleptic judgement. The establishment of numerical limits or ranges for fish products depends upon their delineation by sensory means. However, once the reliability of the relationship between the objective measurement and the organoleptic assessment is established, then the subsequent evaluation of a product using these ranges can be considered objective and need not involve the senses with their physiological and psychological influences.

Objective Chemical Methods

Attempts to duplicate by chemical means, the olfactory sensory perception of the volatile products (carbonyl and basic nitrogen compounds, volatile acids, hydrogen sulphide and occasionally indole) and the volatile odourous compounds associated with the development of spoilage were made. A procedure was devised which was based on the aspiration from a press juice of the compounds volatile at room temperature into an alkaline solution of potassium permanganate which oxidized them. The amount of the reduction of the permanganate is used as a measure of the amount of volatile substances originally present in the sample, and hence of the odour. The substances determined are called Volatile Reducing Substances (VRS) and the concentration is expressed in micro-equivalents of reduction per 5 ml of press juice, the sample usually used. The most recent modification of the apparatus employs the closed circuit principle, in which air from the pressure side of a small pump (a Dyna pump, needing no lubrication is used) is forced through the sample into the alkaline permanganate reagent and thence by the vacuum or suction side of the pump re-circulated through the solutions. A circulation time of 40 minutes is ordinarily used for a determination, but this time may be shortened as desired to speed up the operation.

Methods

Unfrozen fillets, freshly cut from raw fish were immersed for 2 minutes in the 5 per cent salt solution, removed from the solution, wrapped in groups in waxed paper and then stored in a mechanical refrigeration maintained at 5.5°C (42°F). Individual fillets from each lot are periodically drawn for analysis.

Evaluation of the Condition of Fish

Fillets were removed from their packages with sterile forceps and while thus held were judged organoleptically. They were then wrapped in a gauze and squeezed in a previously hot water-immersed hand orange juicer, with two circular perforated iron discs replacing the conical portion of the juicer. The press juice was sampled for total viable bacterial numbers and then examined for the content of volatile reducing substances (VRS), total volatile basic nitrogen (TVN) and trimethylamine nitrogen (TMN).

The TVN and TMN were measured by Conway's micro-diffusion technique, using 1 ml of press juice as sample, 2 ml of saturated potassium carbonate as alkalizer, 1 ml of 2 per cent boric acid, containing mixed methyl red bromocresol green indicator, as base absorbent and 0.7 ml of neutralized formaldehyde, for the determination of higher amines, as trimethylamine. The volatile nitrogen trapped by the boric acid solution was titrated with approximately 0.02 N sulphuric acid solution (A thick greasy, but water-soluble preparation of starch dispersed in and reacted with glycerol, known as glycerite of starch was used to seal the ground glass plate on to the Conway unit). The results are expressed as mg of nitrogen per 100 ml of press juice.

The VRS were measured in 5 ml of press juice in the closed circuit re-circulating apparatus. The re-circulating pump used was a type not requiring any lubrication, called "Dyna pump". The approximately 2 litres of air in the apparatus are blown through the sample by the pressure side of the pump and then removed after passage through 10 ml of 0.02 N potassium permanganate in N NaOH solution in the reaction flask by the suction side of the pump. The usual aeration time is 40 minutes, but this may be shortened to adapt it to more practical operating conditions. The alkaline permanganate solution is acidified with 5 ml of 6 N sulphuric acid and treated with 3 ml of 20 per cent KI in 0.1 per cent sodium carbonate solution. The liberated iodine is titrated with about 0.025 N $Na_2 S_2 O_3$ solution in 0.2 per cent sodium carbonate, 0.1 per cent sodium borate. A blank for unreacted permanganate is always set up at the time and left at room temperature without aeration. The difference between the titrations for the control and test runs is a measure of the amount permanganate reduced, which is expressed as micro-equivalent of reduction per 5 ml press juice.

Total Viable Bacterial Count

The press juice was diluted with sterile physiological salt solution and 0.01 ml of the desired dilution was pipetted with an 0.1 ml pipette on to the surface of tryptone glucose extract agar in a petriplate. The liquid was then spread uniformly over over the surface of the agar with a sterile curved glass rod. After incubating for 48 hours at room temperature of about 27°C, number of colonies was counted and the total count per ml of original press juice was determined.

Eye-Fluids for Objective Tests for Fish Quality

The study has been confined to haddock (*Melanogrammus aeglefinus*) and cod (*Gadus morhua*) regarding the eyes of fresh fish and their relationship to fish quality. Although examination of lens and outer membranes of fish eyes yielded no satisfactory means of measuring the quality of fish, physical and chemical measurements of eye fluids apparently correlate closely with the quality of fish tested (cod and haddock).

Eye fluids from cod of given quality produce values for refractive index and optical density that differ from those for the eye fluids of haddock of the same quality. It may be assumed that values for other species of fish also differ as far as numerical values of these measurements are concerned. It is therefore probable that each species of fish to be examined for quality by these criteria will require separate standard curves.

A quick simple method of clarification of the eye fluids however, can be developed. The correction factor may be used for temperature in order to provide for the use of a hand refracto-meter in place of the usual laboratory instrument.

Measurements of refractive index and optical density have given results that promise to be satisfactory as criteria of fish quality. Colourimetric measurements made of pyridine extracts or Biuret test mixtures of eye fluids, in which the optical density of the resultant solution is also measured, have shown very good correlation and are apparently a modified means of measuring those substances in the eye fluids that have caused the original optical density measurement to correlate with quality. Appearance, taste and other organoleptic measurements and measurements of trimethylamine-nitrogen have been used as the criteria by which the newly developed tests have been evaluated.

From the primary results, it is concluded that refractive index and optical density measurements of fish eye fluids appear to offer satisfactory criteria of the quality of fresh fish. The tests are quick and easy to perform and should be readily adaptable for use outside the laboratory.

Methods

Fresh eviscerated haddock and cod used in the storage tests were purchased only if they have been caught less than 24 hours earlier, and were held at 2.2 to 4.4°C.

On each test day, the sample was removed from the storage, the eyes were cut out from each fish and pierced, and eye fluids (both aqueous and vitreous humor) were expressed into a container. Each sample of eye fluids was centrifuged at approximately 3400 rpm in a clinical centrifuge for 5 minutes and the supernatant liquid was then passed through sentered glass filter for clarification.

Refractive Index

A few drops of each sample of clarified eye fluids were placed on the prism of an Abbe refractometer and the refractive index was measured at 20.1 to 19.9°C. Three readings were made on each sample.

Optical Density

Two ml of clarified eye fluids was added to 8 ml of a buffer solution at pH 7. After thorough mixing, the resulting solution was measured in a colourimeter at a wave length of 420 millimicrons.

Biuret Colourimetric Measurement

The Biuret reagent consisted of 10 per cent potassium hydroxide to which 25 ml of 3 per cent copper sulphate per litre was added. 2 ml of the Biuret reagent was added to the sample used for the measurement of optical density. After thorough mixing of the sample and Biuret reagent, the optical density of the resulting solution was read at a wave length of 420 millimicrons.

Pyridine Colourimetric Measurement

Two ml of clarified eye fluids was added to 8 ml of pyridine. After thorough mixing, the sample was allowed to stand for 10 minutes and then was centrifuged for 5 minutes. The supernatant liquid was poured into a colourimeter tube, and the optical density of the extract was measured at 420 millimicrons

Vacuum Distillation Procedure for Estimating the Quality of Fish

The volatile acids and volatile bases content are two indices of fish spoilage that are used widely. Several methods are used for the determination of each index. All these methods have been found to be time consuming. The determination of volatile

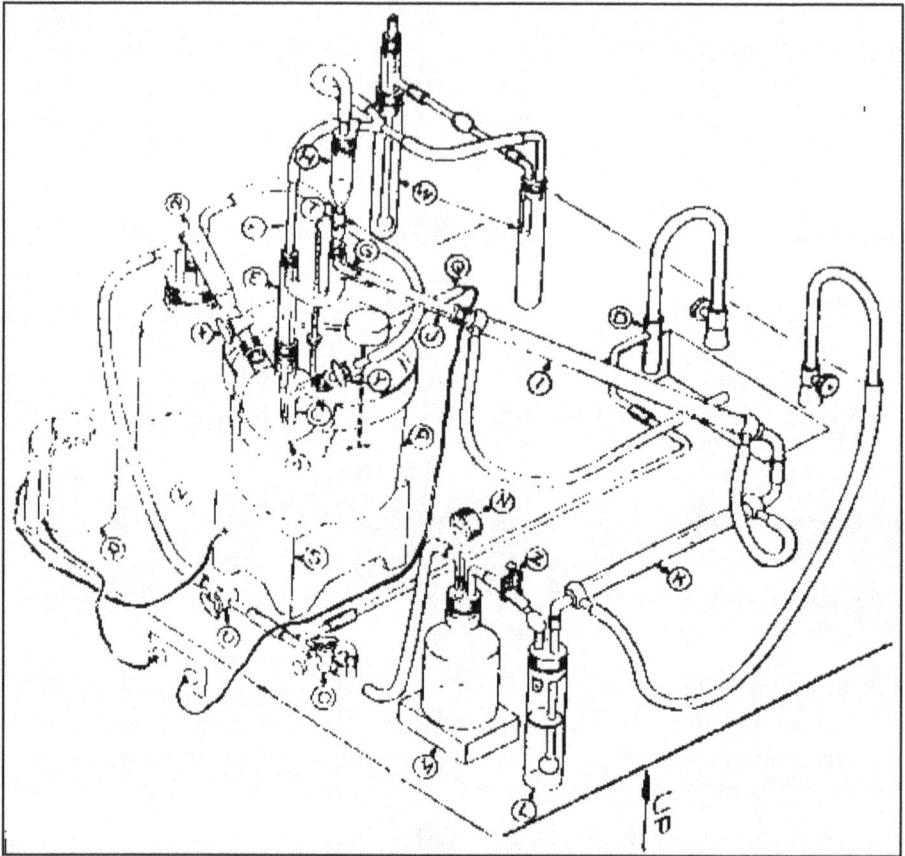

Isometric Drawing of the Vaccum Distillation Apparatus

acids, as used officially in the USA utilize a steam distillation procedure. The determination of volatile bases, as used officially in Japan, utilizes a vacuum distillation procedure or an alternative aeration procedure. A steam distillation procedure for the determination of volatile bases. has been used extensively in the USA.

Recently, Tomiyama, DA Costa and Stern developed an apparatus and vacuum distillation procedure suitable for the determination of both volatile acids content and the volatile bases content of fish flesh. The method was found to be accurate and to yield high per cent recoveries of both volatile acids and volatile bases. The primary advantage claimed for the procedure was the rapidity with which determination is made, approximately 5 minutes for the determination of volatile bases and 10 minutes for the determination of volatile acids.

Methods – Preparation of Sample

A representative sample of flesh is taken and passed through a grinder three times. 85 gram of the comminuted flesh are added to 200 ml of water in a blender top and the mixture homogenized for 3 minutes. Four hundred ml of magnesium sulphate solution are added to 200 gram of the homogenate in a large flask. The mixture is shaken well and then filtered through Reeve Angell 235 filter paper. The resulting filtrate is clear and at pH of 2.0. Fifty ml of the filtrate are equivalent to 5.0 gram of flesh.

The magnesium sulphate solution is prepared by dissolving 60 gram of magnesium sulphate in distilled water, making the solution to a volume of 100 ml with additional distilled water. Then 2 ml of 6N sulphuric acid are added to the solution.

The apparatus utilizes a 500 ml round bottom three neck flask as the distillation vessel (A). A tube passes through the stopper in the left neck of the flask and is connected to a small funnel (B) by a wide bore stopcock (X). Another tube (C) leads from the bottom of the flask through the stopper in the right neck of the flask to a waste trap (V) with a connecting stopcock (Y). The waste trap is connected by a stopcock (U) to the vacuum system, which may be an aspirator pump (D).

A modified Claisen-type distilling (E) of almost equal diameter as the central neck of the distilling flask, extends a short way into the flask beneath the stopper in the central neck. The vertical neck of the tube extends only half inch (1.27 cm) above the joint of the side neck of the tube. The mouth of the vertical neck is stoppered. Passing through this stopper, a capillary tube (F) extends to the bottom of the flask. The purpose pf the capillary is for the admittance of air, as an agitating agent during the distillation.

The side neck of the Claison-type tube extends only 1 inch (2.54 cm) above the joint of the side arm leading to the cooling system. The mouth of the side neck is stoppered. Passing through the stopper is a small bent tube (G), the tip of which projects into the side arm of the Claison-type tube, distilling tube. A small funnel (H) is attached to the other end of the tube by a short piece of rubber tubing and a pinch cock

The side arm of the Claison-type tube enters the cooling system through a two-holed stopper in the mouth of a 300 mm Liebig condenser (I). A thermometer (J) is placed in the second hole of the stopper with the bulb extending approximately one half inch below the stopper.

The adapter of the condenser (I) is bent in a vertical plane to an approximate right angle with the body of the condenser. A second 300 mm Liebig condenser (K) is modified by drawing the mouth into a bent tube and by bending the adapter tip. The adapter ends of the two condensers are connected by a short piece of rubber tubing in such a manner that the second condenser (K) is approximately at right angles to the first condenser (I) in an approximately horizontal plane.

The rubber mouth of the second condenser (K) is attached by as short piece of rubber tubing to the bubbling tube. The bubbling tube passes through one hole of a two holed rubber stopper in the mouth of a 38x200 mm test tube which serves as a receiving vessel (L) and extends below the surface of the trapping liquid in the receiving vessel. A celluloid disc rides on the bubbling tube. The purpose of the disc is to prevent escape of the trapping fluid from the receiving vessel by breaking up large bubbles which may develop during the distillation or when the vacuum in the system is released.

A tube, a bulbular flash-trap passes through the second hole of the rubber stopper in the mouth of the receiving vessel and extends only a short distance into the vessel itself. This tube is connected to a flash-trap-bottle (M) by rubber tubing and a screw type pinch cock (Z). The flash-trap-bottle, which has a vacuum gauge attachment (N) is connected to the vacuum system. An air inlet tube, controlled by a stopcock (O) is placed in the vacuum line for maintenance of the proper pressure line in the system.

All air drawn into the apparatus during and following the distillation is first bubbled through NaOH solution (W) to remove carbon-dioxide.

A controlled, constantly agitated water bath (P) is used as the source of heat for the distillation. The distillation vessel (A) is immersed in the bath to the necks of the flask. Both necks and the side arm of the Claison-type distilling tube as well as the mouth of the first condenser (I) are insulated to prevent premature condensation of the distillate.

Procedure

The stopcock (U) and (Y) and the screw pinch cock (Z) are closed. A receiving vessel (L) is prepared and attached to the system. If the volatile acid number is to be determined, 10 ml of 0.01 N, NaOH are placed in the receiving vessel. If the volatile acid bases content is to be determined, 10 ml of N/28 sulphuric acid are placed in the receiving vessel.

The stopcock (X) is opened. One drop of silicone anti-foam agent and 50 ml of the sample solution are added to the distilling flask (A) via the funnel (B). If the determination is of the volatile acid number, the distillation is then begun. If the determination is of the volatile bases content, 5 ml of a 10 per cent NaOH solution are added to the flask (A).

The distillation is started by closing the stopcock (X) and slowly opening the screw pinch cock (Z), thus permitting a vacuum to be drawn in the system. As the distillation begins, the temperature as shown by the thermometer (J) in the mouth of the first condenser (I) rises rapidly. As the distillation continues the temperature slowly falls. When the temperature reaches a steady state 3 or 4°C above the room temperature, the distillation is completed.

If the determination is that of bases, the distillation procedure is at an end. If the determination is that of acids, 20 ml of distilled water are added to the distillation vessel without breaking the vacuum by means of the funnel (B). The distillation is permitted to continue until the temperature, as shown by the thermometer (J) has again reached a steady state, a few degrees above room temperature.

Following the distillation procedure, the screw pinch cock (Z) is closed and 20 ml of distilled water are flushed into the apparatus through the funnel (H) and bent tube (G) assembly of the side neck of the Claison-type tube. This water washes the cooling system and is collected in the receiving vessel. The vacuum is then released by allowing air to enter the system through this same assembly.

When the vacuum has been dissipated, the receiving vessel (L) is disconnected and removed from the apparatus and the trapping liquid is titrated with either standard acid or standard base, depending upon the determination. The titration is carried out in the receiving vessel itself, using the bubbling tube as a stirring rod. In the determination of volatile acid number, neutral red is used as the indicator; in the determination of the volatile bases content methyl red is used.

Water is then run into the distillation vessel (A) through the funnel (B) until the flask is full. Stopcock (X) is closed and stop cocks (U) and (Y) are opened. The water in the flask is siphoned through the tubular system into the waste trap (V). The operation is repeated at least twice or until the flask is clean. The condenser system (I,K) and the Claison-type distilling tube are thoroughly washed by immersing the receiving vessel tip of the second condenser (K) in water while the siphoning operation is in progress. The entire washing procedure requires about two minutes.

The temperature of the water bath is maintained throughout the distillation and washing procedures at 75°C. The vacuum is maintained at a constant 29 inch (739 mm) of Hg.

Chapter 6

Methods of Retarding Spoilage

Fish must be preserved by storing it in time of abundance for use in time of shortage. Therefore all methods or any technical measures to retard the spoilage of fish to keep the products;

(a) Nutritional quality be retained.

(b) Food is free of pathogenic and spoilage micro-organisms and their toxins

(c) Food is free of chemical compounds, causing problems.

(d) To extend the shelf-life of fish and fishery products by; 1) Icing and chilling, 2) Drying, 3) Smoking, 4) Salting, 5) Boiling and steaming, 6) Freezing, 7) Chemical reagents and additives, 8) Packaging and 9) Radiation.

Drying is one of the oldest methods of food preservation than any others. The requirement is to reduce the free water content by heating and evaporating, which is able to depress the growth of micro-organisms. Water is essential for micro-organism's growth. The method, which is commonly practiced for traditional dried fish, the fish is washed and gutted and splitted, then hangs for cool air or hot air blow, which is able to drive the moisture out. It is found that in 15 per cent moisture or less mould will cease to grow. The concept of heat treatment is to inactivate pathogenic organisms. There are three ways to propagate heat energy; convection, conduction and radiation. Convection heating means bodily transfer of heated substances, while, conduction heating take place by transfer of molecular activity through one substance to another. Radiation heating is transfer of heat energy in the same manner as light, and with the same velocity. Heat transfer by convection must be accompanied by some conduction heating. Conduction heating is very slow compared to the usual cases of convection heating.

The second law of thermodynamics states that heat energy flows only in one direction, from hot to cold bodies.

When a can of food is sealed at 180°F and placed in a steam pressure vessel which is brought to 15 pounds per square inch with steam, the steam chamber is the reservoir of high heat energy and the can of food is the reservoir of lower heat energy. Heat is then transferred from the hot body to the cold. The mechanism of heat transfer in canned food during such thermal processing may be divided into several rather definite classes. To a certain extent, it is possible to place food into heat transfer classes by knowing their physical characteristics. The heat is transferred by conduction from the steam to the can, and from the can to the contents. The can contents will either heat by developing convection currents or heat by conduction.

Salting is the most popular method for preserving many of fishery products. There are a bacteriostatic and bacteriocidal action that both delay the growth and kill many groups of micro-organisms, especially putrefactive rod if the salt concentration is 10 per cent. But 15 per cent of salt will able to kill putrefactive cocci and halophilic group is able to grow on higher salt content

In foods containing salt as preservative, the salt has been ionized collecting water molecules from each ion. This process is called hydration. The greater the concentration of salt, the more water employed to hydrate ions. A saturated salt solution at a temperature is over that has reached a point where no further energy is available to dissolve the salt. At this point (26.5 per cent sodium chloride solution at room temperature) bacteria, yeasts and molds are unable to grow. It has been postulated that there is no free water available for microbial growth at that point.

Chapter 7

Refrigeration in Retarding Fish Spoilage

Refrigeration is not only a means of storage, but serves throughout all stages of food processing, commencing at production and culminating in the product, at retail level. It comes into processing either as an integral operation or as the primary function of a processing procedure, and in some instances must be considered in relationship to other methods of conservation. Processing and/or storage can well call for heat transfer between produce and refrigerating media temperature controlled environmental conditions or air conditioning, involving the use of refrigeration to attain desired hygrometric levels.

Long term preservation of "dead" and sometimes "live" produce may be affected by pre-storage processing in immersion freezers, air blast freezers, fluidized bed freezers, indirect contact freezers of the plate type or the scraped-surface heat exchange design, or cryogenic freezer units etc. The type of freezer employed varies to a large extent with the produce, the form in which it is to be processed or retarded and with processing requirements. Furthermore, it can be multi-purpose, for example contact plate freezers used for processing meat or fish are equally efficient for packaged pre-cooked meals.

Processing Procedures

The provision of refrigeration facilities to cater for all facets in food processing is governed by many factors-.It may be as an aid or expedient to one or more processing stages- extending to conditions where refrigerating operations are vital to processing and often constitute the major operations. Processing alone can be, but seldom is, the only service application. Storage, cold or frozen, is generally required. This does not however rule out the desirability of refrigerated storage in all processes, where the operation is not an integral part of the manufacturing cycle. Here requirements for the application may be prior to processing, interposed between process operations and or for safe keeping of the finished product.

Processing methods vary considerably, being influenced by the product handled. In some cases both temperature and humidity control of the environment in which the process is carried out is necessary, whilst in others humidity control alone is sufficient. Cooling or freezing of the product itself may in addition be needed. Forced or gravity circulation of cold air during pre-cooling and chilling operations is again largely determined by the characteristics of the product.

A greater divergence exists in freezing. Not only are there broadly defined methods, but considerable design variants occur in each classification. Furthermore, a combination of methods may be employed to expedite processing. Even with the same produce, methods differing from current practice can prove a valuable alteration. Other considerations center around the volume of goods requiring refrigeration and this can vary widely within the same trade.

Ice-Freshwater Ice

One of the most important facts to bear in mind concerning the use of ice for preservation of fishes that spoilage is only retarded, not stopped. Under proper conditions of handling, fish such as prime halibut may be preserved in ice for 8 to 12 days before a noticeable lessening in quality occurs. Cod similarly iced will keep for over seven days. On the other hand, bottom fish, left on deck for 12 to 18 days and iced improperly in a deep hold without the use of shelf boards may show appreciable loss in quality after only 4 to 5 days.

Ice when properly used in adequate amounts, aids in preservation of fish in two ways; (1) The temperature of the fish is lowered to approximately 32 to 36°F, which slows the bacterial and enzymatic changes; and (2) the melting of the ice bathes the fish in clean cold water and with proper stowage, washes away considerable slime, blood and bacteria. The resulting contaminated water accumulates in the bilge of the boat, and at intervals is pumped overboard.

Every pound of ice, on melting, absorbs 144 BTU of heat from its surroundings. The absorption of heat is sufficient to lower the temperature of 19 pounds of fish 10 degrees F (assuming that the fish has a specific heat of 0.760 and that no external heat was absorbed in the process). In actual practice, the heat transfer from the boat hold and air equals or exceeds the heat transfer from the fish. Dunn (1946) showed that on a 6-day trip, the heat gained by a trawler hold with a capacity of 100 tons of fish and 30 tons of ice caused about the same amount of ice to melt as was required to cool the fish. This transfer of heat demonstrates the desirability of insulating the hold, which would make possible the saving of a substantial quantity of ice during long trips.

Every fisherman soon learns how much ice to "take on" in order to carry him through a trip. The expected duration of the trip, the temperature of air and sea water, the insulating value of the sides and deck head of the vessel, and the expected quantity

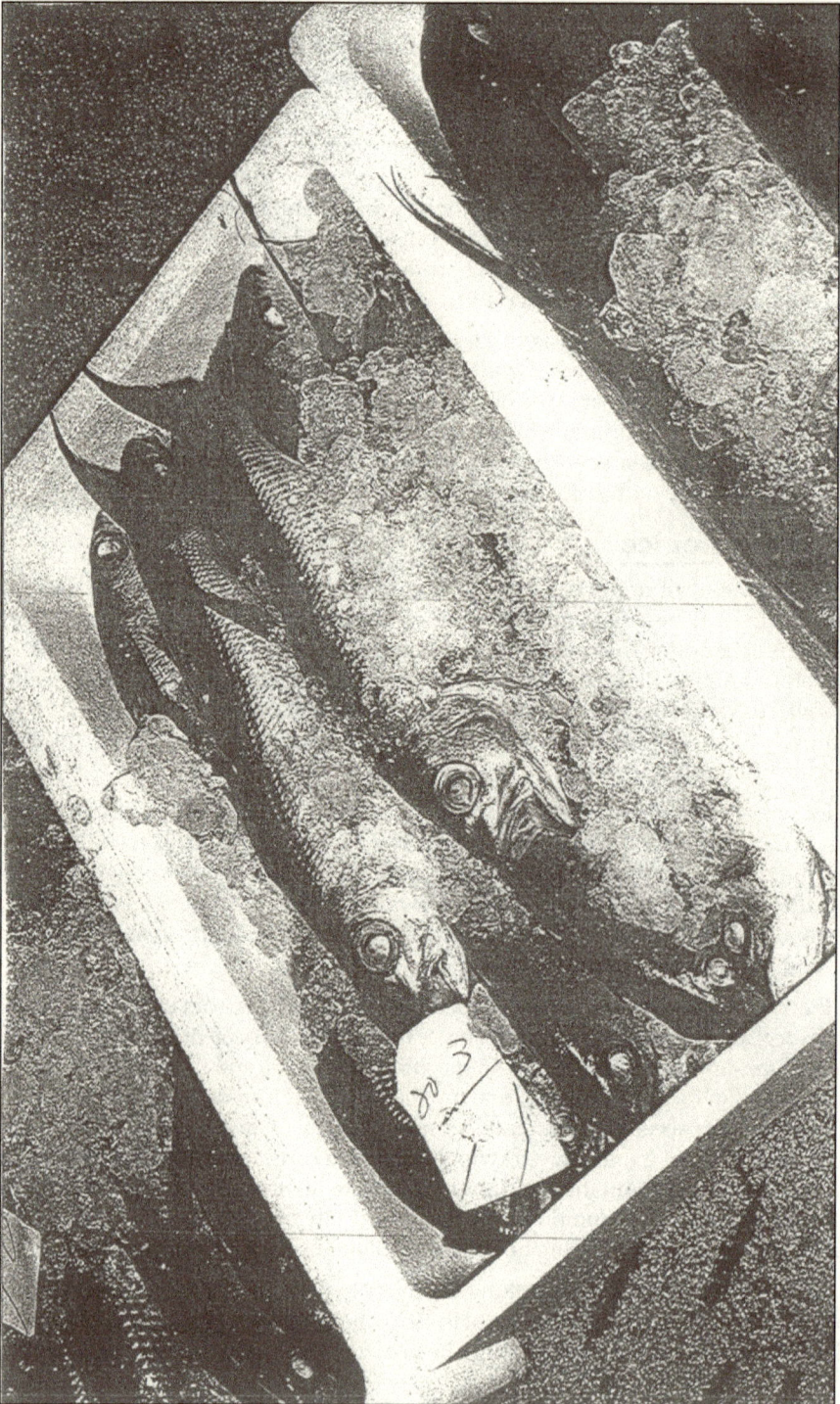

Fish Preserved in Crushed Ice

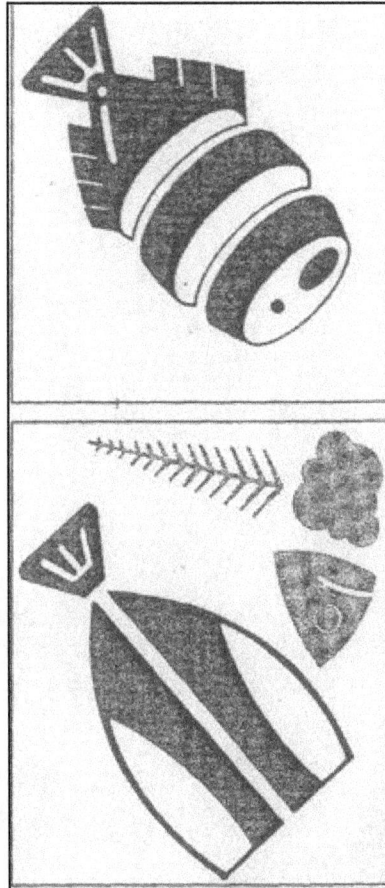

Freezing Plant

of fish to be obtained are all factors to be considered in estimating the amount of ice to be loaded. Ice is cheap compared to the other expenses in fishing operation. Hence no fisherman should cut short his estimated need. The exact ratio of weight of ice to weight of fish to be carried varies commonly from 1:4 to 1:1. In un-insulated holds of wooden vessels a ratio of 1:2 is common. Studies have shown that more rather than less ice should be taken by fishing vessels, because it was found that additional ice (compared to present practice) should be allowed at the sides of the vessel and adjacent to the wing boards of each pen. Any exposure of fish at these joints due to melting of ice contributes substantially to quality losses.

To correctly ice the fish in the hold, three things should be accomplished; (i) the fish should be placed with sufficient ice around them to cool them as promptly as possible and to maintain their temperature as close to the melting point of the ice (32°F) as is practical for the duration of the trip; (ii) the ice and fish should be arranged to allow accumulated water, blood and slime to drain through the mass into the bilge; and (iii) the fish should not be subject to great pressure from the weight of fish and ice placed above; otherwise, the physical damage, as well as the shrinkage or loss of weight by the fish will be excessive.

Correct icing requires considerable care and experience, and every vessel has a separate problem depending on the construction, hold and pen layout, and the relative heat transfer from the water and air outside the hold. Kanke (1946), in discussing the correct icing of fish at sea, has pointed out that from 50 to 60 per cent of the profit of a trip may be lost if the quality of the catch is reduced through inadequate or incorrect icing. Ample ice should be placed on the floor of each pen, 8 to 12 inches deep, for a trip of 8 to 12 days. A like amount should be placed at the skin of the vessel and on top of the fish. A smaller amount should be used at the wing boards (the transverse partitions) and sides of each pen to keep the fish away from contacting the board. For eviscerated fish, the gut cavity or poke of the fish should be well filled with ice, taking special care to pack the ice in the gill cavity and around the nape. Preferably, each fish should be surrounded by ice or the fish placed in alternate layers such that the ice is in actual contact with the greater portion of each fish.

The practice of using a bed of ice, the layering 10 to 12 inches of fish, followed by a thin layer of ice and another thick layer of fish results in most inefficient cooling. Two or three days may be required, in this case to lower the temperature of the fish to 36°F.

In some instances, improperly iced fish does not cool appreciably throughout the entire period of storage. Under proper conditions however, not over 3 to 6 hours should be required to lower the temperature to 36°F of fish weighing about 2.5 kg. Fish should be placed on a rounded layer of ice so that the melt water drains away from the fish to the sides of the pen. In placing the fish on the ice, the belly cavity should be turned in such a manner that there will be adequate drainage from it. A good icing job has been done if, at the end of the trip, sufficient ice remains on the bottom and at all sides so that the entire load has been maintained at a temperature not higher than 36°F (or 32°F, ideally).

Salt Water Ice

Fish do not freeze at one point in the temperature scale, as water does at 32°F. Rather they begin to freeze at about 30°F and gradually harden as the temperature drops. At 23°F, the fish have passed through the zone of maximum ice-crystal formation, but are still not solidly frozen. The lower freezing range of fish means that an ice melting at a lower temperature than 32°F can be utilized to lower the holding temperature further.

Tests on the effect of temperatures close to that of freezing on the storage of fish showed that a reduction in the temperature of the fish from 37 to 31.5°F increased their keeping time in ice as much as did a temperature reduction from 77 to 37°F. The lower temperature of the fish was obtained through the use of salt-water ice (about 3 per cent sodium chloride), which has a melting point of approximately 28°F. Salt-water ice is best prepared by the flake-ice method in order that the salt may be distributed uniformly through the ice. In a pilot plant study, sub-cooled flaked salt-water ice and ordinary crushed freshwater ice were used in icing similar lots of fish, both held under otherwise similar conditions of storage. The flesh temperature of the salt-water iced fish ranged from 30 to 32°F, which was 6 degrees lower than the temperature range of the fish iced with crushed freshwater ice. The fish stored in salt-water ice were superior in quality at the end of the test.

Operational Environment

The ice maker is designed for indoor or outdoor operation. Ambient air should not be below 40°F (4°C) or above 100°F (39°C). Water temperature should not be below 35°F (2°C).

Sea water is pumped over the inside and outside surfaces of the hollow-core type 316 stainless steel Seafarer evaporators. The refrigerant, circulating in the hollow evaporator core removes the heat from the sea water and the sea water ice is formed with its salinity retained.

The corrosive effect of sea water is minimized by use of cupro-nickel stainless steel and plastics in those areas that come in contact with sea water. Regular maintenance is required.

At regularly timed intervals, cycles are reversed and sufficient heat is re-circulated through the evaporators to enable ice to slip, by force of gravity, downward into a revolving auger which reduces the ice to irregular shaped pieces and discharges them into storage bin.

Dual refrigeration system increases the chance of always having ice. Requires a reasonable maintenance. No moving parts in the freezing zone, means no expensive repairs. Regular preventive maintenance helps eliminate costly downtime.

An independent power system (marine diesel generator) is required to be installed in the boat to operate the ice-machine. A smaller unit (1 ton ice/24 hour) will require a 7.5 kw generator.

Water flow to cool the machine and condense the refrigerant is controlled automatically by the regulating valve assembled in the machine. Water consumption will depend entirely on the temperature of the water.

A centrifugal pump should be located below the water line of the boat. A sea water strainer should be in the water line, to protect the pump and the ice maker from foreign matter. A pressure regulated by pass should be installed in the pump discharge line to protect the pump by ensuring a flow through the pump in the event the ice maker is shut down for any reason. By using a centrifugal pump in this way, no trouble should be experienced, from the polar regions to the equator.

The pump must have a positive head. The deeper the sea suction, the less chance there will be for the pump to ingest floated impurities such as, oil slicks.

The circulating water pump is a magnetic driven pump which eliminates the conventional shaft seals. The pump housing, cover and impeller are made of glass filled delrin. Being magnetically driven, the pump housing is completely enclosed, eliminating the possibility of leakage.

The pump motor is encapsulated in epoxy. This epoxy encapsulation allows the pump to run totally submerged.

Advantages of Sea Water Ice

Sea-water ice is colder than freshwater ice, since sea water freezes at 28°F and freshwater at 32°F. The colder temperature aids the preservation of the fish because the catch is kept colder, but not frozen.

Salt water ice is softer than freshwater ice. Therefore it does not bruise the fish and is easier to handle. Salt water ice can even be shoveled by hand, because it does not bridge.

When salt water ice is used to preserve fish on board the trawler, the catch is kept in its own environment. Most freshwater ice is manufactured in large ice plants, dock side. The freshwater is not purified and contains bacteria alien to the fish, which can spoil and contaminate the catch.

Fishermen must wait in line to purchase freshwater ice from a port-side ice-plant, sometimes detaining fishing boat several hours and losing precious hours of fishing. Many a times, fishermen will be at sea for two days without finding a good school of fish and realizing a profitable catch. Then on the third day, a good school of fish will be found, but if freshwater ice is being used to preserve the catch, the third day's profitable fishing will have to be limited because the freshwater ice has melted and the catch will spoil, if not quickly brought back to port. Valuable fishing time must be lost because the trawler must return to shore due to lack of ice.

Another problem which frequently occurs is when fishermen realize a good catch during the first day out, but have to return to port because of a threatening storm or other inclement weather conditions at sea. If the trawler was loaded with freshwater ice before embarking, all the remaining ice will have to be discarded. With sea water ice making machine on board, ice is made only as needed. Consequently there is always ice when needed and ice is not wasted, if it is not needed.

Premium prices have been realized because of the high quality of fish or shrimp preserved with sea water ice while at sea.

Ice-Making Machine

When a process calls for plate ice, in particular the fishing industry, the needs of

A.P.V. Parafreeze Ice Machine Prototype

such plant, should cover (a) the ice making capacities from 6-50 tons per day, (b) the storage of ice with holding capacities from 20-250 tons, and (c) the ice handling conveyors for discharging at rates up to 50 tons per hour. The ice is harvested on specially designed vertical aluminum plates in which evaporating refrigerant builds up a plate of ice as water is sprayed over the plates. A reverse hot gas defrost cycle causes rapid expansion of the ice on the surface of the plate and, as the sheet of ice falls off, it is broken into pieces approximately 2x3 cm. Ice builds up at a rate, dependent on the evaporation time and the ice thickness is variable from 3 mm to 1.3 cm. The ice falls from the base of the machine

Cylindrical ice silos are available with all the ice handling equipment to store ice in bulk for rapid discharge onto fishing vessels or transport for land use.

Freezing - Air Blast System

Forced convection systems, commonly termed, air blast freezers, may be of the batch, semi-continuous or fully continuous types, all of which use rapidly circulating air at sub-zero temperatures as the heat transfer medium and can incorporate various means for transporting products through the freezing atmosphere. The sphere of application is extremely wide; many plants now being designed specifically for the continuous freezing of fragile products. The method also finds wide employment in the pre-freezing of packaged goods prior to storage.

The method of refrigerated processing is incorporated in a great diversity of plant designs. These include tunnel freezers in which the produce is passed through freezing tunnels on shelved trucks furnished with several tiers of wire mesh shelves which simplify loading and unloading, while permitting the tunnel to be made shorter for a given capacity.

There are also continuous conveyor freezers employed to secure in-line freezers. These includes units fitted with continuous straight belts with fluid box attachment and may be applied to the freezing of fish, shell fish and shrimps, as well as fish fingers etc. Single belt conveyor giving long dwell periods in low temperature air occupy a large volume in the production area. To oviate this, resort may be had to multi-level operation. Transfer between levels may be by short belts to avoid damage of the product and to maintain controlled passage through the freezer.

Spiral Design

Spiral freeze systems operate by continuous air blast freezing, giving high output, while occupying minimum space. The gentle freezing action safe guards the quality of such products as prawns, crabs, shell fish, fish and fish fillets.etc.

In these spiral freezing systems, the belt and its stationary supporting bed are arranged in either an ascending or descending spiral path around a driven drum. This imparts a continuous drive input to the belt, thereby providing a very low tension level and low drive power requirements.

The spiral system is enclosed in a chamber insulated with 127 mm polyurethane fabricated panels. An under floor heating mat is supplied to prevent frosting. In feed and outlet transfer points are provided immediately external to the chamber. This position can be arranged to suit site conditions and ensure trouble-free transfer of products from and to conveyors.

The freezer is obtained complete with a purpose built air cooler design for easy defrosting by hot gas or water, or a combination of both. Air circulation is achieved by the use of axial fans which are fitted with peripheral electric heaters to prevent ice formation between the blade tips and the casing when the fans are inoperative. A variable speed drive fitted to the freezer enables residence time in the freezer to be adjusted to suit any particular product.

Para-freeze spiral freezers are obtainable with belts in 14 widths of between 305 and 965 mm (12-18 inches). Freezers with 305 mm and 455 mm belts are supplied as packaged units, which only require bolting together at site and connecting to a refrigeration unit.

Directly from a production machine or processing line the product is loaded on to a flexible belt which conveys it into the freezing zone. Following a spiral path, the continuous moving belt lifts the material up through the freezing zone. After a set period, determined by the product freezing time, the frozen product is brought out at the desired location. There the belt passes over a discharge roller, ensuring that the material is gentle rolled off undamaged.

The belt returns to infeed via a belt take-up station which compensates for length variations due to temperature changes etc. The belt is driven by friction between its inside edge and the driven drum, the speed of which can be infinitely variable from an electric drive unit placed outside the freezer. Freezing air is circulated by the fans through the cooling coils and horizontally across the product on the belt before it is recirculated. These machines are now widely used in the fishing industry.

Boxa Freezing

A completely automatic in-line freezer with capacities ranging from 1 to 16 tons/hour for the freezing of fish, boxed meat poultry, ice cream, prepared meals and catering packs has recently been developed. Built on a modular principle, allowing capacity increase as required on site with minimum disturbance and relatively low capital outlay, the produce travels through the tunnel on individual carriers powered by a simple hydraulic mechanism which is identical on all sizes of the tunnel.

Product to be frozen is automatically fed on to a carrier, which is then passed through the tunnel using a simple hydraulic pusher mechanism. During the freezing cycle the product remains on its individual tray of the carrier, which passes through the freezing zone. This ensures a constant freezing performance, low power consumption and a uniformly frozen product.

Arriving at the discharge point, the frozen produce is automatically fed off the carrier, which then moves on ready to receive unfrozen material. At all times the carriers remain in the tunnel. The product is only handled twice during the whole process; once on feeding and once when discharged.

The hydraulic mechanism is identical for all sizes of tunnel and incorporates a maximum of eight hydraulic rams, all of uniform design, and a nominal 7.5 hp hydraulic pump and motor. This is the only power source required for the complete operation of the tunnel mechanism. The hydraulic system is completely automatic and simple in design and operation, being controlled by a solid state logic pack.

Fluidized Belt Freezing

Low temperature air is used in fluidized belt freezing, but here the air, instead of flowing counter or cross-current to the direction of the product, as in air blast freezing, is caused to flow upwards through a bed of material. It thus supports, conveys and freezes the produce without any product damage. Fluidized belt freezing is applicable to food particles of a size sufficiently small to be impervious when closely packed

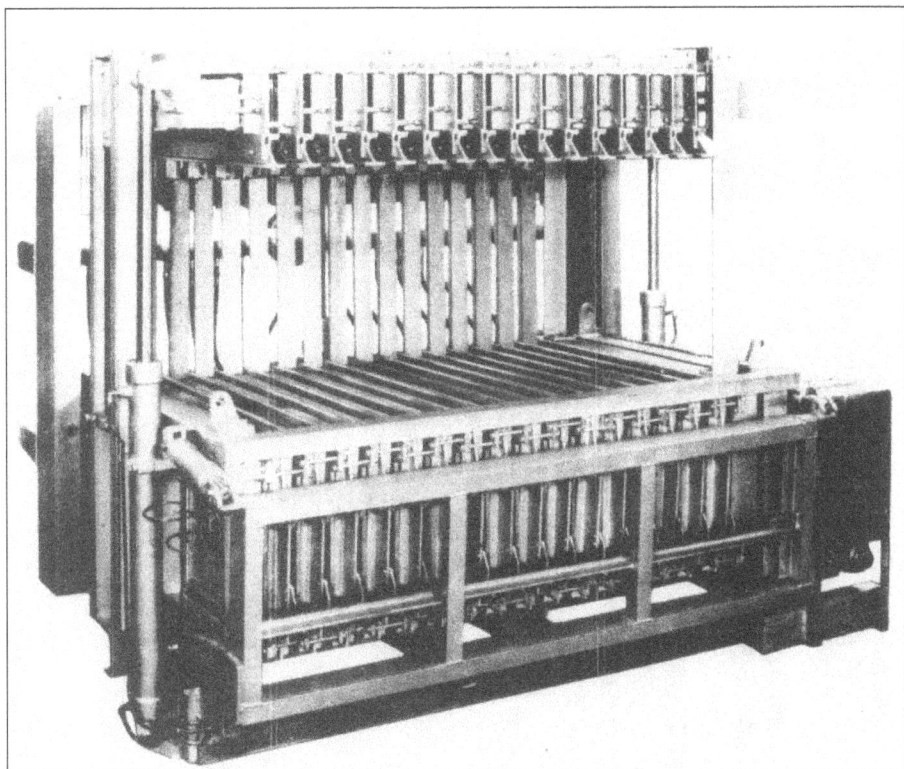

A.P.V. Parafreeze Plate Freezer

and sufficiently large to float on an air cushion as distinct from very small particles which tend to become entrained in air currents.

Processing is suitable for completely light weight bodies, such as, peas, broad beans, sprouts, diced vegetables, mushrooms, potato chips, and all berried fruit. Heavier bodies, such as, carrots, fish fingers and cakes and sausages can be satisfactorily frozen, whilst the application has also been extended to fresh cream sponge, pies and meals-on-tray.

Contact Plate Freezing

With this method, freezing is effected by conduction, the product being brought into intimate contact with metal plates through which sub-cooled brine or evaporating refrigerant is circulated. Freezing is very fast and the plates can be incorporated into plant giving continuous automatic production. Application ranges from ice cream to the freezing of small irregular-shaped produce, such as, small roasting chicken. Contact plate freezers are made with plates vertically or horizontally disposed.

The automatic plate freezer is of horizontal double contact design and has application to a wide range of products, such as, fish packed products, ice cream in packs or cups, packages spinach and sausages and many other products packed in trays, cartons, polythene and paper packs.

A.P.V. Parafreeze Horizontal Plate Freezer

A.P.V. Parafreeze Spiral Freezer

Food products including fish and meat are frozen to minus 20°C (-5°F) at the center of the package in from 20 minutes to one hour, depending on the product, the incoming temperature and the thickness of the package. The freezer automatically loads any flat packages between 18 mm and 75 mm (1/2 inch and 3 inch) in thickness at up to 200 packages per minute depending on size. Packages up to 125 mm (5 inch) thickness can be accommodated at reduced output.

Both plates are in contact with the package under light pressure, which results in a flat uniform pack for easy over-wrapping. A choice of seven standard plate sizes allows maximum flexibility in selection freezer capacities. All units are equipped with quick-lower and fail-safe devices as standard. The primary refrigerant can be controlled as required by an individual float valve with direct mounted accumulator, or by a refrigerant pump and central low pressure receiver.

The freezer and all equipment is enclosed in an insulated stainless steel cabinet. It is designed to operate in normal ambient conditions. These units are available, if required, with automatic unloading and station changing is extra.

As regards to automatic freezer of large capacities, the units are specially designed batch freezers for the smaller processors of packed and cooked foods. These are available in three sizes, are convenient, easily-moved packaged units, ready for installation on any convenient site.

Vertical Plate Freezer

The vertical top loading and unloading freezer vary from 12 stations producing block 101 mm thick to 30 stations producing blocks 51 mm thick. Nominal capacity vary from 4 to 11.75 tons per day.

The vertical freezers enable without the use of trays, the bulk freezing of products into large blocks of uniform dimension. Each standard freezer is specially designed to produce blocks measuring 535 mm x 1070 mm.Facilities for economical stacking on the Euro pallet are also available.

All vertical freezers are fitted with aluminum alloy freezing plates and are designed for operation of pump circulation refrigeration systems using either ammonia, refrigerants R 12, R 22, R 502 or secondary systems using brine. In use the top loading and unloading model is simple to load, as the forks can be lowered to any desired position between the fully raised and lowered points which allows careful placing of products, if required, between the freezing plates. If this is unnecessary, the forks are fully lowered, thus allowing produce to be poured into the stations. Such products as whole fish, fish offal are ideally frozen in the vertical freezer.

A special vertical freezer, fitted with top press assemblies has been developed. This makes it possible to freeze blocks which have accurate dimensions within limits of plus minus 2 mm and parallel square sides.

Heat Exchangers

When considering refrigeration in processing for certain applications, the rotary scraped surface heat exchange plant should not be overlooked. Such equipment can be used not only for cooling and freezing, but also for heating of viscous products, with or without fibrous content, through a wide range of temperatures. The cooling medium can be by means of water, chilled water, refrigerated brine or direct expansion of ammonia or other refrigerants. The working principle is that of bringing a small body of product into contact with large heat exchange surface which is rapidly and continuously scraped,thereby exposing the surface to the passage of untreated product.

The above equipment can be used in conjunction with the following.

Water and Glycol Chilling

The para-freeze zero thermal storage units for economic chilling of water and glycol coolants are available in a range to suit any requirements up to 23,500 kg of ice storage, giving ample and instantaneous supplies of chilled coolants what ever the demand is needed.

The units can be operated on either the fully flooded or the direct expansion system. They consist of high performance ice bank plates submerged in a reservoir of coolant and arranged to form a series of liquid passes. As the refrigerant flows through the plates, ice accumulates on the external plate surfaces to a predetermined depth.

When the chilled coolant is required for duty, it is circulated through the thermal storage unit to the process secondary heat exchanger. Re-entering the unit after its cooling function, the warmed coolant flows through the ice filled formed by the series of plates. Angled baffles on the plates induce turbulence in the flow and ensure that maximum mass heat transfer is obtained between the ice surfaces and the coolant. Since the built-in agitation system has no moving parts, the unit is practically maintenance free.

The main advantage claimed for the zero thermal storage unit is economy, because it is designed to transfer heat through ice; accurate temperature control is un-necessary and the formation of ice on the refrigerated surfaces provides buffer thermal storage, making it unnecessary for the instantaneous load to be matched by the compressor. In addition to the immediate availability of chilled coolant on demand, the system provides for economical operation under any conditions.

For example, where coolant is needed for a limited period, during the day, a storage unit can be sized so that the ice accumulated during off-peak times will supply sufficient mass heat transfer for the entire day. Alternatively, with a duty requiring a large supply of coolant over a long period, off-peak ice accumulation can be augmented during processing by running the refrigeration plant as required. Ice melting is not uniform throughout, naturally being highest at the warm coolant inlet. Where the ice forming surfaces become bare near the inlet, a high heat transfer rate occurs directly between the refrigerated surface and the coolant due to the absence of the insulating effect of ice.

Ice depth is controlled by thickness gauges either manually operated or arranged for automatic setting. When the ice depth reaches the pre-ser thickness figure, the compressor cuts out and restarts automatically when the ice depth diminishes.

The units are equally applicable to glycol coolants. For such duties, the freezing plates are surrounded by open mesh fences to aid ice retention. Calculations assume a 14 per cent propylene glycol solution, chilled to a working temperature of minus 3°C (26°F) and glycol specific heat of 0.96. The packaged units come ready for connection to services and can be obtained for operating with R 12, R 22 or anhydrous ammonia refrigerants as required.

Refrigeration in Fishing Vessels

The principal requirement of all commercial fishermen is to land their catch in the best possible condition and command the highest price possible in the fish market.

Fish flesh deteriorates from the moment of death and retardation of this spoilage can be achieved by making full use of the science of refrigeration.

The first tentative experiments with on-board fish cooling were made in the 1920's, when a number of vessels were fitted with freezing systems primarily designed to handle halibut caught on the Newfoundland banks, which as a prime quality fish justified the additional cost of equipment. Subsequently increases in standards of consumer acceptability resulted in the introduction of the ice cooled fish hold and to economize in ice loss, particularly on steam trawlers where there was substantial heat gain in the fish hold through the engine room bulkhead. Simple refrigeration plants became standard equipment from the 1930's onward.

Ice

The most commonly used refrigeration effect is still obtained from the use of melting ice. Each pound of ice absorbs 144 BTU (0.042 kw) of heat in melting through intimate contact with the product and has the desired effect of reducing heat content and hence retarding spoilage rates.

In its various forms, ice to the fishermen is money spent. Whether purchased from a shore installation or made on board in a marine type of ice maker, the end result is melt water in the bilges. However, the importance of allowing ice to melt over the catch must never be minimized and as market prices will reflect, such effective use of ice will be well rewarded.

Normally, ice obtained from a shore ice plant will be made from fresh bacteria free drinking water. In recent years, the use of clean sea water made ice has become more popular and acceptable. Several models of on board marinized ice makers are available, sized from 1 ton per day production upwards. A typical unit to produce 4 to 6 tons of ice per day of freshwater/sea water ice would require a 25 HP refrigeration compressor and its associated equipment

Preservation of Ice

A worthwhile addition to a fishing vessel using ice, particularly when operating in tropical waters, where the melting of ice can be a serious problem, is to fit a small refrigeration plant capable of cooling the insulated ice storage space (normally the fish hold) to a temperature where ice melting will be contained until the vessel reaches the fishing grounds. A 5 HP compressor with its refrigeration plant working in conjunction with a set of evaporator grids on the deck head of the fish hold, is the usual installation. Fish hold using this system operate between 25°F (-4°C) and 28°F (-2.2°C).

It is normal practice to shut the refrigeration plant off when the catch is onboard to enable the melt water from the ice to cool the fish.

Insulation

A well insulated fish hold is very much necessary. The hold lining must be water resistant to allow for the constant melting of ice and washing down after a voyage. Quite often not sufficient attention is paid to the insulation of the fish hold of a vessel.

Expanded polyurethane slab or foam are in common use, covered by marine plywood. This gives a degree of rigid protection to the insulation. To meet the fire regulations in UK, the practice is to cover the fish hold (on top of the marine ply) with a metal skin, usually sea water resistant aluminium.

Refrigerated Sea Water

The refrigerated sea water method of fish preservation has come to the fore front in recent years. The system consists of a set of built-in tanks in the vessel, occupying what normally would have been the fish hold, constantly circulated by sea water cooled in a flooded shell and tube evaporator, which in its twin is part of a refrigeration system. This method of heat extraction, particularly when handling the fish shoal, such as, mackerel and herring, has proved very effective. Catches of 100 tons and up to 500 tons are handled in this way and the vessels are usually custom built for the fishery.

Storage Temperatures

The effective retardation of spoilage by ice, chilled sea water is limited to approximately 14 to 16 days from catching to eating. The next step in long term preservation of fish using refrigeration is low temperature freezing and storage.

While defining the standard for freezing fish, which will produce an acceptable product, deep freezing is a general term establishing a process where the product is reduced and maintained to 0°F (-18°C) or lower; it does not specify the time of temperature reduction. Quick freezing with which the industry is particularly concerned as the correct process for cooling fish is defined in the United Kingdom as a process whereby the whole of the fish is reduced in temperature to 23°F (5°C) in no more than 2 hours and the fish is then retained in the freezer for a further period of time, until the core of the product is reduced to minus 5°F (-20.5°C) or lower. The accepted United Kingdom standard for the long term storage of fish is minus 20°F (–29°C).

Methods of Heat Extraction

For shipboard application there are three distinct processes used for quick freezing fish, namely, contact freezing, immersion freezing and blast freezing.

Contact Freezing

Early installations of horizontal contact plate freezers involving essentially adaptations of machines which had been in use for many years on land installations. With the advent of vertical plate freezer, developed by J and E Hall Limited in association with, Torry Research Station, large tonnage throughput installations onboard ship became practicable. The decision as to use of horizontal or vertical plate freezer will depend to a large extent on the type of fish being caught and the market being served. Essentially the present practice tends towards the use of vertical plate freezer for processing whole, head on gutted fish, giving frozen blocks of whole fish of 42" X 21" X4" (1067 mmX533 mmX102 mm) weighing 45 to 50 kg.

The blocks when frozen have a skin temperature of minus 29°C and a core of minus 20.5°C. For more specialized fish products for example, fillets, prawns etc and particularly in cases where variable block thickness is required, the horizontal plate freezer is superior.

Immersion Freezing

There are limitations to the extent to which immersion freezing can be used, principally due to the fact that the method involves use of a salt solution and, therefore, must be limited to products where there is a small or at any rate limited salt absorption. Immersion freezing is primarily used in various forms for crustacean, such as shrimps and prawns, where freezing to minus 15°C is carried out by purpose built immersion freezers filled with a sodium chloride brine solution, usually with corn oil or dextrose added to give a glaze.

Particularly for the Mexican market an immersion freezer of GRP construction, heavily insulated with polyurethane slab has been designed which incorporates a stainless steel freezing coil and brine agitator. Due to the limitations of power supplies on small vessels the agitator has been designed either for drive from an electric motor or from a small water turbine. In the later case, the circulating system is designed so that the immersion refrigeration condense- sea water pump is used to supply water both to the condenser and to the agitator. These units, primarily simple in concept, have proved to be well adapted to artisanal/cooperative fisheries.

Possibly the biggest area for the application of immersion freezing techniques, although these cannot strictly be regarded as quick freezing systems are onboard tuna vessels, where the standard design involves a number of separate tanks being arranged as part of the ship's structure, each being cooled by a set of direct expansion cooling pipes. In these vessels, the tanks are filled with sea water to which salt is added, pre-cooled to 30/40°F (0/4°C) and the fish is caught and put directly into the tanks. The cooling process on these vessels are relatively slow 12 to 14 hours possibly to 23°F (–5°C), but the nature of tuna is such that it can tolerate a slower cooling temperature than most other types of fish.

Blast Freezer

The blast freezer, as far as the United Kingdom is concerned has not been extensively adopted as a practical method of fish handling onboard ship. The reason for this may well be in the fact that for northern water fish, the plate freezer has proved to be highly satisfactory and, therefore, the greater amount of space occupied by a blast freezer for an equivalent tonnage throughput, together with handling problems has militated against its extensive use.

However, for vessels operating in warmer waters, there are classes of fish which are more readily amenable to blast freezing than contact plate freezing, both with respect to the size of fish and the ability of the fish flesh to withstand compression in a plate freezer. Blast freezers in general, need to be considered almost as part of the ship's structure and the arrangement of accommodation and equipment designed around specific areas into which blast freezer can be built. There is a penalty always to be paid with the blast freezer in that compared with a contact plate freezer, that there is a substantial fan heat load which has to be handled by the refrigeration plant and frequently a greater throughput in an equivalent area can be obtained with a contact freezer due to the shorter time cycle and ability to split the throughput over a number of separate units.

Frozen Fish Storage

The accepted U.K. temperature for the long term storage of quick frozen fish is minus 20°F (- 29°C). The option exists as far as the storage compartment is concerned

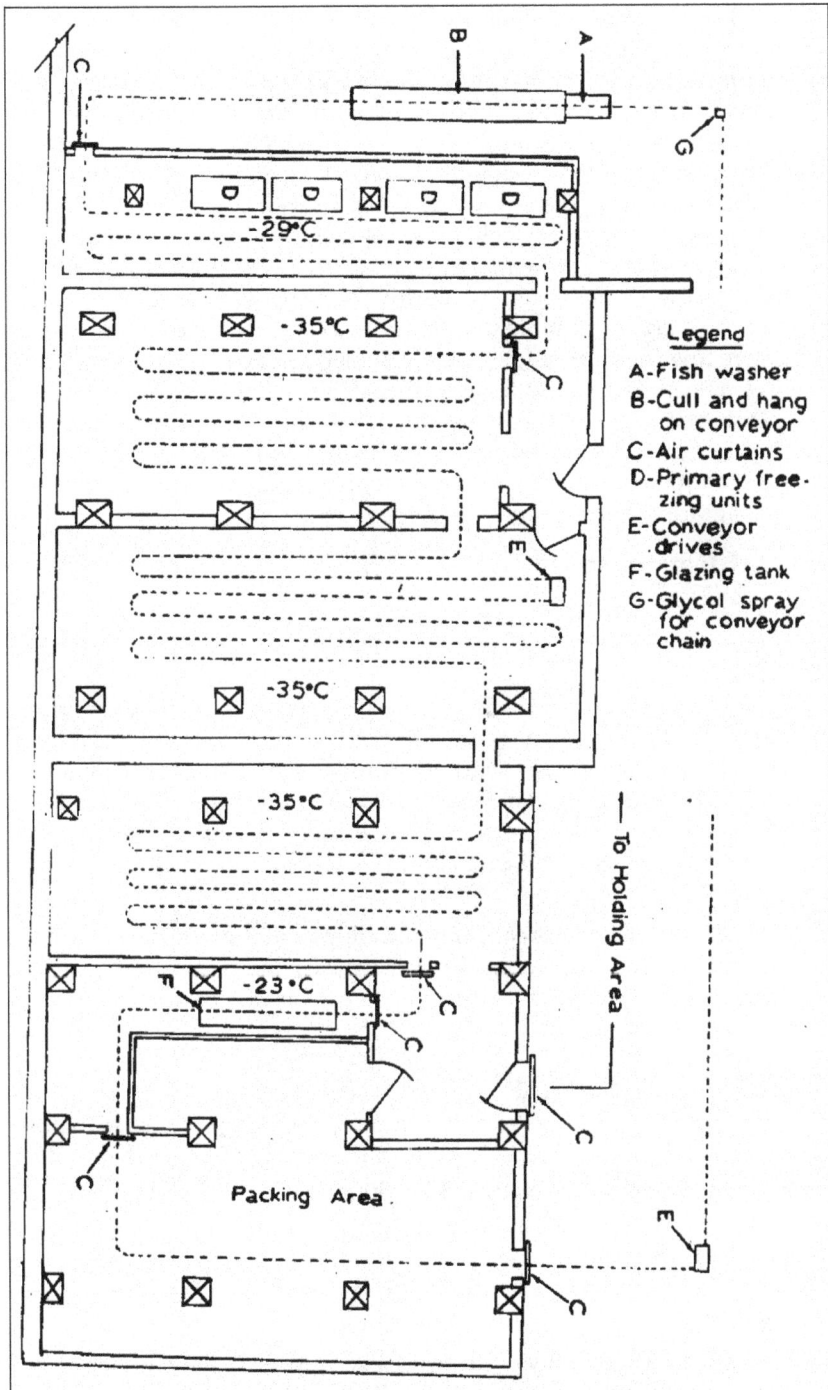

Legend

A-Fish washer
B-Cull and hang on conveyor
C-Air curtains
D-Primary free-zing units
E-Conveyor drives
F-Glazing tank
G-Glycol spray for conveyor chain

General Layout–Conveyorized Blast Freezing

to utilize extended surface cooling grids or plates or an air cooler with fans. The preference is for the former method. Experience has shown that in a properly managed vessel there is not a large ingress of moisture in the fish hold once it has been cooled to temperature and the frost build up over an average period of four to six weeks is not such as to significantly reduce the cooling effect of the grids or plates. Preference is expressed from time to time by operators for air cooled systems. It is however, believed that these have significant disadvantages which are not always appreciated. In the first instance, particularly on smaller vessels there is a limited electrical supply, generally at a non-standard voltage, which raises problems in selection of fan motors for operation at low temperature. Also the frost build up which is spread evenly over a large surface of grid is far more significant on a small extended surface air cooler and therefore, means must be devised for cooler defrosting. This involves either additional electric power for heaters and the associated problem of satisfactory cooler drains or the introduction of spray water piping into the hold for defrosting purposes. There is a risk of uneven temperatures resulting in product spoilage in the fish hold unless specific attention is paid to adequate air distribution either by the introduction of ducting or grating beneath the cargo through which air circulates.

Choice of Refrigerants

The choice of refrigerants is still not large.

Ammonia

Although extensively used on land for many years, has never been particularly popular as a marine refrigerant at least in U.K. although it is still used by many European operators, and it is also extensively used by the American fishing vessels. Although, it had its attraction from the point of view of ready availability, a generally satisfactory performance as a refrigerant over quite a wide range of operating temperatures, these must in all cases be weighed against toxicity and explosion hazard.

Further, the inability to use copper based alloys with ammonia invariably raises problems with heat exchange equipment in a way of condensers, which must either be designed on a throw away basis or a risk of extensive repairs being required at regular intervals must be accepted.

Halogenated Hydrocarbons

To all intents and purposes therefore, the refrigeration equipment for marine use is now dominated by the halogenated hydrocarbon range of refrigerants or they are more generally referred to using the original trade name as the "Freon".

For the general purpose applications using reciprocating compressors and screw compressors, whether it be for cargo refrigeration, provision chamber work, or air conditioning, the choice lies with three of these gases, Refrigerant 12, Refrigerant 22 and Refrigerant 502.

Refrigerant 12 was initially used almost exclusively on the grounds of general suitability, availability and cost, as until the 1960s, Refrigerant 22 was available only to a limited extent and was substantially more expensive.

In fact, Refrigerant 12 as a refrigerant performed satisfactorily in single stage applications for temperatures down to + 5°F. It has a slight disadvantage compared with R 22 in that it has a greater specific volume and therefore, for equal refrigeration capacities, a greater swept volume is required using Refrigerant 12 than is the case with Refrigerant 22.

Also as the operating temperature is reduced and therefore the suction pressure, this latter value falls off more rapidly with Refrigerant 12 to below atmospheric pressure, which can result in air being drawn into a system through glands, stop valve seals etc. Against this, Refrigerant 12 however, has lower operating temperatures which can result ib greater compressor reliability and latterly, it is seen that introduction of Refrigerant 502, which offers a compromise solution between the performance of R 12 and R 22 in having similar operating pressures to R 22, but such lower discharge temperatures as the illustrations show.

It is however, important always to bear in mind about the cost. All figures for costs are difficult, particularly as they may relate to the availability at any one place or any one time. But in general terms, if the price of R 22 is taken as 1, the price of R 12 as 0.6 and the price of R 502 is 2.0.

Selection of Refrigerating Machinery

While selecting refrigeration machinery, sufficient space must be provided for the equipment when the vessel is designed. Even for a wet fish hold installation, the refrigeration equipment occupies 30 to 50 cubic feet of engine room space and absorbs 3 to 5 kw's of power. On moving to quick freezing systems, it must not be forgotten that as far as power is concerned, one ton per day throughput and storage is very approximately equivalent to a load on the ship's generators of 6 to 7 kw.

It is recommended that for the small vessel installation where a wet fish hold of 30 to 50 cubic meter is maintained at 32°F and also for the small shrimp boat type of installation with a throughput of 0.25 to 0.5 tons per day, a simple direct expansion, Refrigerant 12 system be selected, taking the advantage of the lower operating temperatures associated with R 12 as a refrigerant.

The original plate freezer installations for large tonnage throughputs in the United Kingdom used Refrigerant 12 and a pump refrigerant circulation system. Due to the concern regarding possible losses of gas a backward step was taken and Trichloroethylene was introduced as a secondary refrigerant. At the same time the increased tonnage throughput justified the use of Refrigerant 22. In this case, the Trichloroethylene was circulated through direct expansion or flooded evaporators and then through plate freezers and fish hold cooling appliances. The pattern has now been reversed and the more recent large tonnage plate freezer vessels have reverted to pump circulation systems again using R 22, obtaining economics in size of plant. The adequately designed R 22 direct expansion pump circulation system can operate for many years with nominal loss of refrigerant provided correct maintenance procedures are adopted.

To give examples of the two different approaches on plant design the current largest throughput block freezer vessels have a throughput of 60 tons per day and incorporate 3-8 cylinder R 22 compressors having 120 kw driving motors for the freezing duty and an additional 65 kw motor compressor set to handle the hold of 42000 cubic feet capacity maintained at minus 29°C.

The alternative approach for a 24 ton per day vessel using direct expansion pump circulation systems incorporates 2-65 kw motor compressor sets and 1-40 kw motor compressor set.

For both systems compound compressors are used and additionally both for the Trichloroethylene as well as for the R 22 systems, hermetically sealed, glandless pumps have been employed.

The Compression Machine

Compression machines provide dependable, economical refrigeration at all evaporative temperatures. The wide variety of refrigerants that can be used, together with the low space requirements and versatility of drive, makes the compression machine well suited for marine operation.

If ammonia is used as the refrigerant, the compressor should be located in a separate well-ventilated compartment, with a sprinkler system in order to minimize the danger to personnel in event of ammonia leakage. Even though ammonia compressors furnish a supply of efficient refrigerant, with little or no oil return problems, their use is some what limited on small fishing vessels, where space is at a premium.

Freon 12 or Freon 22 compressors can be located in the engine room of the vessel, thereby making possible the utilization of excess space. In designing a Freon compressor system the designer must take care to properly size the evaporators and the suction lines in order to ensure satisfactory oil return. Suitable oil separators should also be used.

A compression machine on the fishing vessels can be driven by (i) a direct current motor, (ii) a diesel-engine coupled to the compressor by a clutch or a flexible coupling, or (iii) a connection off the main engine.

For small brine-cooling installations, operation of the compressor by a separate diesel engine or by a connection off the main engine provides satisfactory operation. In larger installations the use of diesel-electric generating unit and direct current motors is favoured because of small amount of space used and the versatility of operation provided.

Capacity- control for variable-load operation is obtained on many compressors by using suitable unloading device or by reducing the speed of the unit. In general, however, the capacity reduction is limited to 50 per cent of the maximum compressor capacity. Therefore, on brine cooling and freezing systems, where loads less than 50 per cent of the maximum capacity are to be experienced two or more compressors must be connected in parallel, resulting in increased space requirements per ton of refrigeration.

Changes during Freezing of Fish

When fish which have been frozen and held in cold storage for an extended period of time are thawed and examined, a number of changes are found to have been taken place, which differentiate them from unfrozen fish. The flesh of thawed fish will consist of two phases, solid flesh plus a fluid known as drip which was not reabsorbed by the flesh after thawing. The texture of the thawed fish will be soft and additional drip can be expressed by applying a small amount of pressure, the surface may have become desiccated by loss of moisture. The colour of the surface may have altered in some way. After cooking, the fish may be found to have acquired an off-flavour, or the normal flavour may merely be lacking. The cooked fish may be tough or fibrous.

These changes from the condition of the original, unfrozen fish are sometimes said to be due to the freezing process. Actually, almost all of such changes are caused, not by the freezing of the fish, but rather by the subsequent cold storage of the frozen product. If fish are rapidly frozen and then immediately thawed and examined, very few changes are noted, and those changes do occur are of a very minor in nature Such a fish, frozen and then immediately thawed generally exhibits a very small amount drip, and the texture of the flesh is usually slightly softer than that of the fish that never been frozen. A careful examination may reveal a sort of honeycomb appearance caused by spaces being left where ice crystal had formed. Such a condition would be noted if the fish had been frozen very slowly, as may occur with large fish. In case of fish frozen in high velocity air without a protective covering some evidence of surface desiccation would also be present.

Properly frozen and immediately thawed fish can not be distinguished by the average person from that never were frozen. Even an expert might be unable to distinguish such fish after they had been cooked. Although the changes in fish brought about by the freezing process itself are very minor, nevertheless changes do take place to the manner in which fish freeze.

Temperature Changes during Freezing

Fish do not possess a well defined freezing point like water or other pure chemical substances. Rather they freeze over a range of temperatures, and while most of the water in the fish is frozen at 18°F, a small part remains unfrozen even at temperatures far below 0°F.

A fish first starts freezing when its temperature is lowered to about 30.3°F. Freezing begins at the outside of the fish and progresses inward, the center of the fish being the last part to freeze. If thermo-couples are placed at the outside and center of a fish and temperatures recorded as the fish freezes, it will be found that a much longer time elapses before the center freezes than does the surface layer. This is due to the fact that the cooling medium is in contact with the external surfaces of fish. Therefore, the zone of frozen tissue progresses from the outside surface of the fish to the center, so that the tissue at the center does not start to freeze until long after that at the outside surface has frozen. The increased time required to freeze the center of the fish in a particular type of freezer is a function of the thickness of the fish. The fact that the moisture in fish freezes over a range of temperatures rather than sharply at a single temperature gives rise to a curve when temperature is plotted against the freezing time. The time during which most (about 85 per cent) of water is freezing has been termed the critical zone, and it has been suggested that the shorter the time required to pass through this zone, the less damage is done to the fish.

Texture Changes Resulting from Freezing

It has long been known that the fish which are frozen at a very slow rate show a greater difference from the fresh unfrozen fish than do fish frozen at a rapid rate. Thus slow freezing fish tissue yields upon thawing a somewhat greater quantity of drip than does quick frozen fish tissue. The size of ice crystals which form in slow freezing fish are larger than those in quick-frozen fish. Several theories have been proposed to account for these differences caused by different freezing rates.

Reuter's Theory

One of the earliest and still one of the most comprehensive investigations on the effect of the rate of freezing on changes in fish tissue was carried out by Reuter (1916). In all colloids, the time required for inhibition and the reversal of this process is much longer than in bodies having a crystalloid structure. The separation of water will be more complete, the more slowly it is cooled down to the point where it solidifies. Conversely, the separation of fluid will be lessened in proportion to the rapidity with which the temperature of solidification is attained. By using liquid carbon dioxide, small pieces of muscle could be so rapidly frozen that no changes were produced. Freezing took place within one or two seconds, and sections only showed the effect of freezing on their edges, the greater part of the tissue presenting the appearance of normal unfrozen muscle. No exudation of fluid into the connective tissue, and no aggregation into columns within the separate fibers could be seen.

The conclusion from the above observations is that in very rapid freezing, the water of the muscle albumin freezes in an individual molecular state. In less rapid freezing, a number of small columns of fluid are formed in each muscle fiber, and if time permits, these fuse to form a single column. In still slower freezing, the fluid ruptures the sarcolemma and escapes into the connective tissue, forming large space filled with ice.

Thus Reuter believed that the difference in the condition of quick- and slow-frozen fish was primarily due to colloidal differences caused by different rates of lowering of the temperature below freezing. Fish which pass through the critical zone most quickly undergo the least change in their colloidal state, and when thawed such fish more closely resemble fish which has never been frozen than does fish which, in freezing has passed through the critical zone more slowly.

Cell-Puncture Theories

A some what different theory was advanced shortly after the work of Reuter (1916). He had shown that, in extremely slow freezing, the sarcolemma (the thin layer of tissue covering the muscle fibers) is ruptured to allow passage of moisture, which then freezes between the muscle fibers. The idea of mechanical damage of the fish tissue was greatly expanded by subsequent workers. A theory was built up to account for the formation of drip in thawed fish solely on the basis that formation of ice crystals within the cells ruptured them and allowed moisture to pass through and then freeze outside. It was postulated that upon thawing, only a portion of the moisture returned to within the cell. The remainder was held in the fish tissue like moisture in a sponge and would drip out or could be expressed by mild pressure. Since slow-frozen fish has larger ice crystals than do those that are fast frozen, it was assumed that these lager crystals did correspondingly greater damage to the cell walls resulting in formation upon thawing, of larger quantities of drip. This theory seemed to be confirmed by photographs of sections of fish tissue. For many years belief in this theory was wide spread, especially by authors of popular articles in which an attempt was made to explain the advantages of quick freezing to the lay public.

Recent Theories

Recent work seems to have disproved the theory of mechanical damage to cells by crystals formed during freezing, or at least to indicate that such changes are of secondary importance. Most of the photo-micrographs used to support this theory were taken after the tissue had been prepared by the usual histological techniques involving use of a fixing agent followed by imbedding with paraffin and cutting of sections with microtome. Lebeaux (1947) has shown that when photo-micrographs are taken with the polarizing microscope where none of these histological techniques is employed and where pictures of the tissue can be taken while the fish is still frozen or even while the freezing process is going on, there is no evidence for the cell-puncture theory. Although large crystals were formed when fish tissue was frozen, no damage to the cell wall took place. The cell walls were found to stretch and deform during freezing, but no actual rupture occurred. Apparently, damage to the cells previously demonstrated by photo-micrographs had occurred, not as a result of freezing, but rather as a result of the histological preparation.

A new chemical technique for investigating tissue alteration caused during freezing has recently been proposed by workers at Torry Research Station, British Food Investigation Board (Love, 1955). This technique is based upon a determination of the quantity of desoxypentose nucleic acid (DNA) occurring in the press juice of the thawed fish. DNA occurs only in the nucleus if the cells are unaltered. Hence, it is reasoned that its concentration in the press juice will be a measure of the extent of liberation of nuclear material into the interstices. Concentration of DNA has been shown to vary in a rather irregular pattern with rate of freezing. The technique offers a considerable promise for a better understanding of the nature of such changes can be reached.

Effect of Ice Crystal Size

There can be no doubt that the size of ice crystals which form during freezing of fish ia function of the rate of freezing. The crystal size was determined by the thickness of fish or portion of the fish frozen and by the rate of freezing, but not by freshness of the fish. There is no proof, however, that any extensive damage to the fish (*e.g.* marked in crease in drip after thawing) is due to the larger crystal obtained by slow freezing.

In cases however, where extremely slow freezing has resulted in formation of very large ice crystals, the thawed fish may have a porous texture with tiny holes appearing in the flesh, in some cases approaching a honey-comb like appearance.

Quick Freezing

Fish as well as prawns contain about 80 per cent water. The physical change which occur during freezing fish comprise formation of ice with expansion of volume and desiccation starting from the frozen fish with the consequent damage of muscle cells and concentration of minerals damaging the proteins and irreversibly altering them. During freezing, protein water gel is completely altered because the water separates out as pure ice, leaving the proteins more or less dry. Formation of ice is initiated when the temperature of fish is lowered to about 1°C. There is an increase in the volume of fish when it is frozen. At 3°C about 70 per cent of the water is frozen, at minus 5°C about 85 per cent, at minus 25°C about 95 per cent and at minus 40°C almost all the water in the fish is frozen. Thus the largest part of water freezes between minus 1°C and minus 5°C and this zone is called the zone of maximum crystallization and it is the rate of cooling during this temperature interval which determines the size of ice crystals.

Quick freezing means generally that the temperature of every part of the product falls below the zone of 0° to minus 5°C as rapidly as possible. Rapid freezing results in the formation of small ice crystals. As the freezing time is shorter, less time is allowed for the diffusion of salts and evaporation of water. Besides decomposition is prevented during freezing, bacterial built up will be considerably reduced too by quick freezing. On the other hand, slow freezing results in the formation of large ice crystals. These can cause the tissue of fish to become more porous and it may even become spongy. Thus the rate of freezing has been considered very important for obtaining high quality product. Microscopic examination reveals that the slow freezing result in greater destruction of the tissue than quick freezing. Theories of the effect of freezing led to the conclusion that large crystal formed by slow freezing are able to penetrate the cell walls resulting in large drip loss when the fish is thawed than when the fish is quick frozen.

When the frozen block is thawed, it loses weight in the form of thaw drip. Thaw drip is defined as the exudates of tissue fluids that flow free from the muscle during thawing of frozen fish or muscle. Drip leaches along with it soluble proteins, vitamins and minerals. Drip was regarded, occurring as a result of cell damage caused by the freezing. It was popularly believed that quicker the freezing rate, the lesser will be the thaw drip, there by bringing about rebuilt quality of the product. It was established that freezing of shrimp meat as blocks in lesser time, does result in lesser thawing losses than slowly frozen product. Quick freezing definitely brings about again the quality of the product

It is widely understood and practiced that quick icing, quick processing and quick freezing if all done efficiently will definitely improve the quality of the product. A block of prawn, if it can be frozen very near to zero time will yield the least drip, which all strengthens the theory that quick freezing definitely is better as far as the quality of the product is concerned.

Consumers look certain factors in the product that he buys, like nutritive value, degree of spoilage, intrinsic composition, damage, deterioration during processing, storage, distribution, sale and presentation to the consumer, hazards to health, satisfaction on buying and eating, aesthetic considerations etc. So for quality for profit, quick freezing and good manufacturing practices will be the only answer.

Advantages of Quick Freezing

In the early development of the frozen fish industry it was believed that one of the major factors in producing a frozen product of proper texture was that the product must be quick frozen. This belief rested upon certain hypotheses which have since proved to be untenable. When the texture of a large fish such as, 150-pound halibut which was frozen in still air is compared with that of a one pound fish fillet, which was quick frozen in a plate freezer, a very definite difference in favour of the texture of the quick frozen product is found. This difference is large only in case of such extremes. Differences in texture, immediately after freezing of the same type of product frozen by standard commercial quick- and slow-freezing techniques are small, sometimes indistinguishable.

This does not mean that slow freezing is as good as quick freezing. Quick freezing possesses several very important advantages over slow-freezing, which are in no way connected with actual changes brought about by the freezing process itself.

Rate of Chilling

Fish to be frozen must first be chilled to the freezing temperature before they will start to freeze. When large quantities of fish are placed in a freezer, it takes some definite period of time for their temperature to be lowered to the freezing point. The center of large fish will require longer period of time for this cooling to take place before freezing can commence than will the outer portions. During this cooling period, bacterial spoilage can proceed.

In cases where the initial temperature of the fish is high and either where abnormally large quantities of fish are involved, or where the refrigeration capacity of the freezer is too low, this chilling period may be quite long- from a few hours up to several days. In the latter case, extensive spoilage can occur while the fish is in the freezer, but before freezing commences. Such spoilage has occurred to an extensive degree in commercial operations, where tuna or salmon have been frozen aboard the fishing vessel in brine at too slow a rate because of inadequate refrigeration capacity or overloading of the brine wells with excessive quantities of fish. In commercial operation ashore, fish have spoiled where excessively large batches have been placed without pre-chilling, in freezer rooms without adequate air circulation. Spoilage in such cases occur locally at the center of a stack of cartons or fish. Quick freezers are far more efficient than slow freezers in rapidly lowering fish temperatures to the freezing point, and they thus minimize spoilage.

Refrigeration Capacity

At periods, when large quantities of fish must be frozen in a short time, the availability of quick freezing equipment may have a very real advantage from a production stand point. Moreover during periods when unusually large quantities of fish must be frozen, the possibility of over loading slow freezers is greater than the quick freezers. In some instances, fish within the slow freezer are packed together so tightly that the circulation of air is greatly restricted, resulting in some of the fish not being frozen when removed from the freezer.

Some times, fish are slow frozen by merely placing them directly in cold storage rooms containing fish which are already frozen. This practice may lengthen considerably the period of time required for the product to freeze and possible product spoilage in extreme cases. Also if a sufficient amount of refrigeration is not available to compensate for the increased load of warm fish, a drastic rise in the temperature of the room may occur. This rise will damage the rest of the frozen fish within the room by their being held a too high a storage temperature.

Freezing Under Pressure

A third advantage of some types of quick freezing equipment rests in the possibility of freezing the fish under pressure. Packaged frozen under pressure in plate type of quick freezers possesses greater storage life, owing to the elimination of air voids from the packages. This can be a very important factor in favour of such plate freezing when dealing with certain species of fish which are subject to rapid oxidation. Quick freezing also prevents buckling of the packages.

Choice of Freezing Equipments

Quick freezing equipment is always to be preferred to slow-freezing installations and whenever at all possible the quick freezing installation should be employed. In general, freezers employing a freezing medium other than air (*e.g.* brine), those using contact of the product with plates from two sides and under pressure, and those utilizing blasts of air are classified as quick freezers. Stacking fish, in storage-type rooms in still air results in slow freezing. The practice whereby fish are frozen by placing them on shelves made up of pipes or plates containing the refrigerant medium (sharp freezers) but without rapidly circulating air may result in a relatively quick freezing or a slow freezing, depending upon the product being frozen and the manner in which it is stacked in the room. If small packages of fillets are placed directly on such sharp freezer shelves and the shelves are not overcrowded, it is possible to attain a relatively rapid rate of freezing. On the other hand, when such rooms are overloaded with large fish, a very slow rate may result. Quick and slow freezing are only relative terms with no sharp line of demarcation separating them. Similarly freezing equipment does not always fall into one or other of these categories. The use of same equipment in one way with a certain product may result in quick freezing with another product handled otherwise, in slow freezing.

Thawing and Refreezing

From a theoretical point of view it would seem that fish thawed slowly might return to the condition of the unfrozen state more closely than if they were thawed rapidly. A slow thawing might give the water from the melting ice crystals in the tissue a longer time to be reabsorbed than a more rapid thawing. This question remained unsolved. Apparently, the rate of thawing is not a major factor in the determination of the completeness of re-absorption of fluids by the tissue.

A distinction exists between the rate of thawing and the temperature of thawing. While it is true that a sample of fish thaws more rapidly at a high temperatures than at a low one, it is possible to thaw samples at different rates at the same temperature. Frozen fillets of cod were thawed in air and in water at 34°F and at 80°F, and the total free drip was determined in each case. When the quantity of free drip was compared in samples thawed at the same temperature in air and in water, the rate of thawing made very little if any difference in the amount of drip forming in the thawed fish. On the other hand, the temperature of thawing made a considerable difference. Free drip formed at 34°F averaged 5.5 per cent as compared to 11.4 per cent at 80°F.

When fish is frozen, thawed, refrozen and thawed again, the drip is greater after the second thawing than after the first, but the difference is only a small one. Successive refreezing and thawing makes for less and less increase in drip as the process is continued. Accordingly, it is entirely feasible to thaw whole frozen fish, cut fillets, package and refreeze without any important deterioration in quality of the refrozen fillets resulting from the second freezing.

Salt Absorption

Whole fish are often frozen in brine. When this is done, salt is absorbed by the fish during the freezing process. The extent of salt absorption depends upon a number of factors, one of which is the species of fish. Non-oily species like cod and haddock absorb salt more rapidly and to a greater extent, especially after freezing has been completed than do oily ones, such as, tuna and salmon. Investigations on the factors controlling salt penetration into whole fish have shown that the temperature of the brine is an important factor governing salt absorption. About three times as much salt is absorbed in the ¼ inch outer layer of haddock when frozen at 15°F, as when frozen at –6°F. Another important factor is the brine concentration. Under certain conditions, 1.2 per cent salt was absorbed when the fish were frozen in 22 per cent (eutectic) brine. Under identical conditions, except that a brine of 15 per cent concentration was used, 0.72 per cent salt was absorbed. The amount of salt absorbed increases with the length of immersion time and brine continues to be absorbed even after the fish have been completely frozen. In one test, haddock werefrozen in 23 per cent brine at 5°F, and the amount salt absorbed in the outer ¼ inch layer of meat was measured. After 1, 2, 4, and 24 hours, the amount of salt found was 0.5, 0.9, 1.2, and 2.1 per cent respectively. After 10 days in brine, 10.6 per cent salt had been absorbed.

In handling oily fish like salmon, tuna, the rate of absorption of salt is much slower. Particularly after the fish has been hard frozen, little or no salt penetrates into the flesh. The salt content of chum salmon frozen in brine and held after freezing for two weeks in the brine was only 1.1 per cent. It is common commercial practice aboard tuna clippers to freeze and hold frozen tuna in brine for several weeks. Salt uptake of tuna is relatively slow at least compared to the rate of salt absorption by haddock under similar conditions. Salt absorption is however, sometimes a considerable problem even with oily species of fish. In recent years the average size of tuna taken has decreased. This decrease in size has resulted in a greater salt absorption (due to greater ratio of exposed surface to weight for small fish) to such an extent that some small tuna being landed contain so much salt that, even after usual thawing and pre-cooking products, the ultimate canned tuna retains salt in excess of what is desirable from the flavour stand point.

Other Changes Caused by Freezing

Fish become dehydrated during the freezing process. Products protected by packaging materials or frozen in still air do not dehydrate to any significant extent during freezing, but in cases where whole fish are frozen in blast freezers, excessive surface dehydration usually occurs. This dehydration occurring during freezing is termed "freezer burn". Sometimes freezer burn is erroneously attributed to exposure of fish to too low a temperature. Actually this condition results only from loss of moisture from the surface of fish and is not in any direct way associated with freezing temperature. Of course, very cold air may have a lower relative humidity which in turn will cause more moisture to be lost from fish frozen in a blast of such air than if warmer air having a higher relative humidity had been used.

Cryogenic Freezing

Nitrogen is an inert, colourless, odourless, non-toxic, sparingly soluble, tasteless gas, which in its free state forms 79 per cent of the atmosphere. By itself it will not support life or most microbiological action. In its liquid state it evaporates at minus 196°C. This extremely cold liquid can be brought in direct contact with food, thus enabling high heat transfer, giving high refrigeration efficiency. It is therefore refrigerant. The liquid nitrogen (LN) process enables the lowering of the product temperature to an incredible low range of – 35°C or below easily and quickly, something unlikely in mechanical systems.

Refrigeration potential of liquid nitrogen = Latent heat of vaporization (47.51 KCAL/kg) + Specific heat (0.25 KCAL/kg/degree C rise in temperature).

Liquid nitrogen chilling and freezing of food has a great advantage of producing a rapid drop in temperature, so foods spend the minimum time in the bacteriological danger zone of + 60° to + 5°C. It also retards the oxidation enzyme activity, and metabolic deterioration. Nutritional and vitamin properties of food are thus preserved intact.

Liquid nitrogen is safe. Ozone layer depletion resulting from the use and leakage of chlorofluorocarbon (CFC), the compound used to produce cold in mechanical refrigeration system is a serious hazard to mankind. The use of liquid nitrogen for refrigeration, while producing more intense cold than any other commercial refrigerant, has no ill effect on environment and mankind. Liquid nitrogen is the answer as the hour approaches for the drastic reduction and ultimately complete suspension of use of CFC for refrigeration, thus leaving the ozone layer in the atmosphere in tact to fend off the harmful rays.

Ease, speed, quality of freezing and flexibility of operation has made liquid nitrogen the prime choice of food processors world-wide.

Liquid nitrogen freezing reduce dehydration and drip loss. Dehydration may be a serious factor when dealing with high value and delicate seafood. Mechanical refrigeration (air blast IQF) can cause moisture loss of 3 per cent to 4 per cent and in some cases even higher. Liquid nitrogen process reduces this loss to 0.1 per cent to 1 per cent, an important economic consideration for specialist food processors. The liquid nitrogen process subjects the food product to liquid nitrogen spray for a short time, while in mechanical systems, the product is in a blast of much higher volume of cold air for a longer time; hence the increased loss of product moisture.

Osmosis and ice crystal growth cause damage to cell walls and both are governed by the time taken to freeze the product. The quicker the product is frozen, more rapidly the effect of osmosis is arrested, resulting in formation of smaller and more uniform ice crystals, which do not damage the cell walls. So the loss of inter-cellular fluid (drip loss), along with it the loss of protein and vitamin is minimized, when quick freezing is achieved using liquid nitrogen as compared to mechanical refrigeration. Typical freezing time for shrimp using liquid nitrogen is less than 10 minutes as against 30 minutes to three hours for mechanical air blast units.

Reduced bacteriological deterioration, minimum deterioration as food spends minimum time in the bacteriological danger zone of $+ 60°$ to $+ 5°C$.

The nutritional and vitamin properties of food are preserved intact by reducing the rates of oxidation, enzyme activity and metabolic deterioration.

Ensures maximum market value and market acceptance of frozen product by retaining original colour and better texture.

Liquid nitrogen can freeze marine products (shrimps, value added fillets and steaks) by individually quick freezing (IQF) food to a core temperature of $- 30°C$ or below.

Methods of Application and Freezing Equipments

Basically there are three methods of liquid nitrogen application for providing refrigeration in the food industry.

(a) Spraying of liquid nitrogen on the product, equipment used is a tunnel freezer.

(b) Vaporizing the liquid in a forced draught, and blow the cold vapor over food, equipment used is a spiral freezer or batch freezer

(c) Immersion of the product in liquid, equipment used is immersion freezer.

Spraying of liquid nitrogen directly on the food in a tunnel freezer is the most efficient method of using liquid nitrogen for freezing a wide variety of food product. The freezing tunnel is the best answer for quality IQF processing, especially for the frozen shrimp export industry.

Freezing Tunnel and its Operation

The cryogenic freezing tunnel consists of an insulated stainless steel chamber placed on a steel frame.

Food product is carried through the tunnel on a stainless steel mesh conveyor, operated on a variable speed drive, speed depending on the product thickness, desired product temperature and throughput required.

The liquid is injected on to food product on the conveyor. Liquid nitrogen on contact with the food evaporates, drawing the necessary latent heat from the food, causing freezing in a short time (freezing zone). About 50 per cent of cooling is achieved in this zone.

On evaporating the liquid forms a cold dense vapor which travels towards the feed end of the tunnel, effecting counter current heat exchange with the incoming product, thus pre-cooling the same and in some cases crust freezing before it reaches the liquid nitrogen spray. The turbulence fans pushes the cold nitrogen vapor on the product to increase efficiency of heat transfer. This part of the tunnel is the pre-cooling zone and the balance 50 per cent of the cooling is achieved in this zone.

The equilibrium zone, to which a small portion of the cold nitrogen gas is admitted, protects the product from contact with ambient air which would otherwise warm up the colder product surface thus not allowing the cold from the surface to penetrate to the product core.

The operation is simple. Set desired temperature and belt speed, automatic microprocessor control liquid nitrogen spray system, ensure adequate freezing and optimum liquid nitrogen consumption. Short cool down time and reduced equipment thawing requirement of liquid nitrogen freezing system ensures greater productivity.

Low output standby power source can operate tunnel in the event of power failure. Typically the power requirement for equivalent capacity liquid nitrogen and mechanical system are 5 HP and 100 HP, respectively.

Comparable mechanical system costs 3 to 4 times more capital outlay. Simple mechanical design, does not incorporate mechanical compressors, hence does not require of any specialized refrigeration engineer.

Liquid Nitrogen Freezing System Layout

Liquid nitrogen for the cryogenic installation, is stored in vacuum insulated storage tank (VIT) outside the user's processing plant/building, but as near as possible to the point of use. This tank is kept topped up by regular deliveries of bulk liquid nitrogen

Vacuum insulated storage tanks (VITs) are low pressure vessels for storing liquid nitrogen and delivering it as liquid nitrogen, as and when necessary. VITs come in various capacities (up to 15000 litres) and the right sized vessel depending on the user's monthly liquid nitrogen requirement will be installed at the processing plant.

Liquid nitrogen is carried to the point of use, that is, the tunnel, spiral or cabinet freezer by an insulated pipe line. While best result is obtained from use of vacuum insulated transfer lines, polyurethane insulation of recommended thickness also provides the necessary insulation. Polyurethane pipe sections may be used for this purpose.

Chapter 8
Preservation of Fish in Cold Temperature

Fresh Fish Preservation

There are no doubt that more edible marine products are discarded as unfit for human food purely as a direct result of bacterial spoilage than from any other known factor or combination of factors. There are probably many thousands or hundreds of thousands of bacteria associated with spoiling of fish, which differ from one another, sometimes very radically and sometimes only one or two very minor ways.

The most significant and fundamental contributions to the knowledge of fish spoilage originate from the knowledge that fish will spoil much less rapidly in sea water chilled with ice (about – 1°C) than when stored in crushed ice at 2 or 3°C. It was shown that fish would spoil about twice as fast at 2.5°C as it -1.1°C. The reason for this was the fact that fish, unlike meats from warm blooded animals is naturally contaminated with psychophilic bacteria, some of which will grow at -7.5°C.

The lowest temperature at which fish may be stored without their muscle tissues freezing is about -1°C and it is this temperature which must be attained and fairly carefully controlled if one is to store fresh fish ideally. Unfortunately it has not proven easy to find conditions under which this is practicable, though various methods have been proposed. These have included scattering salt over ice used for icing fish, making flaked ice with a melting point of about -1°C from sea water or eutectic block ice from 2 per cent or 3 per cent solutions of sodium bicarbonate or disodium phosphate, and storing fish or fillets, which are well protected by moisture proof wrapping materials in circulating air at -1°C.

Clear sea water (or 3 per cent salt brine) cooled with crushed ice or by other means will have a temperature of about -1.7°C and might be of value in fish wells similar to the refrigerated brine wells used in tuna boats. In salmon trolling boats in which fish were stored in tanks containing circulating sea water which was mechanically refrigerated proved highly successful.

Preservation and Transport of Oil Sardine

Sardinella longiceps has a long and stout head, the body narrow and tapered towards the tail, covered with scales. In fresh condition the fish has an admixture of blue, green and brown colour on the back. Flanks are silvery with iridescent pink. During peak season the catch comprises of medium sized fish ranging in size from 12 to 15 cm, which belong to the one year class weighing 40 gram each. The fish has an oil content, the value reaching as high as 17 per cent on wet weight basis in November.

As the fish is landed along the coast line, and the potential markets are generally thousands of km away, expeditious transportation of fresh fish is not easy. The conventional method of preservation by icing also is not as effective as with other fishes because of its high oil content.

During ice storage as also freezing and thawing, the belly flaps of sardine break and the viscera protudes, reducing the consumer preference even though the fish is of prime quality, and the organoleptic properties are not affected. The incidence of belly bursting in certain cases ranges up to 25 to 30 per cent, depending on maturity, fat content and nature of stomach contents.

With an oil content as high as 17 per cent (wet weight basis) during peak seasons, with high degree of unsaturation, due to the presence of a good amount of poly-unsaturated fatty acids, the incidence of oxidative rancidity is very high.

Investigations have shown that sardine can be preserved for 10-12 hours in bamboo baskets when packed with 1:1 ice and fish and for 14-18 hours in tea chests. By providing an inner lining of gunny and polythene or bitumen coated craft paper, the storage life could be further increased up to 18-20 hours. When the tea chests are insulated with 2.5 cm thick thermocole covered with polythene, the storage life goes up to 55 to 60 hours. Frozen and glazed sardine when transported in thermocole insulated tea chests/plywood boxes can reach destinations as distant as would require 4 days journey in a safe condition. The fish at the end point however, will be in a thawed state.

Belly bursting is a serious problem encountering in freezing of sardine. This phenomenon is maximum in small size, sardine with low fat content and minimum in bigger sardine with higher fat content and can be mostly overcome by a dip treatment in 15 per cent brine for 30 minutes prior to freezing.

The problem of oxidative rancidity in freezing and preservation could be effectively controlled and the storage life extended considerably by dipping the fish in a 0.05 per cent hydroquinone solution for 5 minutes or in 0.1 per cent agar agar solution.

There is no significant difference in organoleptic and bio chemical characteristics between individually quick frozen and block frozen sardine. Oil sardine lose its acceptability and shelf life if the pre-freezing ice storage lasts for more than 3 days.

Brine Cooling of Fish Aboard a Fishing Vessel

The design or selection of the equipment for a brine –cooling or a freezing system to be used on a particular fishing vessel is based on the amount of fish to be handled, the freezing or cooling time required, the brine temperature desired, the size of the vessel and certain factors peculiar to the particular fishery in which the vessel to be engaged.

Among these, the more important factors should be considered in regard to brine tanks, evaporator and refrigeration machines.

Brine Tanks

In order to permit effective cooling of the fish, the brine tank must be so designed that an adequate, constantly renewed film of cold brine is maintained in intimate contact with the entire surface of the fish. To accomplish this, it is necessary that the ratio of brine to fish be sufficiently large to prevent packing of the fish and that a system be used employing movement of fish through the brine or circulation of brine around the fish.

The movement of fish through the brine while they are held in a perforated cylindrical basket offers one of the most positive methods of ensuring rapid and uniform cooling. A perforated cylindrical container with an allowance of 10 kg of fish per cubic feet (320 kg/sq.m) moving through the brine at a speed of 30 feet (9.14 m) per minute presents ideal cooling conditions. An allowance of more than 10 kg of fish per cubic feet results in "packing" thereby reducing the cooling and freezing rate considerably. Movement of fish through the brine at speeds of less than 30 feet per minute may result in slower cooling of some of the fish, whereas, movement at speeds greater than 30 feet per minute will not increase the heat transfer between fish and the brine sufficiently to warrant the extra power required to drive the basket mechanism. A brine tank capacity of about 3.5 kg of brine for 0.5 kg of fish contained in the cylindrical basket will prevent an excessive rise of the brine temperature when the fish are initially cooled.

A method of brine cooling or freezing such as that described above for cooling and freezing ground fish on the trawler has many other applications. Its use in cooling and freezing shrimp would prevent the shrimp from "fusing together" when frozen in sodium chloride solution. Other applications would be the cooling or freezing of herring, mackerel, salmon and other fish that do not exceed 6 inches (15.2 cm) in thickness. The longer cooling time required by fish over 6 inches thick, due to the limited transfer of heat due to the increased thickness, does not warrant the rapid movement between brine and fish that is necessary for cooling or freezing smaller fish. The extra space required by the baskets and the driving mechanism together with the additional power required by the mechanism for the basket drive, would be decidedly disadvantageous if this system were used for cooling and freezing thick fish, such as, large tuna. The time required for loading and unloading the baskets is also a factor which must be given adequate consideration. This system, while being ideal for low temperature cooling of fish which is to be removed after reaching the desired temperature to a refrigerated hold, does not seem to offer many advantages for cooling and subsequent storage in large quantities of fish in the brine tank.

In brine cooling tanks in which the brine is circulated, an attempt is made to maintain a continuous flow of cold brine around the fish either by using a brine agitator in the brine tank, as in the cooling and freezing of shrimp or by circulating the brine within the tank by a brine pump, as in the cooling and freezing of tuna. In cooling and freezing shrimp, the small size of the shrimp, the proper agitation of the brine and the use of a low-temperature salt-sugar brine solution permits rapid and uniform cooling. In cooling and freezing tuna however, the large ratio of fish to tank volume (about 25 kg/cubic feet) and the limited circulation of brine within the brine tank might well result in slower cooling, causing increased salt penetration into some of the fish. A smaller ratio of fish to brine tank volume, increased agitation of the brine and close control of brine density to reduce the buoyancy of the fish will result in quicker and more uniform cooling, thereby reducing the rate of salt penetration. The rate of salt penetration can be considerably reduced by the use of lower brine temperature.

Evaporators

The refrigeration effect necessary to cool the brine to the required temperature can be provided either by using refrigerated surfaces in the brine tank or by circulating the brine through a brine cooler located in the refrigeration room.

In the most brine cooling and freezing installation aboard fishing vessels, bare pipe coils are used to furnish the necessary cooling surface. The size of the particular coil needed depends largely on the particular refrigerant used. If ammonia is used, oil return to the compressor is not a major problem. If however, Freon 12 or Freon 22 is used, the coils should be sized properly in order to obtain a high enough velocity to ensure satisfactory oil return. Bare pipe coils, 3.2 cm diameter, when used indirect expansion ammonia system offer a heat transfer rate ('U' factor) of approximately 15 Btu/hour/sq. feet/degree F (73.2 kg.cal/hour/sq. cm/degree C) if immersed in an agitated brine solution. High capacity brine pumps circulating the brine within the tank directly over the coils will increase the 'U' factor considerably. A flooded-coil system in lieu of a direct-expansion system will further increase this heat transfer rate.

In some installations, bare pipe cooling coils are located directly against the sides of the brine tank. A space of 2 or 3 inch (5.1 or 7.6 cm) between the sides of the tank and coil would permit brine circulation around the entire periphery of the coil, resulting in slight greater heat transfer between the coils and the brine.

Bare pipe coils, finned pipe coils or refrigerated plates, located in a separate portion of the brine tank and arranged so as to permit the circulation of brine over the coil at a high velocity, offer the advantage of an increased 'U' factor, resulting in a reduction in the area of the coil surface required.

In an installation in which the refrigerated surfaces are located in a separate compartment in the brine tank, or in any other installation in which the fish do not come in contact with the coil surface, the use of finned pipe coils with a fin spacing of not less than 1 inch (2.54 cm) or if refrigerated plates should be considered, because of the high heat transfer afforded per linear foot.

The formation of "slush ice" on the cooling coils, which is due to lowering of the temperature of the film of brine surrounding the cooling coils below that of its freezing point, will reduce the coil efficiency considerably. The formation of slush ice can be eliminated by maintaining a close differential between the coil temperature and the freezing temperature of the brine (not more than minus 12.2°C) and by ensuring positive brine circulation around the entire surface area of the coils.

Circulation of the brine in the cooling tank, through a brine cooler offers quick, efficient cooling at high evaporative temperatures. A flooded type brine cooler, which offers a high heat transfer rate has certain disadvantages peculiar to marine operation, the principal one of which are, danger of the brine freezing and rupturing the tubes because of lack of circulation caused by the accumulation of the foreign materials in the tubes or by brine-pump failure, decreased performance due to difficulty in maintaining proper ammonia level because of excessive rolling and pitching of the vessel; and deterioration of the tubes due to corrosive action of the brine. In existing installations, these disadvantages can be reduced by use of an adequate brine-straining system, automatic control to isolate the cooler in event of failure of the brine pump, a salt-sugar brine with low freezing point, a proportioning-type level control, equipment having 20 per cent larger capacity than required and heavy gauge wrought-iron tubes, galvanized on the inside.

The use of direct expansion-type brine cooler should be seriously considered to brine-cooling applications. The advantages of increased space and higher initial cost over that of the flooded type cooler are offset by the reduction in danger due to brine freezing, the absence of troublesome level controls and the protection afforded against corrosion (this is especially true of a Freon-type cooler, where copper alloy tubes are used).

Refrigeration Machines

There are two principal types of refrigeration machines, the absorption machine and the compression machine.

The absorption machine which operate on steam from the boiler of the ship, offers a low initial cost, an economy of operation at low evaporative temperatures and a minimum of moving parts. Also its satisfactory operation at variable loads and evaporative temperatures makes, possible the utilization of one machine for variable capacity operation in lieu of two or more compressors. A problem in existing absorption machine has been the low performance and difficult operation, owing to the lack of suitable automatic controls. To obtain higher performance and ease of operation the following should be considered.

1. The maintenance of the proper liquid level in the generator by use of a proportioning-type level control.

2. The proper proportioning of the steam flow to suit the load condition by use of a steam-regulator valve that is actuated by the absorber-suction pressure.

3. The proper proportioning of the aqua-flow (from absorber to generator) to suit load conditions, so that a constant ratio of about 4 kg of aqua for each half kg of ammonia flowing through evaporator can be maintained at all time. This proportioning can be done by means of a rheostat on the aqua-pump motor, operated by an air-supply proportioning type relay that is actuated by the float mechanism in the brine cooler or accumulator.

4. The prevention against freezing of the brine in the brine cooler (flooded type) by solenoid valves, that are located on the liquid line and the vapor line and that are activated by a brine thermostat and a pressure switch on the brine-pump discharge line.

5. The control of the head pressure by a temperature-control valve installed in the discharge line of the circulating pump.

6. The use of suitable liquid-level sight glasses in the generator, absorber, receiver and brine cooler.

7. The use of thermometers, pressure gauzes and suitable alarms to warn personnel if the system is not operating properly.

In order to obtain continual satisfactory operation of the controls mentioned above, the operational engineers would have to be instructed in the proper servicing and operating techniques, in much the same manner as with other new piece of mechanical equipment.

The absorption machine should be located in a separate vented compartment that is provided with a water sprinkler system. The space requirement for absorption equipment of about 24 cubic feet (0.68 cubic meter) per ton of refrigeration for a 30-ton machine and brine coolers seems high Such space requirement should be compared with the space required by the compressors that would be used to achieve satisfactory operation at the minimum and maximum load involved.

A waste heat boiler, utilizing exhaust gases from the main propulsion diesel engine equipped with an automatic fuel system for maintaining a constant steam output at variable engine speeds would provide a cheap supply of steam to the absorption refrigeration unit.This possibility should be considered in selecting an absorption system for a new vessel that is to be equipped for brine cooling or freezing fish.

Changes taking Place in Cold Storage of Fish

The popular concept that frozen fish may be fish of low quality, is not due to the freezing process itself. If only fresh, unspoiled fish are frozen, the resulting product resembles the original, fresh fish more closely than fish preserved in any other way Rather, the low quality fish is due either to the fish having already been of low quality before it was ever frozen, or to its having been stored improperly or for too long period of time.

The adverse changes occurring during storage of fish involve alteration in texture to some extent with all the species of fish and results in such changes as toughening and separation, when the fish thaws, of a fluid called "drip". The second type of change is caused by oxidation of oils and pigments in the flesh and results in the development of off flavour and in the discolouration or fading of colours. The second type of change occurs to a much greater extent in some species of fish than in others.

Changes in Texture of Stored Fish

One of the most important changes which takes place when frozen fish are held in cold storage is the alteration in texture. Unfrozen raw fish has a firm gelatinous consistency, and application of pressure does not result in expression of any fluid from the tissues. Frozen fish held in cold storage for an extended period of time and then thawed consists of solid phase plus drip. Upon applying pressure to the solid flesh, considerable additional drip is expressed.

Similarly the texture of stored frozen fish may be different, when it is thawed and then cooked; from the texture of unfrozen cooked fish. Cooked fish which has never been frozen and stored, is of a moist, flaky texture, firm but not tough. Only a small quantity of fluid separates from such fish during the cooking process. When frozen fish which has been held in cold storage for long time is thawed and cooked, considerable "cook drip" separates during the cooking process. The cooked fish itself may be extremely soft rather than flaky. In other cases it may be woody, fibrous, stringy or tough. All these changes in texture involve, either directly or indirectly, the moisture in the fish. In some cases, a surface desiccation or drying during cold storage occurs, resulting in actual loss by evaporation of a part of the water. In other cases there may be a change in the manner in which the moisture is held in the fish flesh. Lean fish tissues contains about 80 per cent moisture held by colloidal bonds to about 16 per cent protein. Alterations which occur to the protein during cold storage of frozen fish result in lowering the capacity of the protein for holding moisture. When the fish thaws, not all of the fluid is reabsorbed.The resulting tissue is spongy, and additional moisture can be expressed, the texture may be fibrous and tough.

These changes in the moisture relationship to the fish tissue occur to a very small extent in fish which is frozen and immediately thawed. They become progressively worse as the time which the fish is stored or the temperature of storage is increased.

Some of these changes in the moisture in the fish result in alteration of the appearance as well as the texture.

Desiccation-Adverse Effects of Moisture Loss

When frozen fish is inadequately protected against moisture loss by improper packaging or glazing, or when storage conditions are not optimum, considerable moisture may be lost. The loss of moisture has several bad effects upon the product. In the first place, there is the economic loss in the weight of fish (5 per cent or more under average conditions)

More serious even than this loss in weight of the product is the very detrimental loss in quality of the fish, which in extreme cases may mean that it becomes completely unmarketable. Loss of moisture taken place to the greatest extent at the surface of the fish, and this results in "freezer burn". Cut surfaces, such as fillets or steaks become dry and lose their glossy appearance and on further desiccation become pitted with tiny holes, so that a honey comb effect results. At the last stage the surface becomes chalk-white in colour. With whole fish, loss of moisture may alter the colour of the skin, and the skin dries out to a dull surface.

When badly desiccated fish is cooked, the texture is "woody" or fibrous, especially in portions at the surface, where dehydration begins and becomes more extreme. The fish is generally dry rather than moist and the flaky texture is lost.

Minimizing Desiccation

Loss of moisture during storage of fish can be completely eliminated by enclosing the fish in a hermatically sealed container or covering which adheres closely to the surface at all points of contact with the fish. This can best be accomplished by completely covering the surface with a glaze of ice or by freezing the fish in block form and then dipping the frozen block in water to form an ice glaze. Such methods are used both for whole fish, and to a lesser extent, for fillets and steaks. Such glazed fish are protected completely against moisture loss so long as the ice glaze remain intact. Any loss in moisture occurs from the ice glaze rather than from the product. This loss in glaze through evaporation may eventually reach the stage where the fish is exposed. Desiccation will then take place unless the glaze is renewed.

Ordinarily, it is considered undesirable to glaze consumer-sized cuts of frozen fish, and some kind of packaging material must be employed. Moisture loss occurs through all types of packaging materials to a greater or lesser extent except through hermatically sealed metal or glass containers. When a highly moisture-vapor-proof wrapper is employed, the moisture loss may be held to a few tenths of a per cent per year under good storage conditions, and this loss, both from an economic and quality stand point, is almost negligible.

When fish is packaged in such a manner that large air spaces occur within the package, desiccation may occur within the package. Moisture is transferred from the surface of the fish at the air pocket to the inner surface of the packaging material where it collects as frost. With large air voids, this moisture transfer may occur in an extensive degree resulting in serious freezer burn. In a similar manner, when fillets are frozen in paper board boxes (wax chip board) over-wrapped with packaging material, and a number of them cartooned together, moisture may pass from the fish to the inside of the over-wrap and eventually through the over-wrap to the carton. While little or no over-all moisture loss takes place from the carton, appreciable freezer burn may occur to the fish.

Although it is theoretically possible to prevent all desiccation by using ideal packaging methods, this is difficult or impossible to attain under practical, commercial conditions. Where moisture is being lost from the fish, the loss can be held at a minimum by using some of the precautions in operation of the cold storage rooms.

Temperature Differential

When frozen fish is stored in a cold storage room, which is refrigerated in the usual manner with pipes containing the refrigerant, the pipes are always at a lower temperature than the air and the fish. Moisture vapor that travels from the ice of the (relatively) warmer frozen fish and is deposited as ice on the colder pipes. If the pipes were made somewhat warmer than the fish (*e.g.* if the fish temperature were 0°F and the refrigerant in the pipes were suddenly replaced with brine at 20°F) then ice from the refrigerating pipes would sublime and be deposited upon the fish.

The rate at which the moisture from the fish is transferred to the refrigeration pipes is dependent in part upon the temperature difference between the fish and the pipes. With no temperature difference, no such transfer would take place. By keeping the temperature differential as low as possible, the tendency for transfer of moisture from the fish to the pipes is held to a minimum.

Lowering of this temperature differential is achieved by increasing the ratio of exposed pipe surface to volume of cold storage content. Economic considerations will limit the amount of piping which can be feasibly employed in a cold storage room.

Humidity

Moisture is transferred from fish to refrigeration coils by first subliming from ice to moisture vapor. The moisture vapor may be considered to dissolve in the air just salt dissolves in water. In a similar manner, as water will dissolve only a certain maximum amount of salt at given temperature and then becomes saturated, the air will dissolve only a certain maximum quantity of moisture vapor at a given temperature. When the air is saturated with moisture (at any given temperature), the relative humidity is 100 per cent ; when half this amount of moisture is present in the air (at the same temperature), the relative humidity is 50 per cent and a considerable capacity for taking up additional moisture vapor exists. Thus desiccation of frozen fish in cod storage can be minimized by keeping the relative humidity as close to 100 per cent as is feasible. A few installations have been made whereby the relative humidity is kept high by injecting steam into the cold storage rooms; but such systems are not entirely satisfactory. High humidity is best achieved by keeping the temperature differential between pipes and product at a minimum. Since this means increasing the area of refrigeration pipes, there is a limit to the maximum relative humidity that it is feasible to maintain by this method. Ordinarily a 85 per cent relative humidity is as high as is economically feasible.

Air Circulation

Transfer of moisture dissolving in the air adjacent to the frozen fish depends also upon circulation of air within the room from the fish to the refrigeration pipes. Any means of cutting down the air circulation will retard desiccation. Complete blocking of circulation of air from to refrigeration coils is accomplished by means of jacketed cold storage rooms as proposed for fish by Young (1933). Several such commercial installations are in use on the west coast and give excellent results. In this system the cold storage room is sealed inside a second room very slightly larger. The small space between the two rooms contains air circulated from the refrigeration pipes by a blower fan. This refrigerated air, circulated in the jacketed space, adsorbs heat coming from the outside and thus maintains the proper storage temperature in the inner storage room. The air which surrounds the cold storage room never comes in contact with the fish, and hence no transfer of moisture from fish to refrigeration pipes is possible.

In standard cold storage rooms, air circulation can be restricted by proper stacking of the frozen product. When the frozen fish is first placed in cold storage, however, stacking must be made in such a way as to encourage air circulation if the fish is at a temperature higher than that of the storage room; otherwise there would be danger that the fish would not be lowered to the room temperature for an extended period. Proper initial stacking calls for a 2-inch to 3-inch space at floor level and provision of air space by use of "dunnage" sticks or spacers between every second layer of cartons of packaged fish. This provides space for air circulation of about 3/8 inch between alternate layers of cartons. A space of six inches is generally left adjacent to walls and ceiling of inside rooms and at least twice this amount (sometimes 18 inches under refrigeration coils) for outside rooms.

Fluctuation of Storage Temperature

Storage temperature can never be held absolutely constant. Some fluctuation always occur during the on off cycle of the refrigeration compressors. Greater fluctuation will occur when doors are open frequently or when large quantities warmer fish are deposited in the cold storage rooms. At one time it was believed that these fluctuating temperatures were harmful to the quality of the fish by causing large ice crystals to form within the fish tissue which might result in alteration in texture. But experiments on fish and on similar other frozen commodities have shown that no such damage occurs.

There are, however, other disadvantages of fluctuating temperatures. If the storage temperature averages 0°F, but fluctuates from -5° to + 5°F, the fish being stored at part of the time at + 5°F, are subjected to the greater damage (such as oxidative changes) caused by the higher storage temperature than would be the case if the temperature never rose above 0°F. Fluctuating storage temperatures also are favourable to causing somewhat larger quantities of frost to collect inside packages of fish where air voids are present.

Drip in Frozen Fish

Drip is the clear, sometimes slightly cloudy fluid which is not reabsorbed by the fish tissue when frozen fish thaws. The fluid consist of water in which dissolved protein and other nitrogenous constituents as well as minerals. The nitrogen content averages about 1.5 per cent. The quantity of drip depends upon a number of factors including the species of fish involved and the length and temperature of storage prior to thawing. Drip may be less than 1 per cent or more than 20 per cent of the weight of the fish.

Factors Controlling Drip

The quantity of drip forming varies greatly with the particular species of fish involved. Where the moisture content of a species is unusually high or the protein content very low, the quantity of drip is apt to be high. Thus the drip in the oysters is considerably greater in quantity than is the case with most other species.

The rate at which the fish is frozen, is one factor determining the quantity of drip formed, slow frozen fish usually containing more drip than quick frozen ones. Thus in the case of "sole" fillets an average of 5.1 per cent drip was found from quick frozen fish stored for 6 months at – 5°F as compared to 7.8 per cent for slow frozen fish handled in an identical manner.

The rate of thawing appears not to be an important factor in controlling the amount of drip. The temperature at which the fish is thawed is of more importance, a greater quantity of drip forming at the higher thawing temperatures. Thus it was found that fish thawed in air, and water at the same temperature gave no significantly different quantities of drip. In one experiment, in which fish were thawed in air at 34° and 80°F, the amount of drip was 4.9 per cent and 10.1 per cent respectively.

The most important factor in determining the quantity of drip forming when fish thaws is the condition and length of storage. The longer the fish is held in the cold storage and the higher the storage temperature, the greater is the quantity of drip.

Cook Drip

When thawed fish is cooked, a fluid separates during the cooking process, which is termed as "cook drip". A method has been described for measuring its quantity. The cooked drip is usually greater in amount than that which forms when the fish thaws before cooking. It is generally a clear fluid, while still hot. Upon cooling, protein may precipitate to form a viscous or gelatinous paste.

Disadvantages of Drip

The appearance of thawed fish containing much drip is so different from that of unfrozen fish as to make the frozen product unappetizing. Since the drip is ordinarily discarded, any dissolved protein, mineral and flavour components will be lost. Usually the texture of thawed fish from which much drip has exuded is undesirably dry after cooking, and it may be of a woody or tough texture. Accordingly, an unusually large quantity of drip may be an indication that other more serious alterations in texture have taken place.

Retarding Drip Formation

Aside from keeping the storage temperature low and the storage time short, there is no good way of markedly reducing the quantity of drip which will form when the frozen fish is thawed. One method often employed to reduce the apparent quantity of drip is to brine the fillets before freezing. Lean (non-oily) fillets are ordinarily dipped for about 20 seconds in a salt brine of about 6 per cent sodium chloride content. Oily fillets (of species with which rancidity may be a problem) are generally dipped for about 10 seconds in brine of the same strength. Thus Stansby and Harrison (1942) reported 5.1 per cent drip from control (unbrined) sole fillets stored for 6 months at -5°F as compared to 1.7 per cent for brined fillets otherwise handled in an identical manner.

Alteration in Texture

As frozen fish remains in cold storage, the texture becomes progressively tougher and the amount of drip increases after the fish is thawed. Some changes must be going on within the fish during cold storage to account for these alteration in texture. The theory has been advanced that the changes responsible for these texture alterations involve a denaturation of the fish protein.

Snow (1950) reported a lots of solubility of the principal fish protein, actomyosin, as the storage period or storage temperature of the frozen fish increases. This change in solubility is attributed to to protein denaturation since the solubility of denatured protein is lower than that of the native protein.

Because such changes in protein solubility run parallel to increasing toughness of the fish as they remain in cold storage, it has been postulated that the principal cause of toughness is the denaturation of the fish protein.

Changes in Colour and Flavour of Stored Fish

When fish are held in storage, changes in colour and flavour develop. Since these changes, though quite different with respect to the final quality of the product are caused by the same factor, oxidation, they will be treated together. Oxidation is caused by chemical combination between oxygen and some constituent of the fish tissue, such as the oil or one of the pigments. This process may be a purely chemical reaction, or it may be one involving certain bio-catalysts, such as, enzymes.

Colour Changes in Stored Fish

Changes of colour in fish occur shortly after they are removed from the water. In very freshly caught fish, that portion of the blood remaining in the tissue is bright red in colour. Thus in skinned fresh fish fillets, a pattern of red blood is especially noticeable at the skinned surface where blood vessels still retain some of the blood. If the fish is held in ice for a few days before the fillets are cut, the red colour of any remaining blood, instead of being bright red, is a dull red. With the fish held still longer, the colour may be light brown or even dark brown. These colour changes result from a change in the state of oxidation of the bright red pigment, hemoglobin which occurs in the blood.

Most fish contain two types of flesh, a dark and a light meat. The light meat is usually white or light grey. The dark meat which occurs in the tissue at certain localized areas, such as, along the lateral line and at the tip of the nape varies with the species and may be a light yellow or light grey. Pigments in these dark flesh areas are especially subject to oxidation (probably because of localized concentration of bio-catalysts) which results in a darkening of the colours to deep yellow, dark brown, or in some cases to almost a black. These colour changes take place in the frozen fish after they have remained in cold storage for an extended period of time. Such changes, when extreme, may result in so unsightly an appearance as to render the fish unsalable.

Some fish such as salmon have a pink or red light meat. The pigments which cause these colours are slowly oxidized when the frozen fish is held in cold storage, resulting in a fading of pink or red colour. In extreme cases, caused by extended storage periods, the pink or red may completely disappear.

The natural oil occurring in fresh fish varies in colour depending upon the species. With some species the oil may be almost colourless; in others it is light yellow, with a few species notably salmon, it is some shade of red or pink. The colour of the oil is caused by colour pigments dissolved in it. These pigments, similarly to the pigments in the flesh are subjected to oxidation. The colourless or light yellow fish oils change, when oxidized to deep yellow, orange or brown. The red or pink oils of salmon first fade in colour, as a result of oxidation, and then alter to yellow or brown.

The oils in frozen fish may oxidize and darken while present in the tissue. A more rapid oxidation occurs when a part of the oil is forced out of the tissue during freezing and oozes to the surface. This occurs especially with fish frozen whole. The oil working its way to the surface owing to the pressure built up by expansion of water as it changes to ice during the freezing process, collects on the outside surface of the skin of the fish and alters to form a brown colour. Such fish is said to have "rusted". This development of a brown rust colour has been shown to involve not only oxidation of the oil, but also reaction with nitrogenous constituents of the flesh. A similar condition may occur, but to a less noticeable extent with frozen dressed fish such as fillets or steaks. With certain wrappers used for such forming of frozen fish, the oil may be absorbed by the wrapper. This exposes a large area of oil to oxygen of the air and may cause a rapid oxidation and discolouration of the pigments in the oil.

Flavour Changes in Fresh and Frozen Fish

Fresh and frozen fish is often described as having a "fishy' flavour. It is not always recognized that this so called fishy flavour is not the normal flavour of the freshly caught fish. Rather it is an altered flavour, which has developed through some faulty or too prolonged handling. In the case of frozen fish, this may be due to the fish having been stored in ice for too long a period of time before it was frozen or it may have resulted from improper protection or excessively long holding of the frozen fish during cold storage after freezing.

Quite different from the rather unpleasant "fishy" flavour characteristic of improperly handled fish is the pleasing taste of freshly caught fish. This palatable flavour is fully retained if the fish is promptly frozen and under favourable cold storage conditions, the fish will reach the consumer in a condition closely resembling its high quality immediately after it was caught.

Unfortunately, the changes which result in a transformation of these pleasing fresh flavours to unpleasant fishy ones occur very rapidly with certain species of fish, and these changes can be avoided only by prompt and careful handling. The alteration in flavour takes place in two stages; loss of characteristic flavour followed by development of abnormal flavour.

Loss of Flavour

Each species of fish has its own characteristic flavour. This flavour is not the "fishy" one which develops upon protracted storage, neither is it a bland, flat flavour or absence of taste. Rather, it is the delicate flavour which distinguishes one species from another, when they are eaten.It is not known about the chemical compounds present in fish which give rise to these characteristic flavours. It is also not known that the compounds which give the fish these individual flavours, which are very unstable.Shortly after the frozen fish is placed in cold storage these delicate flavours slowly disappear. Such changes take place early in the storage of the fish and considerably earlier the development of any "off" flavours. Eventually all or nearly all, of the flavour characteristic of the particular species disappears.

While no systematic study of the cause of the loss of these naturally occurring flavours in fish has been made, it is probable that such loss is due to an oxidation of the chemical constituents causing the flavour.

Samples of fish stored in evacuated tin containers retained the characteristic flavour longer than did samples stored with greater access to air.

Development of Abnormal Flavour

Shortly before the disappearance of the last trace of the normal flavour, in fish held frozen in cold storages an abnormal flavour starts to develop. This flavour, which is described as a rancid one, is characterized by a bitter taste which persists in the mouth long after the fish has been eaten. In the initial stages of rancidity, this flavour may not be objectionable to all persons eating the fish. As a matter of fact, frozen fish so often have become slightly rancid at the time they are consumed, that many persons are unaware that this flavour is not a normal one. As the rancid flavour develops, it becomes more objectionable until a point is reached where the fish become unmarketable.

This rancidity development in frozen fish is caused by an oxidation of the fish oils to produce carbonylic compounds. The reaction whereby the fish oil becomes oxidized may be a purely chemical changes (auto-oxidation) or one which catalyzed by bio-catalytic systems occurring in the flesh of the fish.

Mechanism of Oxidation of Fish Oils in Frozen Fish

When fish oils which have been extracted from the fish tissue are exposed to the action of air and light, they oxidize spontaneously according to a typical auto-oxidation type of reaction. Such reactions follow a pattern whereby the oil is at first oxidized very slowly (this stage is known as induction period). Then the rate of oxidation increases at a rapidly accelerating pace until a maximum rate is reached, after which the oxidation rate decreases. Auto-oxidation of fish oils is accelerated by high temperatures, by the presence of light and by the presence of pro-oxidants (certain chemicals, metals like copper and iron). It is retarded by limiting access of the oil to air and by the presence of anti-oxidants (various chemicals, including certain phenolic-type compounds). The reaction of the fish oils with oxygen occurs first as an oxidation of fish oil fatty acids to form peroxide-type compounds which then decompose to form numerous carbonyl types of compounds which are responsible for the rancid odours and flavours.

When fish oil is oxidized in fish flesh under the conditions that occur during development of rancidity in frozen fish, another type of oxidation is involved at least in part. Banks (1937) has shown that in the oxidation of herring oil in frozen herring held in cold storage, the oxidation is catalyzed by the presence of some bio-catalytic substance present in the herring flesh. Khan (1952) has more recently investigated the reactions which he attributes to enzymatic oxidation of the oil in frozen salmon. Tappel (1953) has carried out studies with certain meat products and concludes that hematin is a powerful catalyst in accelerating the oxidation.

The present knowledge indicate that the reaction mechanism whereby rancidity develops in frozen fish is predominantly a bio-catalyzed reaction, although auto-oxidation probably plays a part.

Types of Fish Most Susceptible to Colour and Flavour Changes

Oxidation of fish oils and pigments is a more serious problem with some species than with others. If a species of fish contains only a small quantity of oil (under 1 per cent), rancidity is ordinarily of minor importance. Such fish as haddock and cod can be kept in cold storage for many months with only very slight alteration in flavour. Even with these species however, some rancidity does develop. When salt cod is stored at room temperature, the small amount of oil present in the flesh oxidizes to a considerable extent because of the relatively high temperature of storage. This oxidation leads to development of a rancid flavour typical of salt cod which is sometimes described as a "salt fish flavour". This same "salt fish flavour" develops to a lesser extent in frozen haddock and cod stored for extended periods of time. Because the flavour is not as pronounced as occurs with more oily fish species, it is not often recognized as rancid flavour.

The content of oil in the flesh of the fish is not the only factor involved in the susceptibility of the species to rancidity development. A good example of this is the case of five species of salmon, which contain an average oil content in their flesh as follows; king salmon-16 per cent, red salmon-11 per cent, coho salmon-8 per cent, pink salmon-6 per cent and chum salmon-5 per cent. Of these five species, the pink salmon containing only about 6 per cent oil (next to the leanest) is by far the most susceptible to development of rancid flavours (as well as to change in colour). Pink salmon steaks become rancid and badly discoloured in less than one-tenth the time that the much more oily king salmon reach the same stage of alteration under identical storage conditions. Whether this is due to solely to the fact that the pink salmon oil is more unsaturated than the king salmon oil (iodine value about 165 for the pink salmon as compared to about 115 for the king salmon) or whether there are more active lipoxidase or other bio-catalytic systems, or less naturally occurring antioxidants, is not definitely known. It would appear that some of the latter factors must be of importance.

In general, the rates of the discolouration and rancidity development run parallel for the various species. When rancidity is a severe problem, discolouration is also usually severe. When rancidity develops slowly, discolouration is usually a minor problem. This correlation suggests that the underlying cause for rates of both rancidity development in the oil and discolouration of pigments in the flesh may be primarily a matter of the particular bio-catalytic systems (or their degree of activity) present in the flesh.

Minimizing Colour and Flavour Changes

The most important consideration in retarding changes in fish brought about by oxidation is to keep air (and hence oxygen) away from the fish. When this condition is completely attained, no oxidation can occur, and consequently no flavour or colour change in the frozen fish will develop. Keeping all oxygen away from the fish, however, is beyond the present practical means in handling frozen fish.

The amount of oxygen which will cause discolouration or rancidity to develop is extremely small. Fish frozen in an evacuated hermatically sealed tin can does not of course, contain a perfect vacuum. Even the small amount of air left in such a can will discolour to some extent the surface of the fish adjacent to head spaces. Likewise, when fish is covered with an ice glaze, a small quantity of air (and hence oxygen) is dissolved, both in the glazing water and in the water contained in the tissues of the fish. Even this small quantity of oxygen is sufficient to cause a small (though sometimes quite perceptible) slow development of slight degree of discolouration and of rancidity.

The problem of keeping away from fish during frozen storage consists first of eliminating as much as possible of the air in contact with the fish within the packaging medium and secondly of preventing further air from entering the package. An ice glaze renewed as required is by far the best method of attaining the first objective. A thick glaze, when properly applied, effectively forms a barrier between the fish and the outside atmosphere. No pockets of air whatever are left between the fish and the glaze. Because the ice glaze slowly evaporates under most cold storage conditions, and in some cases cracks and is lost, the ice glaze alone is not necessarily the permanent protection.

A hermatically sealed tin can provides the best solution of the second problem of keeping air from entering the packaged frozen fish. In facts, it provides a complete solution to this problem, since no additional air whatever can enter once the can is sealed.

Frozen fish in a tin can contains, even under best conditions, a sufficient trace of air to bring about some slight oxidation. The best means for preventing all but a trace of oxidation is packing the ice glazed fish within an evacuated tin can. This method is not at present commercially acceptable means for handling frozen fish. For round fish, an ice glaze, renewed at suitable intervals represent the most feasible and usually entirely adequate means of preventing oxidative changes. For dressed fish products, such as fillets and steaks, the products, if ice glazed and then packaged in a suitable moisture-vapor and oxygen-proof packaging material, such as good grade of cellophane, will have excellent protection.

Where an ice glaze can not be used, the method of applying the packaging material to avoid pockets of air within the package is extremely critical. In addition to application of wrapping material in the proper manner, filling of the packages completely and freezing under pressure aids in elimination of air pockets to a very great extent.

Choice of a packaging material which cuts down on oxygen transmission through the package is also of extreme importance. As a general rule those papers having a high resistance to moisture-vapor transmission are also effective in cutting down on oxygen or air transmission through the wrapper.

The packaging material should be grease proof. Wrappers which adsorb oil will form a thin layer of fish oil at the surface where air, slowly passing through the package, will cause rapid oxidation. The following papers are, in the order given, of descending quality as far as grease proofness is concerned; cellophane, pliofilm, vinyl derivatives, polyethylene, glassine, vegetable parchment, "grease proof" papers, waxed papers.

Keeping Fish Temperature Low

The second most important precaution in retarding oxidative changes in frozen fish is to keep the temperature of the frozen fish as low as possible. Some handlers of frozen foods believe that fish can be safely stored at temperature considerably higher than those for other frozen foods. Fish which appear to be "hard frozen" are considered by such handlers of frozen fish to be at their optimum storage temperature. Occasionally, special storage rooms at about 20°F are used for fish, while frozen fruits and other products are held at 0°F or even lower. This practice is completely unfounded. It probably originated from the fact that fish appear to be frozen quite hard at 20°F, whereas fruits frozen in sugar syrups are not hard until a temperature of 0°F or lower is reached. Actually, the hardness of the product has nothing whatever to do with the determination of the proper storage temperature.

The oxidative changes occurring with fish are greatly retarded by lowering the temperature, and fish held at 0°F are oxidized at only about half the rate at 20°F. Similarly an equal diminishing in oxidation rate occurs if the temperature is lowered from 0° to -20°F. Since frozen fishery products are subject to greater oxidative changes than almost any other frozen food, these products should always be stored at lowest economically feasible temperature.

Avoidance of Pro-Oxidants

Certain chemicals, when in contact with fish, accelerates the rate of oxidative changes. These substances include copper and iron, and fish should be processed in such a manner as to avoid contact with these metals, as much as possible. Common salt, when present in large quantities, also has an accelerating effect on the oxidative changes in frozen fish. Where salt dips are used to minimize drip formation in the frozen, thawed fish, it is advisable to use weaker brines or shorter brining periods when dealing with species which give trouble with rancidity or discolouration than with other species. Furthermore, salts, such as, those of calcium and magnesium which sometimes occur as impurities in common salt (sodium chloride) cause a greater acceleration in rancidity development than does the sodium chloride itself.

Hence, only salt of relatively high chemical purity should be used for preparing brine dips for species of fish which are especially subject to oxidative changes.

Use of Antioxidants

Antioxidants are effective in retarding auto-oxidation of oils,.when the oils are present in the tissue, as in the case of frozen fish. The effectiveness of the antioxidant in the presence of moisture from the fish is greatly diminished. Further antioxidants which give protection to oil auto-oxidizing may be of limited or of no value to protect oils which oxidize through the action of a bio-catalyst. Application of commonly used antioxidants, such as ascorbic acid or mixtures of ascorbic and citric acids, propylgallate and nordihydroguaiaretic acid are effective in retarding rancidity development in frozen fish. Certain antioxidants, however, have some value in slowing up to a degree the discolouration of certain species of fish. Even this effect is very limited.

Bacteriology of Frozen Fish

When fish are frozen and stored at the usual storage temperature, the bacteria present in the fish tissue, are most part, inactivated. A portion of them are killed, but those surviving will, when the fish is thawed, grow once more and contribute to spoilage of the thawed fish.

When a frozen fish is stored at usual storage temperature (0°F), those bacteria not destroyed by freezing process are slowly killed. Even after a year's storage, however, the fish are far from sterile, since a considerable quantity of bacteria are very resistant to holding at low temperatures.

Bacteria present in frozen fish normally will be killed, when the fish are thoroughly cooked. If any pathogenic bacteria are present, they will all be destroyed during the cooking process so that ordinarily no danger to health exists. When pre-cooled frozen fishery products, such as, crab meat, fish steaks or breaded shellfish are frozen, the thawed product may be eaten either without further cooking or after only inadequate heating. Under such condition, should any pathogenic bacteria be present, they would constitute a very real health hazard. Accordingly, preparation of such pre-cooked frozen fish products must be made under the best sanitary conditions in order to prevent their being a potential source of harmful bacteria. This handling under the best sanitary conditions is, of course, the only acceptable practice with all foods, whether they are to be eaten raw or cooked.

Minimizing Changes Occurring during Storage

The first prerequisite in obtaining satisfactory storage life of frozen fish is that the quality of the fish before freezing be first class. In the early days, when frozen fish was a new item, a practice developed of selling iced fish on the fresh fish market until spoilage was imminent and then freezing what was left. Such fish of second quality to begin with, certainly would not improve upon freezing and storage and would also have a very short storage life.

Assuming that good, first quality fish are selected for freezing, the second important step in ensuring satisfactory storage life is proper processing. This includes, avoidance of contamination by metals or other substances, use of a suitable brine dip to minimize loss of drip, and use of good sanitary handling practice, particularly for any pre-cooked products and for those to which batters and breading materials have been added.

The third important step is to start freezing the product almost immediately after processing. Also an improved method of commercial freezing should be employed so that the particular product can be frozen as rapidly as is economically possible. Long time intervals between the termination of processing and the beginning of freezing and the use of slow un-orthodoxed freezing methods will greatly affect the storage life of the product.

The next important precaution is application of satisfactory protection against desiccation and oxidation. A good ice-glaze is the best protection both for dressed cuts of fish to be packaged and for whole fish. For packaged fish, the packaging materials must be moisture-vapor proof and resistant to oxygen transfer. The fish must be packaged in such a way as to avoid air pockets within the package.

One of the most important precautions is the use of as low as storage temperature as is economically feasible. The temperature should not be higher than 0°F, and still lower temperature are desirable.

Finally proper operation of the cold storage rooms is important. Such operations includes, provision of high relative humidity, minimizing unnecessary air circulation and keeping the temperature constant as well as low.

Storage Conditions and Storage Life for Packaged Frozen Fish

Category	Type of Fish	Brine Treatment	Protection Required	Storage Life at 0°F
1.	Species most difficult to store	None	Ice glazing before packaging	4-6 months
2.	Oily species difficult to store in 6 per cent brine	10 seconds	Ice glazing before packaging	5-9 months
3.	Non-oily species difficult to store	20 seconds in 6 per cent brine	Packaging	7-12 months
4.	Fish easy to store, mostly Non-oily species	20 seconds in 6 per cent brine	Packaging	Over 12 months

Species	Category	Species	Category
Carps	2	Pollock	2
Cat fish and Bull head	2	Rock fishes	3
Chub	1	Sable fish	3
Cod	4	Salmon Chinook or king	3
Flounder	3	Salmon Chum or Keta	2

Species	Category	Species	Category
Haddock	4	Salmon Pink	1
Hake (white)	4	Salmon Red or Sock eye	3
Halibut	3	Salmon Silver or Coho	3
Herring (Lake)	4	Sardine Pacific (Pilchard)	2
Herring (Sea)	2	Smelt	2
Ling cod	4	Spanish mackerel	2
Mackerel	2	Tuna	2
Ocean perch, Atlantic	3	Whiting	4
Ocean perch, Pacific	3		

Freezing Characteristic of Tropical Fishes, Tilapia

Analysis of tilapia muscle during ice storage showed a regular decline in the nitrogenous constituents (TN, PN, SSN, SN, and WSN) with increasing of ice storage, which may be attributed to leaching by the melting ice. Similar results were obtained in sardine muscle stored in ice and in prawns stored in ice. Free fatty acids (FFA) registered a slow increase (1 to 4 per cent), while peroxide value (PV) increased regularly. Development of FFA was slightly faster after 6 days of storage, when the salt soluble proteins were also found to fall concomitantly. Similar results have been reported in cod held in ice. A comparison of these changes in tilapia from freshwater and brackish water sources revealed close similarity except that the extractability of myofibrillar proteins was found to increase more rapidly in the brackish water.

Organoleptically the freshwater tilapia was acceptable up to 13 days of storage in ice, while the storage life of the brackish water tilapia was limited to only 10 days. Slight rancidity odour was detected in the former after 13 days in ice, while it was perceptible in the latter after 6 days. Yellow discolouration was observed in the brackish water fish iced for 10 days, whereas the freshwater fish retained its fresh organoleptic qualities even after 13 days of storage in ice. While no appreciable changes was observed in the texture of the freshwater fish muscle, marked textural changes were observed in the muscle of brackish water fish after 10 days. No significant differences in the organoleptic qualities of male and female species in brackish water tilapia was observed during ice storage, although the meat of the latter was adjudged to be comparatively sweeter. An examination of the changes in total free amino acids in both the species has brought out certain interesting observations. A higher amount of glycine (41 mg per cent) was found in female species as against 24 mg per cent in the male species although the total free amino acids in both remained more or less the same. The predominant amino acids in the male species were in the order of arginine > glycine > valine > glutamic acid > praline, while in female species the order was glycine > arginine > praline > glutamic acid with traces of valine. The quantities of valine, threonine and serine were higher in male species, while those of praline, lysine and isoleucine were higher in female species. The quantity of total free amino acids decreased slightly with duration of ice storage in both. The levels of glycine, lysine, and praline decreased in female species appreciably during ice storage.

The total bacterial count of the freshwater tilapia showed gradual decrease up to 6 days and then significant increase on the 13th day in ice which compares well with the organoleptic observations. In the case of brackish water tilapia the total bacterial count showed gradual increase being more significant from the 10th day onwards, which again agrees well with the organeleptic observations.

It has been found that the extractability of proteins of fish muscle decreased during the 24 weeks frozen storage at -18°C. The fall in salt extractable proteins was more pronounced in the brackish water tilapia than in the freshwater fish. In the case of fish held in ice for longer periods prior to freezing, the salt soluble proteins decreased at a faster rate during the first few weeks of frozen storage and thereafter the change took place at a slower rate. The amount of sarcoplasmic proteins in tilapia muscle did not changed significantly during the storage period up to 24 weeks at -18°C. No significant changes were observed in the biochemical characteristics of the male and female species of tilapia during frozen storage. The free amino acid content decreased appreciably in both male and female species during frozen storage. The major amino acids, glycine, valine, praline, arginine and histidine decreased during frozen storage, the rate of decrease being more pronounced in the latter. Lysine, alanine and isoleucine registered an increase after a storage period of 24 weeks.

The lipid fraction of the fish has shown slow hydrolysis, the free fatty acids (FFA) registered a value of 5 per cent after 24 weeks of frozen storage. No significant difference in the rate of development of FFA, during frozen storage was observed between samples stored in ice prior to freezing for varying periods. Peroxide value (PV) rose to 7 for the fresh fish, while the increase was from 14 to 46 for the 13 days iced samples after 24 weeks of frozen storage. The rate of development of PV was slightly faster in samples held in ice for longer periods before freezing. Rancid odours were detected in the samples during frozen storage. The longer the fish was held in ice prior to freezing, the earlier the development of rancid flavours. The 6 day iced, 10 day iced and 13 day iced samples showed signs of rancid odours at the end of 24 weeks, 16 weeks and 8 weeks respectively. PV of the fat at this stage was around 30. No noticeable darkening or yellowing of the muscle was however apparent in any of the frozen samples of freshwater tilapia during 24 weeks of frozen storage. Yellow discolouration, however, was observed in the brackish water tilapia after 4 weeks for the 13 day iced samples and after 24 weeks for the fresh and 3 day iced samples. The discolouration was more pronounced in the male species than in the female ones. Rancid odour was detected in the samples much earlier in the brackish water fish.

Changes Occurring during Preservation of Fish by Freezing and Prolonged Frozen Storage

According to organoleptic studies from time to time at interval of one month, the white pomfret subjected to quick freezing without any chemical treatment and stored in frozen storage were acceptable for only two and half months after which yellow discolouration became much more prominent in addition to thick irremovable white membrane upon the eyes and an undesirable odour on thawing. Ice water glaze was little effective in extending the keeping time of pomfret. Pomfrets that were given ascorbic acid glaze and those packed in gunny bags after treatment with sodium

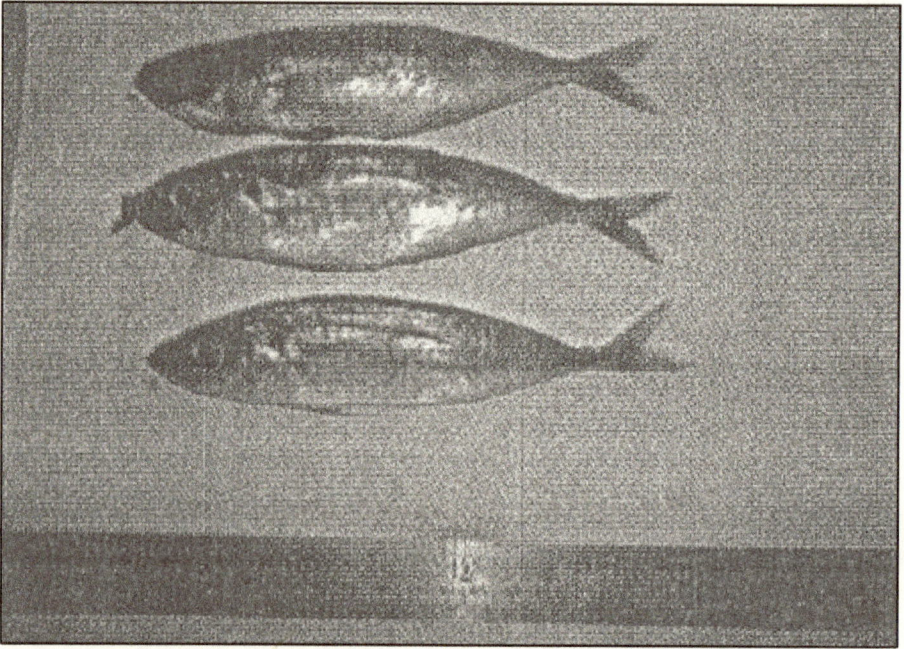

Horse Mackerel (*Trachurus trachurus*)

Preparation of Fatty Fish Mince

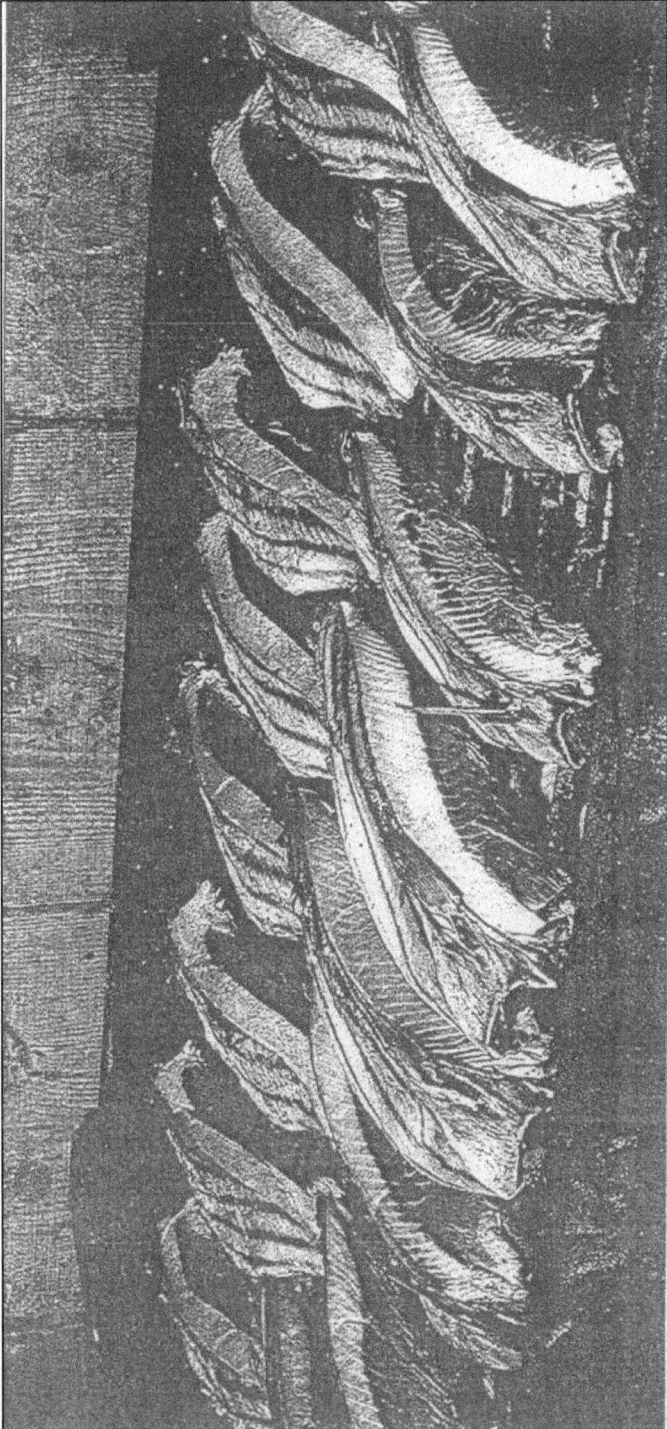

Smoked Fish

Sundrying of Fish in Scaffold

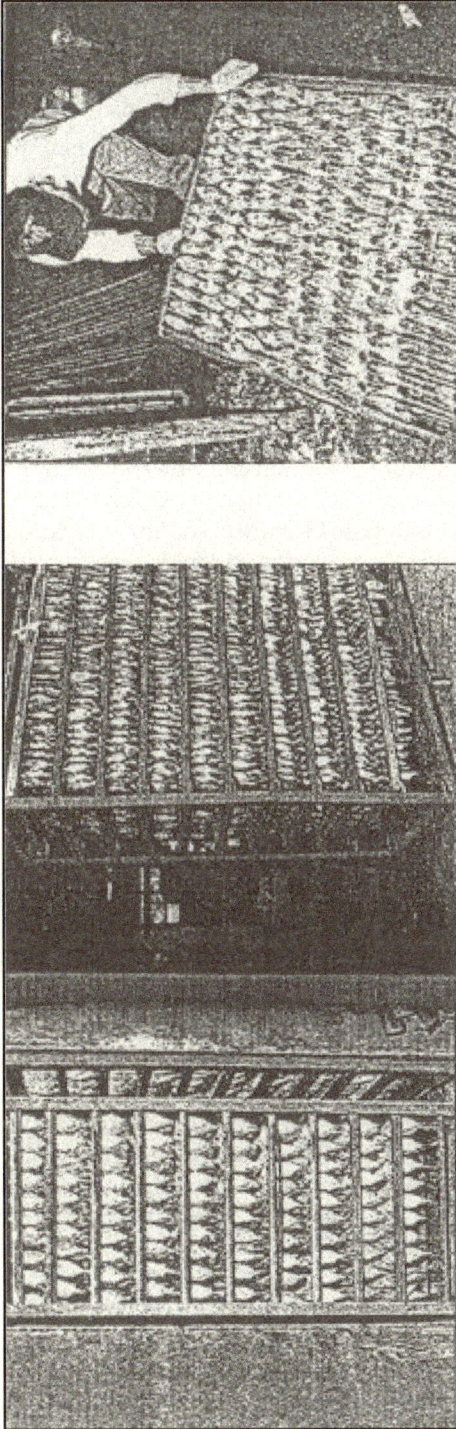

Hot Air Fish Drier

chloride and glucose, had good and acceptable quality up to four and half months Sodium chloride and glucose glaze was the most effective in extending the keeping time of pomfret over five and half months, as no prominent well marked yellow discolouration could be observed and texture was also satisfactory.

Seer fish, with ascorbic acid glaze was acceptable for 5 months only. Sodium nitrite glaze could only preserve fish for more than 4 months in an acceptable condition, but at the beginning of the 5th month it started developing those undesirable characteristics, such as, too tough in texture and thick white membrane formation in eyes. Citric acid glaze, however, extended the storage life of seer fish over 5 months. Mackerels, slow frozen and block frozen could keep well up to 4 months.

Development of yellow discolouration (due to oxidation of fat and unsaturated fatty acids) due to bruising of fish during handling that leads to the rupture of tissue was confirmed that the frozen storage hastened the development of oxidation of fat or rancidity. The process of rancidity was considerably retarded by the application of glazes of ascorbic acid and mixture of sodium chloride and glucose. Since it formed a thin film on the surface preventing the reaction of oxygen with bruised surface in addition to retardation of dehydration. Mixture of sodium chloride and glucose was more effective against oxidation than ascorbic acid in low concentration as shown by the extension of shelf life of pomfrets by 2 months by the former compared to only one month by the latter.

Effect of Freezing, Glazing and Storage on Vitamins and Minerals in the Fish Flesh

The freezing causes some mechanical disruption of musculature of fish, which can cause loss of nutrients.

In pomfrets, seer fish and mackerel moisture content decreased rapidly from its initial value of 75 per cent to 66 per cent after 3 months and then gradually in four and half months to about 63 per cent. Loss of moisture was less in fishes glazed with sodium chloride-glucose solution during four and half months of frozen storage.followed by ascorbic acid treatment. Gradual loss in B-vitamins, such as, thiamine, riboflavin and niacin was noted during storage.

Decrease in moisture content of fish during storage was due to desiccation, as there was nearly 10-15 per cent difference in relative humidity (RH) value of the surrounding environment and that of frozen fish. In the glazed fish losses due ti desiccation were less because moisture from frozen tissue was not lost till the glaze on fish was evaporated. The amount of drip did not appear to be influenced by the rate of freezing, but increased as the period of frozen storage was prolonged. There was considerable loss of B-vitamins, such as thiamine, riboflavin, niacin, pantothenic acid, folic acid and pyridoxine in frozen storage due to thawing of frozen fish stored for long period. Similarly drip of fish exuding on thawing was also found to contain B-vitamins and minerals. Loss is encountered in vitamins and minerals from frozen fish tissue are in the first place, due to various changes that are set in spontaneously by contaminants and then by subsequent freezing and frozen storage over long period. It is common practice to freeze fish when the *rigor mortis* is over. Therefore pH of flesh at the time of freezing is almost neutral and flesh is softened due to loosening. The drip that is developed during frozen storage and liberated on thawing is nothing but emulsion of various nutrients, such as, B-vitamins and minerals. Calcium remains in the tissues in insoluble form.

Chapter 9

Preservation by Reduction of Moisture

Drying

Water is essential for life, and micro-organisms are no exception in requiring plenty of water for their growth and multiplication. Lack of water, or loss of it, can bring to a standstill the activities of the bacteria and moulds that spoil foodstuffs and hence drying can be used as a means of preservation.

Primitive methods of drying fish merely by hanging them, either in the sun and wind, or over wood fires, have been used by man for thousands of years. The process is slow and may take up to several weeks to complete, depending on the size and thickness of the piece of fish, and on the climate. Some traditional dried products are salted as well as dried. They mostly have a strong flavour.

Drying is involved in various ways in processes used by the fish trade. Unintentional drying occurs frequently, for example, when wet fish stands in a breeze on the market or when fish is in cold store, and various measures can be taken, such as suitable packaging or covering, to reduce or prevent unwanted drying. However, many processes, drying is an essential part of the operation, For example, the evaporation of moisture from the product is a complementary part of the smoking process, and is in the production of dried salted fish and fish meal.

The term drying usually implies the removal of water vapor by evaporation, but water can also be removed from fish by pressure or by the use of absorbent pads or by salt.

The rate of drying depends on;

    1. the surface area of the fish;

2. the velocity of air;
3. the thickness of the flesh;
4. the fat content;
5. the temperature of the air and fish;
6. the relative humidity of the air;
7. the initial water content.

Salting

Salting is both a method of preserving fish and a preliminary operation to smoking, drying and marinading.

Broadly speaking, salting is a combination of operations aimed at preserving fish in common salt, beginning with the preliminary work of washing and gutting and ending with packing the salted fish in containers. In a narrow sense, it is the group of operations by which fish is placed in contact with salt (whether as crystals or in solution) and allowed to become permeated by it. These operations include

Salt Curing of Fish

It is washed clean with water.

Salt is sprinkled on the inside and outside of the fish. Salt hardens the meat and extra moisture in the meat is drained off during this process.

After one-day curing, fish are arranged in a flat box.

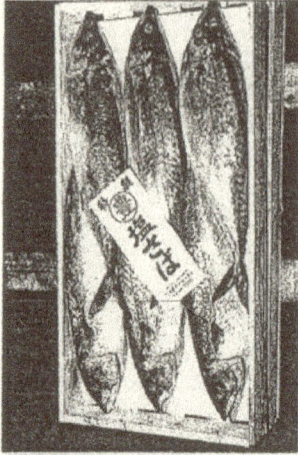

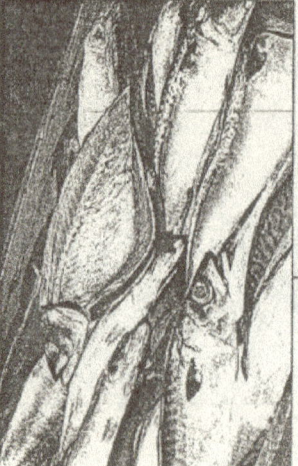

Sending to the market.

Salt Curing of Fish

mixing the fish with salt, laying and keeping the fish in the container in which it is salted, restacking etc.

The main features of salting are the removal of some water from fish flesh and its partial replacement by salt. The passage of salt from fish into water is an example of osmosis; the skin and cell membranes act as semi-permeable membranes which permit outgo of the water and the entrance of some salt. The flow of water through semi-permeable membranes is always from the dilute to the concentrated solution. Thus when fish are placed in concentrated brine, water passes rapidly out of the cells of the fish through the cell wall into the brine. If the fish is placed in dry salt, the moisture on the outside of the fish dissolves some salts and forms a concentrated solution which immediately causes the flow of water from the cells of fish. When sufficient salt has penetrated in the fish and moisture is forced out of the tissues to the point that the inner flesh of the fish loses much of its translucent appearance, it is called "struck through" in the works of the salter.

The preservative effect of salt is often said to be due to the removal of moisture from the flesh, because fish spoilage bacteria cannot grow in the absence of sufficient moisture. Another factor is also apparently involved, salt itself reduces or prevents the activity of these bacteria if more than 6 per cent is present. However, small amount of salt are required by nearly all bacteria for their growth and most of them grow far better in media containing 1 or 2 per cent salt. Salt uptake and water loss are influenced by the fattiness of the fish, the thickness of flesh, freshness, temperature, chemical purity of the curing salt and other factors.

Commercial Methods of Salting Fish

There are two classes of commercial methods of salting fish in common use today; brine-salting and dry-salting.

Dry-Salting

The exact procedure followed in dry-salting fish depends upon the kind of fish and the custom in the particular locality. But, for a general consideration of the subject, the following description is sufficiently detailed: The round, beheaded, eviscerated or split fish are washed and then packed in watertight containers with an excess of dry salt. The proportion of salt to fish varies from 10 to 35 per cent of the weight of the fish, depending upon the kind of fish, the weather and the custom of the fish salter. The fish are usually rubbed in salt just before packing and each layer is sprinkled with salt. After a few hours sufficient pickle has formed to cover the fish; they are not disturbed until completely salted. Then they are either repacked in fresh pickle or removed and dried.

Brine-Salting

Brine-salting is of relatively little importance commercially as compared with dry-salting. The dry-salting method was found to obtain more rapid penetration of salt into the fish and to inhibit decomposition more quickly, therefore it is suitable for the warm climates. The principle of the method is the cleaned fish are placed in large vats, partially filled with concentrated salt solution. A small amount of salt is put on top of the fish floating in the brine. The fish are sometime stirred to prevent the brine from becoming too dilute at any point.

Spoilage of Salted Fish

Salt stop the growth of spoilage bacteria. There are, unfortunately, many types of bacteria found on fish, and they are not equally affected by salt. Some micro-organisms can tolerate much greater quantities of salt in their environment than can others.

Micro-organisms may be divided into three groups according to their sensitivity to salt. Group-I; these that are held in check and even killed by a few per cent of salt. This group includes most of the bacteria that commonly make things go bad. Group-II; those that can tolerate large amounts of salt, even up to completely saturated brine, but which find it increasingly difficult to grow as the percentage of salt rises. Group–III; those that cannot grow without salt. These are so-called *halophiles* or salt-lovers.

Pink Spoilage

Halophilic bacteria is a group of bacteria that has a distinctive pink to rose-red colour, and which is responsible for one of the most troublesome types of spoilage encountered in the dry-salting trade; it is known as *pink*.

Dun Spoilage

Light cures are seldom attacked by pink. There is however, another type of spoilage that is particularly noticeable in the light cures, although it can also occur in the heavy ones, and is known in the trade as dun. The fish surfaces become covered with numerous small spots of a black, brown making the fish look as if it has been peppered. The ' pepper spots' or dun are not caused by bacteria but by molds.

Sun Drying of Bombay Duck

To improve the quality of dry Bombay duck, it is essential that the raw material is brought in absolutely fresh condition. This may be achieved by making two hauls only when ice is not carried out on board or preserving the fish in the holds in ice if more hauls are to be made. The fish must be washed free of mud in clear sea water and other adhering impurities. The fish must be dried on scaffolds gave better products than those obtained by drying them in trays. Optimum rate of drying and quality of dried product were obtained when the fish were suspended at the rate of 50 to 60 per meter length of the rope. The quality of raw material as affected by the delays in different hauls in the fishing trip was reflected in the dried products also.

Reduction of moisture–Heat treatment

Smoking of Fish

Smoked fish is considered as a delicacy, especially in some of the European countries, and it is believed that with more publicity and better smoking methods and packaging, the demand for smoked fish could be increased both in internal and export market. In India however, this method is not widely prevalent.

The best woods for smoking fish are Alder, Oak, Mahagony, Peat, Mango, wild jack, coconut husk, stems of non-resinous hard wood. Wood consists of both combustible and non-combustible substances, the latter being moisture and ash. The combustible substances are complex organic compounds, such as, cellulose, lignin-opentosans, tannic acids, protein substances, resins and terphenes.

Cellulose on heating gives oxymethyl turfural, which is an unstable compound, breaks up into formic acid, levulinic acid and humic substances, which help to give smoked fish its colouration. The other constituents of the normal wood smoke are formaldehyde, methyl ester of pyrogallol, guaicol, vinyl guaicol, creosol, orthocresol, catechol, phenol, engenol etc, which are well known for its bacteriocidal property on pure bacterial culture, especially on non-spore bearing ones, such as, *Bacillus subtilis, Bacillus mesentrieus* survive even when exposed to thick dense smoke for seven hours. Quite different results have been obtained in experiments on the direct effect of smoke on the microflora of smoked fish. Smoke with a temperature below 40°C has a weak antiseptic action, and only with hot smoking, the number of micro-organisms is substantially reduced, but this may be due to the high temperature of the smoke.

Fish is smoked either round (whole, small sized fish), gibbed (gills, visceral parts removed) or dressed gutted as in case of mackerels, sardines, dhoma as fillets (longitudinal pieces) as in pomfrets, seer fish and cat fish and steaks (cross section) as in eels.

Fish is dressed as described above using absolutely fresh fish. The washed fish free from blood, slime, dirt etc. was treated with 8 per cent saturated brine for 30 minutes incorporating 3 per cent sodium proprionate. The salt used should be at least 98 per cent as sodium chloride. After brining, the fish are hung on the rods or hooks and allowed to remain until the surface moisture is removed and sticky to touch. The fish are put in the smoking oven or unit, the smoke fire started so that the temperature should rise within about three hours to about 85°F (29.5°C) or same as that of room temperature. If hot smoke is desired, the temperature of the smoke could be increased beyond 120°F, so that fish get cooked and subsequently dried. Too slow smoking makes the fish dry, too quick smoking deposits too much smoke and make the fish unpalatable. The end product is best recognized when the fish just begins to curl,

In smoking fish, the following are the important consideration; (a) smoke quality, (b) smoke volume, (c) smoke temperature, (d) humidity, (e) smoke/air velocity and (f) distribution.

Smoke Quality

Must be such that it is free from a non-resinous wood. Hard woods are largely used for smoking, because the pitch and resins present in the soft wood are easily volatilized and deposit themselves on fish. Coconut husks also could be used, but has a tendency to flame up.

Smoke volume should normally be delivered by a large diameter pipe or chute to permit condensation of tarry substances.

Smoke temperature should be able to control and it should be at the right temperature.

The relative humidity of smoke should be between 60 per cent to 70 per cent. Smoke velocity should be such that all the material get uniform smoking avoiding air pockets. In smoking kiln's the normal velocity of the smoke is around 100 feet per minute. For uniform distribution of smoke guides or aerofoils can be used.

In traditional smoke houses a standard consistent product to get is difficult. In Torry Kiln, it is possible for even drying the whole batch, even smoking of the whole batch in the same period of drying and controlled warming up of the cure. Smoke of controlled density, temperature passes through a horizontal chamber in which the fish to be smoked are hung. The smoke is generated in a producer hearth lined with fire bricks. The smoke is mixed with air and the mixture then passes over electrically heated grid, whose temperature is controlled thermostatically, and then it is driven by a fan into a duct which runs along the top of the kiln. Partitions in the duct distribute the smoke mixture evenly across its full width. As the smoke is turned downwards at the end of the kiln, it is further evened out. Fish for smoking are loaded on to trucks or trolleys, which are pushed into the kiln. Smoke of the moisture laden smoke is discharged from the kiln through the chimney and is replaced by fresh air and smoke.

There are different types of smoke; (a) distilled smoke, (b) blown smoke, (c) electrostatic smoke, (d) electric smoking and (e) liquid smoke.

The use of chemical agents, such as a liquid smoke or of salt impregnated with chemicals found in wood smoke has been advocated as a substitute for smoking. It is claimed that smoke flavour is better controlled, flavour is improved, greater length of preservation is secured, the requirement for curing is greatly reduced, and that the smoke house may be eliminated. The appearance, texture and flavour were not equal that of fish smoked by common methods.

A well dried product, has a muscle moisture of 8 to 12 per cent and yields range from 22 to 30 per cent of the pre-drying (smoking) weight.

Some of the popular smoke cured fish products are, "Bonga", smoke cured tilapia of Africa; "Pla-krob" smoke cured catfish of Thailand; "Katsubushi" smoke cured mackerels, sardines and horse mackerels of Japan; "Tinapa" boiled smoke cured tuna of Philippines; "Trepang" smoke cured sea cucumber in Philippines, Malaya and Thailand; "Masmin" smoke cured tuna of Lakshadweep; "Bandeng" smoke cured tuna like fish in Indonesia. In western countries cold smoked fish especially considered as delicacy.

Smoking of Eel Fillets

Fresh eels (*Muraenesox talabonoides*) of medium size weighed 2.5 to 5 kg and 120-150 cm length were brought from the fishing grounds and the surface of the fish was rubbed with coarse salt and wash thoroughly with freshwater until the surface became clean and completely free from slime. After removing the bladder and guts and washing, fillets (with skin on) of the appropriate size were prepared.

Fillets of 16x2, 16x2.5 cm, 16x3 cm, 18x4 cm anf 20x5 cm were salted in 20 per cent brine for 30 minutes, dried for 30 minutes and then smoked. The final drying periods required to obtain products with desired level of moisture and appearance.

Salting

Fillets of uniform size were dry salted for 4 hours in the ratio of 1:5, rinsed in self brine, dried for 30 minutes and then smoked.

Fillets of uniform size were dipped in 5 per cent, 10 per cent, 15 per cent and 20 per cent brine for 30 minutes, followed by pre-drying, smoking and final drying.

Fillets of uniform size were kept in 20 per cent brine for periods of 15, 30, 45 minutes, 1 hour, one and half hours, 2 hours and 3 hours period for salt absorption.

Pre-Drying Period

Brined eel fillets of uniform size were dried in sun for 15, 30, 45 minutes, 1 hour and 2 hours followed by smoking. It is preferable to press the cut sides of brined fillets before drying, either with hands or against the sides of the tray to get the surface smooth. It was found convenient to hang the brined fillets on pointed galvanized iron rods for pre-drying and subsequent smoking in kiln.

Smoke Kiln

An asbestos smoke kiln (a vertical box type, traditional kiln) was used, which was less expensive and easy to operate (Size of kiln, 2,45x0.61x0.61 m).

Smoke was generated by burning saw dust in an earthen pot having a narrow hole at the bottom for air passage. The saw dust was filled in two layers, bottom (thick) layer having about 20 per cent moisture and the top layer with 26 per cent moisture in the ratio 3:1. Fire was introduced in the bottom layer near the hole. Uniform dense smoke was obtained by this method.

Various types of saw dust were used. Normally it contained about 10 per cent moisture and water was sprinkled on it to adjust the moisture level in the range of 20 to 36 per cent.

Both green and dried coconut husk were procured, split into small chips and dried to a moisture content of about 20 per cent.

Every kind of wood produces smoke with its own definite characteristics. The most suitable combination to produce best quality smoke is fresh saw dust of different types of wood such as, teak, dajad, mango tree etc and coconut husk were used in different proportions.

Period of Smoking

Brined fillets were smoked for different periods and colour, appearance and odour were observed. Phenol estimation of smoke were determined by exposing 100 ml of water in Petri-dishes in the smoke kiln at different intervals of smoking. Smoke was produced from a mixture of sag saw dust and coconut husk (1:1). The fillets (brined and pre-dried) was smoked for 15 hours and the smoked product was dried in the sun for different periods and moisture and salt contents analyzed.

The fillets of bigger size (20x5 cm) retained more moisture after smoking and took much longer time for final drying (more than 10 hours in sun) to bring down the moisture content to desired level (30-35 per cent). Fillets of 16x3x2.5 cm and 16x2.5x2.5 cm were the most suitable and required only 4 to 5 hours of final drying to get the required level.

Salting makes the flesh firm and it is essential for the development of surface gloss or "pellicle" during subsequent drying and smoking. Salt also contributes to the overall flavour of the finished products and exerts pronounced preservative effect against mold and bacteria at a level of 8 to 10 per cent in the fish. A salt concentration of 9 to 15 per cent definitely affects the flora, killing off or keeping a check many of the putrefactive types normally found in fresh fish. Hence proper salting of fish fillets prior to smoking is of prime importance. Hence salting of eel fillets in 20 per cent brine for 30 minutes is suggested for preparing the smoked eel fillets.

Surface drying (pre-drying) prior to smoking to get "tacky" surface for the development of good colour and gloss is essential. An optimum pre-drying period of 30-45 minutes to get a product with "pellicle" form and maximum smoke deposition was found optimum in 70 per cent relative humidity in the smoke kiln during the initial stage of smoking at a maximum temperature of 50-62°C.

Addition of some extra water in saw-dust is essential for the partial oxidation of sawdust in the commercial type of smoking, since the active constituents of wood smoke result from the destructive distillation and/or partial oxidation of wood and complete combustion of wood gives only carbon dioxide and water. A very attractive smoked product with red colour was obtained by smoking with sag saw dust, but the odour and flavour were considerably acrid and unpleasant. The best results were obtained with a mixture of coconut husk and sag saw dust in equal weights. The fillets smoked by this mixture had a very rich golden yellow colour which turned te reddish during subsequent drying. The product had a very palatable taste without any undesirable after taste.

Smoking for 15 hours is found to be the best for imparting the required level of red coor and bringing down the moisture content sufficiently. During the first 10 hours of smoking phenol was deposited and about 20 per cent water was lost by evaporation during the same period. From 10 to 20 hours, the phenol content did not increase considerably, but about 54 per cent loss of water was observed. This reveals that the rate of absorption of phenol might be higher during the initial 10 hours of smoking and rate of evaporation of moisture might be higher after this period.

The moisture content of smoked eel fillets (smoked for 15 hours) was in the range of 50-60 per cent. Further drying was necessary for the proper preservation of the product. Smoked products having moisture content above 50 per cent were attacked by molds and spoiled within 3 to 4 days of storage at room temperature and those having 37 to 40 per cent moisture had storage life of only 2-4 months, after which they spoiled gradually. Samples with moisture content between 30-35 per cent (15 hours of smoking) remained good for more than 6 months without significant spoilage. However, no mold attack was observed up to 8 months of storage. Below 30 per cent moisture level the product tended to acquire toughness and pronounced shrinkage on the surface. The smoked product with moisture content between 30-35 per cent and salt content about 13 per cent has the best appearance as well as shelf life.

About 86.7 per cent reduction in the initial bacterial load was noted on smoking and drying. In "hot" smoking, most of the organisms are killed and apart from the occurrence of a few resistant micrococci or spore formers, the smoked product as it is taken from the kiln, is nearly sterile. Aldehydes, phenols and fatty acids jointly accounted for most of the bactericidal properties of smoke. About 30.5 mg/100 g steam volatile and 10 mg/100 g steam non-volatile phenols were present in the final smoked product. It is believed that the deposited phenols are responsible for the formation of the colour of smoked products. The analysis for phenols in smoked fish appeared to provide a more useful index of smokiness than the estimations of formaldehyde, total carbonyls and volatile acids.

The proximate yield of fresh fillets and final smoked product from the original weight of whole fish is;

    (a) Fillets–60-62 per cent

    (b) Final smoked and dried product–25-22 per cent

Smoke Cured Fillet from Oil Sardine

So many reasons are attributed for not paying due attention to smoke curing. In the first place, smoke curing, as practiced traditionally yields a product of short storage life. Secondly smoking calls for more careful and critical operations and the carefree approach usually adopted in the conventional dry and wet curing will not yield a product of satisfactory qualities.

Fresh oil sardine (*Sardinella longiceps*) of average length of 14 cm (20 g each) were treated with refined salt, and food grade sodium propionate, butylated hydroxyl anisole (BHA), potassium sorbate.

Fresh fish was headed, gutted and cleaned well in running water and then dipped in brine (1:1 W/V) containing 3 per cent (W/V) sodium propionate and 0.1 per cent (W/V) BHA.

After brining the fish was drained for 10 minutes, spread on a wire mesh tray and smoked for 6 hours at a temperature of 70-80°C. In a smoke kiln coconut husk, shell and saw dust were used to produce smoke. Smoked fish was then removed from the kiln, and allowed to cool to room temperature. Skin, backbone and most of the spines were carefully removed. Twenty five gram lots were packed in clean, coloured cellophane paper. They were stored in glass containers under ambient conditions.

Saturated brine was found to be beneficial for penetration of salt into the fish muscle due to high fat content of the oil sardine. A period of 15 minutes dip was found to be optimum.

Proper draining of the brined fish reduces the content of surface water of the fish, thereby preventing the excessive absorption of smoke. This in turn, ensures an uniformly smoked product having better appearance and flavour. Draining for 10 minutes has been found sufficient for brined sardines.

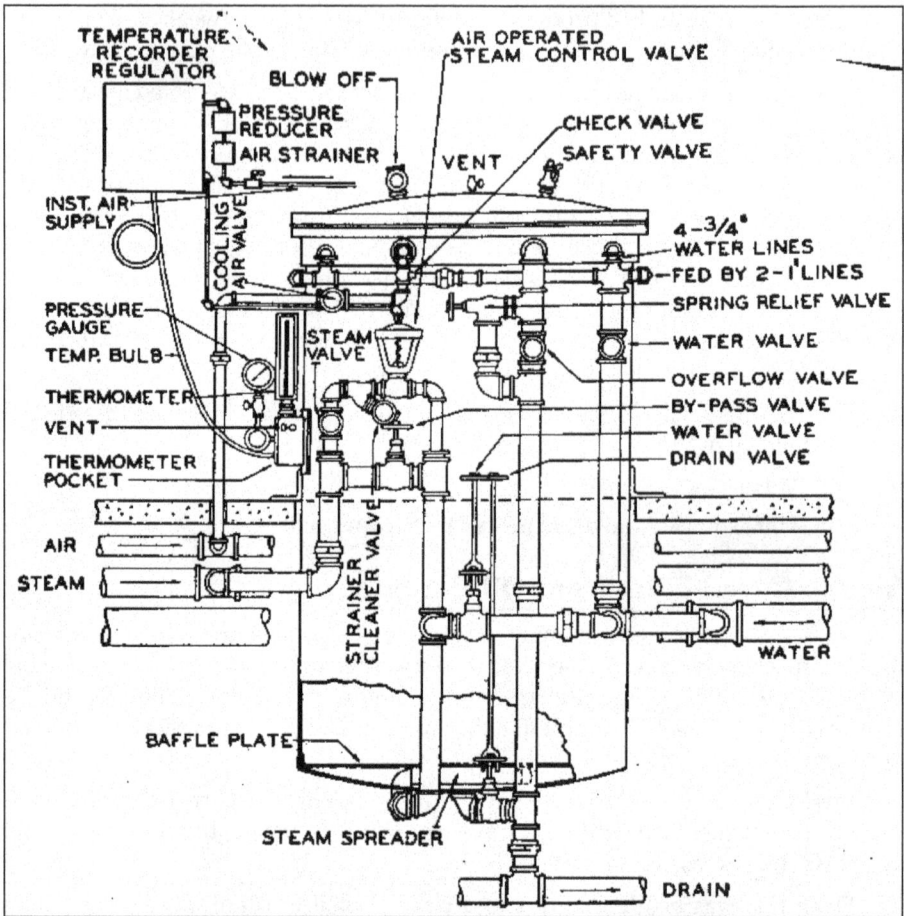

Diagram of Vertical Retort Equipped for Pressure Cooling

By smoking the brined and drained fish for 6 hours at a temperature of 70-80°C, the fish has been found to attain a characteristic smoked colour and flavour. The moisture content under these conditions comes down to 30 per cent. Lower moisture level was found to diminish the reconstitution properties of the product, besides accelerating the oxidation of fat, and also adversely affecting the appearance and texture. The smoked product prepared thus was found to have very short storage life. After three weeks, the product become susceptible to spoilage by fungus, rancidity and the attack of insects.

For improving the keeping quality, sodium propionate was found to be the most effective in warding off the onset of fungus and molds. This is most active when incorporated in the brine at an optimum concentration of 3 per cent (W/V). Higher concentrations of sodium propionate, although slightly improve the storage life, unpaired the taste of the product.

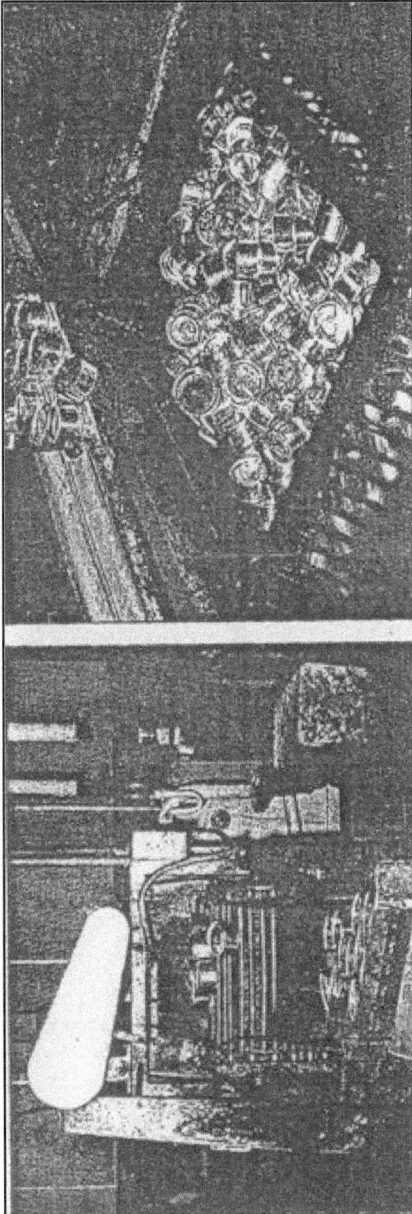

Sterilizing Cans

The high fat content of sardine and the high temperature of smoking favour the early onset of rancidity and the products develop rancidity after 3 weeks of storage and very soon it loses the characteristic smoked flavour and taste. 0.1 per cent (W/V) BHA dissolved in minimum volume of dilute alcohol (1:5) and dispersed in the dip bath by vigorous stirring, the rancidity of the product will be prevented for a period of 4 months.

For the preservation of the attack of mites, the smoked fillets are kept at 125°C for 15 minutes and then allowing it to cool down to about 65°C. After keeping for 5 minutes in the open, smoked fillets were packed in cellophane. The treated product remains free of insects for more than 5 months.

Chapter 10

Preservation by Heat Treatment– Canning of Fish

Historical Outline

It is generally agreed that the original inventor of the art of canning was Nicolas Appert, Frenchman, and that the basic methods of the industry are the result of his work.

In 1795 the revolutionary French Government was at war on land and sea with most of the other European governments in areas as widely separated as the West and East Indies. Few ports were open to the French, who therefore had to depend largely on dried, smoked and pickled foods brought from France. These products were subject to spoilage and their use resulted in the widespread incidence of deficiency diseases, such as scurvy, which greatly weakened the military forces. The French Government therefore offered a prize of 12000 francs to anyone who would develop a new method of preserving food so that decomposition would be reduced and more of the original characteristics of fresh food retained.

Nicolas Appert, a confectioner, brewer, distiller and wholesale caterer was interested in this problem. Appert worked until 1804 before he attained his first measure of success and not until 1809 was his method finally developed. He was awarded the prize in that year after a thorough investigation of his method.

Principles of Canning Seafoods

Micro-organisms in Relation to Canning

The canning of fishery products in hermetically sealed, heat-sterilized containers has for its objective the prevention of spoilage through micro-biological action. Fresh, dried, salted or smoked fishery products may be rendered unfit for use by a wide variety of causes, but to protect canned fishery products careful consideration must be given to micro-organisms which are the cause of putrefaction or spoilage under ordinary circumstances.

Micro-organisms are intermediate between plant and animal life, some almost plants, other animals. They are divided into three general groups, yeasts, molds and bacteria. Yeasts and molds are destroyed at temperatures much below those required for commercial sterilization. Bacteria is mostly concerned with canning of fish and fishery products.

These bacterial organisms are distributed almost everywhere in air, soil and water. Bacteria may lodge on fish while they are alive and in the water, or during the period after they have been removed from the water and before they have been packed in hermetically sealed containers. A large number of species of bacteria may therefore be concerned in spoilage through decomposition.

Micro-organisms require moisture and a favourable temperature for development because they can utilize only liquid food which is absorbed through the cell wall. A moisture content of less than 35 per cent has a direct inhibiting effect. The most favourable temperature for the development of ordinary forms of bacteria is between 70 and 100°F. Live fish are able to resist the growth of these organisms and their temperature is unfavourable to rapid multiplication. After death, however, there is no longer any resistance to growth of micro-organisms, and the temperature of the body tissues rises rapidly. Conditions are then favourable and the micro-organisms on the fish grow and multiply rapidly.

The body tissues must be liquefied so that micro-organisms may absorb the food necessary for growth and multiplication. These organisms throw out secretions known as enzymes, which have the property of breaking down, decomposing, or liquefying the surrounding tissue so that this material may be absorbed through their body walls. Waste products resulting from this absorption are excreted through the cell wall, thus furnishing such distinguishing characteristics of decomposition as "off" odours and tissue breakdown. This process is known as putrefaction.

The cells in animal tissue also contain enzymes which do not cease their activity upon the death of the animal but may digest the cells in which they are contained. This process is known as autolysis or self-digestion and is distinct from putrefaction. Autolysis is especially marked in species which have been feeding when caught. Death prevents completion of the normal cycle of digestion. The large quantities of digestive enzymes manufactured during the feeding process are prevented from fulfilling their function of preparing the food for absorption and attack the walls of the digestive tract, which are destroyed with extreme rapidity, and the enzymes then pass to the flesh which is also softened in a short time. This process is normal and the flesh is not absolutely inedible, but is unsuitable as raw material for canning due to the extreme softness of texture.

The fundamental principle of canning is the application of heat to food in hermetically sealed containers at a temperature and for a period of time sufficient to destroy any yeasts, molds and enzymes, and to destroy or render inactive any bacterial organisms likely to cause spoilage.

Heat-resistant, spore-forming organisms are the most important types and the heat treatment or process must be sufficient to render them impotent. A short heat treatment at a high temperature or a long heat treatment at a lower temperature of about 212°F may destroy the active or vegetative organism but leave the spores unharmed. These forms may return to activity under favourable conditions even after a long dormant period.

The spore-formers are either aerobes or anaerobes. The aerobes require oxygen to develop. If the can has been properly exhausted, that is, if a sufficient vacuum has been obtained, the aerobic bacteria will lie dormant until the processed container is opened and there is little likelihood of spoilage in the sealed container. The presence of aerobic spore-formers in a properly sealed container containing sufficient vacuum indicates insufficient heating but it does not necessarily follow that the product is unfit for food.

On the other hand the anaerobic organism requires the absence of oxygen to develop and multiply. If anaerobic bacteria are present, a complete seal of the container and a sufficient vacuum only make conditions favourable for their development unless the spores are destroyed by heat in canning. Spoilage from this cause is usually indicated by swelled or bulging can ends, accompanied by an extremely pervasive and offensive odour when the can is opened. Spoilage may occur without gas formation and bulged ends, if certain types of anaerobic organisms are present. Such cans are known as "flat sours".

Sterilization

Sterilization in the literal sense of the word, means the absolute destruction of all organisms present, and the maintenance of that condition.

Theoretically, complete sterilization is possible and desirable, but actually there are many factors affecting the cooking process in canning, which work against the attainment of this ideal. Processing is used in canning for reasons other than sterilization or destruction of organisms causing spoilage. In certain products the texture, in others appearance, while in still others the flavour is improved. However, if processing is carried out at too high a temperature or for too long a period of time it may also seriously damage the quality of a product, by causing excessive softening of the texture affecting the appearance through darkening or discolouration, or the taste through scorched or overcooked flavours.

To be effective a process not only must preserve a product, but also must operate within certain limits. A temperature sufficient to destroy spoilage organisms establishes the minimum temperature and time, or lower limit of heating, while quality considerations determine the maximum, or upper limit of processing.

Adequacy of process varies with individual products and depends on a number of factors, the most important of which are given below.

Condition of Raw Material

The first and most important consideration in sterilization processes is the condition of raw material. It is useless to expect the process of heat sterilization to produce a sound, marketable article from inferior or partly-decomposed raw material. Processes are worked out on the basis of the use of fresh raw material in such condition so as to result in a finished article of good edible quality. Variations in the quality of raw material affect the value and adequacy of the process.

Freshness

Due to great number of micro-organisms present in stale or decomposing flesh, sterilization is more difficult than for products that are canned while fresh. Fish that have been handled too much or carelessly, packed too deeply in the holds of fish boats, transported too far or held for too long, usually is heavily infected with spoilage organisms so that the incidence of decomposition is hastened. Every effort has to be made by the industry to regulate conditions of handling and transport.

Maturity

Maturity in fish is associated with the physiological changes of approaching spawning. A softening of the texture of the flesh, which favours increased growth of micro-organisms, occurs when spawning is far advanced. In some species in a spawning condition, ammonia and sulphur compounds in the flesh are more likely to react with the tin plate, thus leading to discolouration of the product or container during processing or later during storage.

Fill of Container

Variations from standards in fill of the container, followed as the best practice, may affect sterilization and give rise to certain difficulties.

Heat Penetration

When the container is overfilled there is a greater mass to be brought to required temperature; therefore, the processing used may be insufficient. An overfill of solid material compresses the mass and decreases the rate of heat penetration by hindering possible convection currents. "Packing" or compressing might occur to such a degree that heat would penetrate by conduction alone.

Vacuum

Overfilling not only retards heat penetration but also may reduce the amount of vacuum since it varies inversely with the volume of head space at the time of sealing, other factors being equal. Such can become "springers" and therefore are unmarketable.

"Cut-out" weight

"Cut-out" weight is the drained weight of the product, exclusive of brine or other liquid, after a specified period of draining, usually 1 to 2 minutes on a screen of 1/8 inch mesh. As such, the fill of the can is the most important factor in maintaining the required "cut-out" weight.

Corrosion

In many products the amount of fill in the can is important in its relation to corrosion of the tin plate. The great advance in mechanical exhaust or vacuum sealing methods has been forced partly by corrosion difficulties. Air in the can should be eliminated as much as possible because the presence of oxygen accelerates corrosion. In under-filled cans, there is greater opportunity for corrosion. Under-filled or "slack filling" is sometimes resorted to in attempts to reduce the process and increases the possibility of corrosion in the headspace of the can.

Strain on Containers

If the container is overfilled, especially with solid material, such as, salmon or shad or in using containers of the smaller sizes, the seams of the can may be strained by physical expansion of the product during processing, leading to spoilage of the product through infection from outside.

Consistency of Product

Starch Content

It has long been known that starchy foods have a very slow rate of heat penetration. Fishery products such as, fish balls, fish pudding and fish pastes must be processed for longer periods than fish which are canned without added starchy ingredients.

A product packed too tightly heats much more slowly than a more loosely packed can of the same product.

Size of Particles

The size of the particles into which a product is divided has considerable effect on the rate of heat penetration. For example, shrimp will heat through to the center more rapidly than salmon, because a can of shrimp contains a large number of small pieces and salmon one or two large pieces.

Composition of Container

Tin or rather tin plate, which is about 98.5 per cent steel base (iron) and 1.5 per cent tin, and glass are the two principal materials used as containers of processed foods. Steel or iron is a much better conductor of heat than glass, while water, an important element in the composition of fishery products, is the poorest of the three unless there is a free flow of convection currents.

Conduction

The power of conduction of any material is expressed in terms of "diffusivity". This term may be defined as the temperature change produced in a unit cube of material in a unit time by a unit quantity of heat conducted across a unit area of the product per unit difference in temperature. Diffusivity is a constant for any given material, and equals 10.8 for iron, 0.37 for glass, and 0.084 for water.

Water, when convection currents are absent, is thus the poorest conductor of heat of the three media mentioned. In terms of diffusivity, the heat conduction of glass is 4.5 times as rapid as that of water. Heat conduction through tin plate is approximately 30 times as rapid as through glass and 130 times as rapidly as through water. Iron heats only from 4 to 6 times more rapidly than water when convection currents in the water are unhindered, and in such case water in glass containers heats approximately as rapidly as the conducting power of the glass. If the conductivity of the container is less than that of the product, the container is the limiting factor to the flow of heat and vice versa.

Thickness of Container Walls

The rate of heat penetration will vary according to the thickness of the container wall. Glass containers, in addition to possessing a lower heat conducting factor, have much thicker walls than containers of tin plate. Additional processing time is required for an equivalent amount of material if it is packed in glass rather than in tin plate.

Size of Container

The size of the container is an important factor in sterilization due to surface-volume relationship. In the smaller sizes of cans the amount of surface per unit of container volume exposed to processing temperature is greater than of larger size.

The distance from the surface to the center of the can also affects the rate of heat penetration. Odd-shaped cans such as, the "quarter square" used in canning sardines have a short surface-to-center distance as compared with the volume of the can. A spherical can would have the greatest distance from surface to center for a given volume.

Initial Temperature

Delay between exhausting or hot filling and processing causes a lowering of the initial temperature, which should be avoided. If the cans are sealed hot they may lose their original high initial temperature by delay before processing, but they would not lose the other principal benefit of a heat exhaust, namely, a good vacuum.

Vacuum sealing or mechanical exhaust has displaced heat exhaust in the greater portion of the sea-food canning industry. In the packing of such products as salmon and tuna, this means that the initial temperature is much lower than with exhaust by heat, and that processes based on heat exhaustion are no longer valid. Therefore, an allowance must be made for a lower initial temperature in sterilization procedures, where the product is exhausted mechanically. A low initial temperature requires a longer sterilization period. A miscalculation of the initial temperature may result in under-processing.

'Coming-up" Time

In the more easily sterilized products such as wet-pack shrimp, in which a slight over-cooking brings about a serious deterioration in quality, the heating received during the "coming up" period must be considered carefully. Delay in bringing to retort temperature or the use of a longer coming-up time than usual may result in the over-cooking of products with a narrow range between the upper and lower limits of processing.

Retort Temperature

Theoretically, it is desirable to use the highest possible retort temperature because of the greater temperature gradient between retort and can. If products of like nature and having the same initial temperature are processed in two different retorts, one at 240°F and the other at 250°F, each will reach the temperature of the retort in which it is being processed in the same length of time. However, cans in a retort at 250°F will approach a temperature of 240°F in a much shorter time than cans in the retort held at 240°F and will hold that temperature over a greater period of time. Where there is likelihood of discolouration or a "scorched" flavour, it is preferable to use the higher retort temperature in actual practice, as the use of higher retort temperatures shortens the time of processing and length of processing appears to more seriously affect quality than temperature of processing.

Mode of Heat Transfer

Heat from the processing medium, steam, water or air enters the can or other container by one of the two means; conduction or convection.

Conduction

Conduction may be defined as the transfer of heat between adjacent stationary molecules, that is, the heat must pass from one solid particle to the next. Therefore, heating is slow and gradual. For example, an iron bar heats along its length by conduction so that one end may be quite hot, while the other is still cool. Dry-pack shrimp, fish pastes and fish cakes are examples of canned fishery products which are heated mainly by conduction. Such processes are longer and the product is more likely to spoil unless the sterilization process is closely watched and carefully worked out.

Convection

Convection may be defined as the transfer of heat by currents. A pail of water heats mainly through convection, since water or other liquids tend to expand when they are heated, which decreases the density. The greater density of colder liquid causes the heated liquid to rise and set up a circulation, thus distributing the heat throughout the mass. Transfer of heat by convection is very much faster than transfer of heat by conduction. Wet-pack shrimp, oysters and ready-to-serve soups and chowders are products heated mainly by convection. Products in which convection currents are sluggish or absent due to a semi-solid nature of the product, such as, salmon, require much longer processing at a given temperature than products in brine or liquid in which convection currents are unhindered.

Time and Temperature Relationships

In sterilization procedures, time and temperature are inseperable. From a study of the death rates of heat-resistant organisms, sterilization was found to be 100 times more rapid at 250°F, than at 212°F. Therefore time is essential and must be considered equally with temperature when specifying sterilization temperatures.

A short process at a high temperature has certain advantages; for instance, more "cooks" per day per retort, thereby increasing the capacity of the plant. Also, there is greater uniformity of sterilizing value between individual cans.

Water in the Retort

The presence of water in the retort has no effect as far as sterilization is concerned, since the temperature of water in a retort is the same as that of steam. The time required to bring a retort to processing temperature will be longer as the heat absorbing capacity of water is great. If any considerable amount of water is present in a retort, the product may be affected in other ways.

A small amount of water in the retort is essential to prevent the formation a sudden vacuum upon the entrance of water, when water is used for cooling.

Air in the Retort

The presence of air in a retort during processing is a potential danger, due to the possibility of non-uniform temperature distribution. Air may be present in a retort unintentionally through insufficient venting, or intentionally as in a steam and air mixture process, such as that which has been used for the processing of packs in glass containers.

In a mixture of steam and air there are certain factors that tend to prevent a uniform mixture, which of course is essential to uniform temperature distribution.

Where two gases occupy the same volume the pressure of the mixture is equal to the sum of the components. This is the well known principle of physics called "Dalton's law". The share each gas exerts in the total pressure is called its "partial pressure". Thus in a retort, the partial pressure exerted by the expansion of imprisoned air is additive to the pressure of the steam. Under such conditions the reading on the pressure gauge is not a true measure of the steam pressure.

Temperature Pressure Relationship

Saturated steam at given pressures yields known temperatures. Many ordinary vented retorts are pressure controlled, that is, the temperature of the retort is regulated by varying pressure. Thus to process at 240°F, the controller is set to maintain a pressure of 10.3 pounds. A pressure-controlled retort must be properly vented so as to contain steam only to have a desired temperature.

Thermal Death Time

Thermal death point is defined as the lowest temperature at which a bacterial culture will be killed in 10 minutes, and is not a constant for all bacteria. Micro-organisms differ widely in their resistance to heat, in fact, they have the widest range of temperature resistance of all forms of life. Certain types of bacteria grow and thrive in cold-storage plants, others in hot springs. Most bacteria, yeasts and molds have an active growing range of temperature about 80°F which may be considered as their optimum temperature, growth usually ceasing at 100°F.

The growth range of certain organisms, known as thermophiles, is at considerably higher temperatures, or from 100 to 120°F. Yeasts, molds and growing cells of bacteria are killed at temperatures of from 170 to 190°F.Spores are more resistant and are destroyed only by prolonged heating at 212°F, although they usually are destroyed readily by heating at 250°F for a short time. It is the heat resistant bacteria which cause trouble in canning, especially the thermophiles. The spores of this type of "heat loving bacteria have withstood boiling water for 20 hours and a temperature of 239°F for 10 minutes.

It is desirable to heat the material at the point of slowest heating of each container to 250°F for at least 3 to 8 minutes or to process it for equivalent times at other temperatures. By equivalent process it is meant that time and temperature are the only variables and other factors such as degree of contamination, rate of heat penetration or size of container remain unchanged.

Vacuum

The degree of vacuum in a container has a direct bearing on sterilization only when the rate of heat penetration in the can is dependent upon the vacuum. The rate of heat penetration is affected when heating is by conduction and the can contents are composed of small pieces, with open spaces which are not filled with liquids, such as dry-pack shrimp. Products are considered to be vacuum-packed when a can containing solid pieces has only a small amount of added liquid and the air surrounding the pieces is evacuated mechanically.

When vacuum-packed products are processed, the slight amount of liquid in the can vaporizes, that is, forms steam and fills all the spaces between pieces. However, if the can was not evacuated and the spaces between pieces are filled with air, the partial pressure of the air will retard the development of steam and consequently the rate of the heating of the pieces. Thus, if a process time and temperature were worked out on the basis of high (28- to 30- inch) vacuum in a can and actually the vacuum were much less, the process might be insufficient for sterilization through the retardation of heat penetration by air in the can.

Vacuum in Canned Fishery Products

The mechanical and chemical reasons for desiring a vacuum in canned food products are; (1) to keep the ends of the cans collapsed; (2) to prevent unnecessary strains on the containers during processing; and (3) to reduce chemical activity.

In order to keep the ends of the cans collapsed under the different temperatures encountered during shipping and storage, cans must have a vacuum.

Cans are subjected to internal strains during processing. Strain is at a maximum when the center of the container attains the highest temperature during the period the container is in the retort. Internal strains are considerably reduced if the container has a high vacuum and sufficient headspace before processing.

Corrosion of the steel base within the tin plate is accelerated by the action of oxygen in air left in the headspace. Excessive headspace or "slack fill" increases the rate of corrosion as the oxygen content is consequently increased. A short heat-exhaust is used to secure a vacuum in preparing the pack. A high vacuum would have reduced chemical activity.

The vacuum obtained in any container is influenced by altitude and the effect on vacuum is similar to the effect of an increase in temperature. If canned fishery products are shipped to localities which are both higher in altitude and warmer in temperature, the reduction in vacuum may be sufficiently severe to cause the ends of the cans to bulge.

Method of Exhaust

Air may be exhausted from a container (a) by heat, or (b) mechanically. The heat exhaust, the oldest type, still used to some extent in packing such products as clam chowder, is to fill cans with a hot precooked product and to seal these containers before cooling. Vacuums of 12 or 13 inches may be obtained if there is no delay between precooking and sealing, but if containers are not filled and sealed promptly the vacuum may be insufficient This is most likely to occur where the product is prepared in batches or in small plants where filling is by hand.

Until recently, in packing canned fishery products, vacuum was obtained by heat exhaust utilizing an apparatus known as an "exhaust box". Several slightly different types are manufactured, but in principle all are steam chests at temperatures of 200 to 212°F. Stem exhaust-boxes are still in use to some extent but are being replaced by vacuum can sealing machines which exhaust the cans mechanically. The size of exhaust boxes and the speed at which they are operated varies in different canneries and in the packing of different products, giving a heat exhaust varying from 3 minutes in the canning of alewife (river herring) roe, to 20 minutes in the canning of mackerel.

The vacuum obtained by steam exhausting depends on the sealing temperature of the contents of the can and the amount of headspace at the time of closure, the higher the temperature and the less the headspace the greater the resulting vacuum. Usually a long exhaust at moderate temperatures (200 to 210°F) is better than a short exhaust at higher temperature as in the case of exhaust by steam under pressure.

Vacuum created mechanically by (1) a suction pump, (2) a steam ejector and (3) a steam blast at closure. A suction pump is probably the most widely used mechanical means of obtaining a vacuum in commercial canning.

Headspace or "Fill" of Can

Solids and liquids expand when heated, displacing a certain amount of air, later contracting on cooling to the original volume.

In an open can, the vapor pressure of the water present, plus the pressure of the gasses in the headspace are at equilibrium with the atmospheric pressure.during exhausting, if the contents of the can are heated there is an immediate increase in vapor pressure from the water. According to Dalton's Law, the combined pressures of vapor and gas cannot be more than atmospheric pressure; therefore the pressure due to gas (air) present must decrease with the resultant expulsion of headspace gases.

If the can be sealed and cooled, the vapor pressure falls to the definite pressure for that temperature. There will then be less pressure in the can and, since pressure and vacuum are inversely related, the lowered pressure will result in the formation of a vacuum. The partial vacuum, due to this lowering of vapor pressure, may be calculated by subtracting the vapor pressure at the time at which the vacuum is determined from that of the closing temperature. For example, the vapor pressure of water at 132°F is equivalent to a column of mercury 4.75 inches in height, while at 72°F, it is 0.75 inch or a difference of 4 inches. Therefore, the partial vacuum due to decrease of vapor pressure after exhaust and sealing is, in this instance, 4 inches.

The amount of headspace affects the degree of vacuum present in a container. During exhaust by heat the gases in the headspace expand. Since the headspace in the can decreases during heating, a portion of the expanded gases is forced out by; (1) expansion of solid and liquid contents, (2) increase in vapor pressure, and (3) expansion of headspace gases themselves, each of which exerts influence upon the partial vacuum in the can on cooling. Some of the headspace gases are forced out in expanding. If the can is then sealed hot and then cooled, the contents will contract with the formation of a partial vacuum in the headspace.

Delay between Exhausting and Seaming

In order to obtain a suitable vacuum when using a heat exhaust, the cans should be raised to a high temperature in the exhaust box and sealed immediately. If the contents of the can are cooled to any degree before the container is hermetically sealed, there is an appreciable loss of possible vacuum. Cans sent through a heat exhaust without covers lose a certain amount of possible vacuum before seaming. The practice of exhausting cans with lightly clinched covers has been followed as a sanitation measure but is also useful as a means of heat conservation

It is essential that all vacuum determinations be made at some uniform temperature, which in usual practice is about 68 to 70°F.

Canning of Oil Sardine

Eventhough methods have been evolved for canning of oil sardine in oil and quality specification laid down by the Indian Standards Institution (ISI), no regular canning and no industry worth its name commensurate with the abundance of this

**Elevator from Unloading Pump to Automatic Scales at a
Sardine Cannery, Monterey, California**

Flume Conveyor Extending from the Wharf into a Cannery at San Diego, California
(The tuna are on a slat elevator in transit to the automatic scales)

fish exists. The seasonal nature of the fishery lasting for a period of 4 to 6 months in an year as well as the erratic behaviour of this resource as reflected in the wide variation noticed in the annual landings have discouraged the entrepreneurs to enter into this industry.

In sixties seven firms were engaged in canning of sardine either fully or partly. Owing to the seasonal nature of sardine fishery, canning was only a minor activity for these firms and they were engaged mostly in packing shrimp and other fishes.

There were wide variations in the quality of the canned sardine turned out by different processors and again with the same processor among different batches. This was largely due to the variation in the quality of the raw material, seasonal variation in the composition of the fish and the variation in the methods employed by the trade.

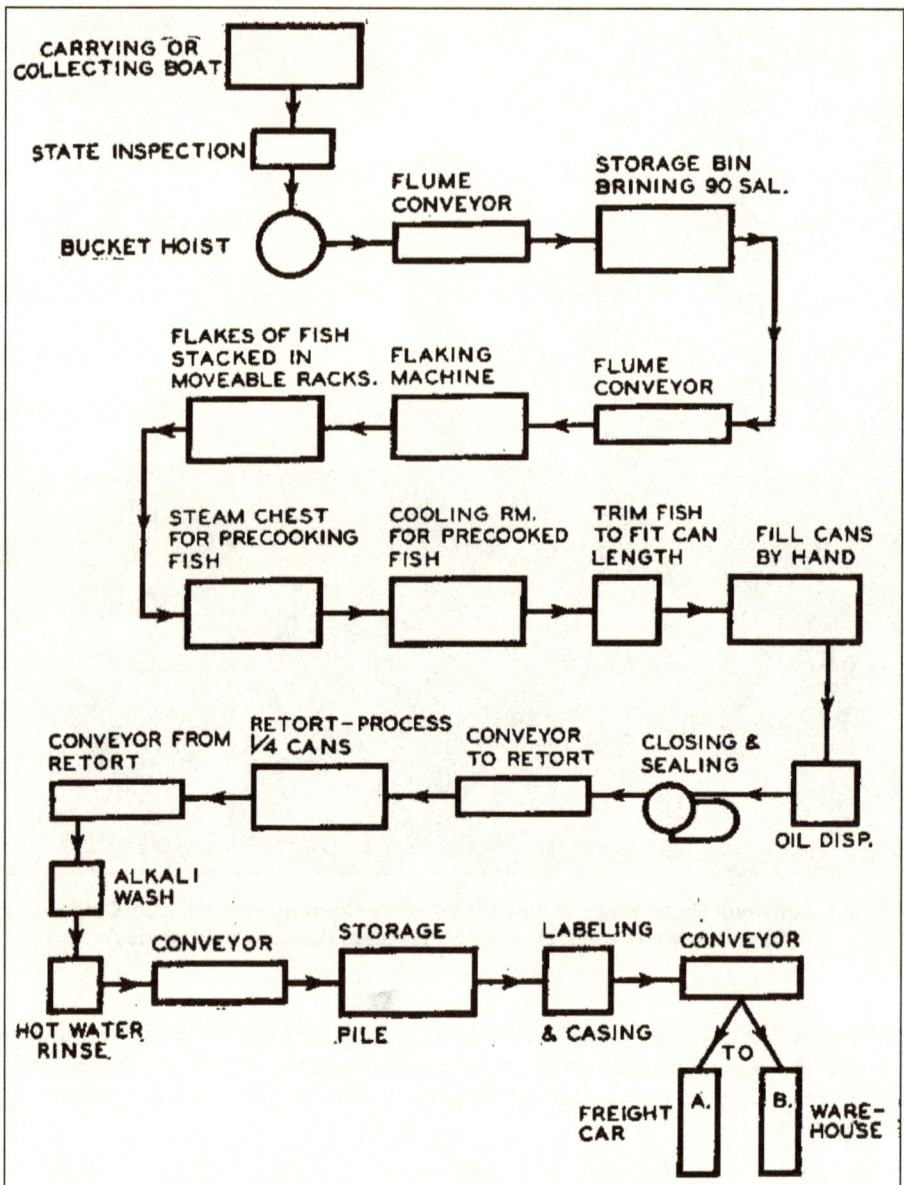

Flow Sheet of Maine Sardine Canning

Besides this, there were various physical and chemical changes, still more significant. Most of the processors procured raw material from important landing centers through agents or brokers and generally there was a time lag from few hours to a couple of days in certain cases, by the time it is ready for processing. During this period, the fish was generally not properly iced with resultant change for incidence of spoilage. Even in properly iced fish, there was loses of nutrients and flavour, development of rancidity and textural changes. Considerable changes occurred in major protein nitrogen fractions, namely, sarcoplasmic myofibrillar and stroma protein. There took place a reduction of 3-6 per cent in sarcoplasmic proteins in about two weeks chill storage, whereas only slight variation did occur in the case of stroma proteins. The yield of myofibrillar proteins was inhibited presumably owing to the presence of free fatty acids. Investigations revealed that the fish stored in ice for more than two days is not satisfactory for canning due to their physical, chemical and organoleptic changes.

The heaviest landings of sardine coincides with the season when the oil content is the highest bringing with it the innate problem of oxidative rancidity and consequent loss of flavour to be faced by the industry. During 4 months of spawning season of the fish (June to September), the problem faced by the industry was the poor appearance of the finished product resulting from belly bursting as the fish is more vulnerable to this phenomenon during this season.

Unlacquered quarter dingly cans (capacity 3 and three-fourth oz) was the container most generally used for packing sardine. Lacquered round cans of 12 oz capacity (301x307) and 8 oz (301x206) were rarely used. The canning medium used in India was mainly refined ground nut oil, though small quantities were packed in tomato sauce.

The fish, called sardine are of different species in different countries and different in appearance, size, taste and composition. They can be classified in three groups; (a) Sprat, obtained only in west European waters, (b) Sea-herring obtained in North Atlantic and (c) Pilchard obtained mainly in the Mediterranean and off the Atlantic coasts of Spain, Portugal, France and South Africa.

The main world producers of canned sardine are Portugal, Spain, Morocco, Norway, South Africa, the USA and Canada.

The methods employed for canning vary from country to country due to local customs, the type of fish available and the medium used. Conventionally sardine is canned after evisceration, and beheading. But Portugal and Morocco are producing canned sardine of the skinless and boneless flesh type. These countries as well as Spain pack pilchard sardine almost entirely in olive oil. In Norway sea-herring, lightly flavoured with oak-wood smoke are used for canning in oil with sauce. South African packs are mainly in tomato sauce though a small portion is packed in oil. Canning medium used in the USA and Canada is mainly soybean oil and to a lesser extent either in tomato sauce or mustard sauce.

Sardine is packed in a number of different types of cans. In the USA aluminum cans are mostly used, the rest being made out of tin plate, size varying from two and half ounce to fifteen ounce. But the most abundant type is quarter size keyless can with a net weight of three and one-fourth ounce.having lithographed lid. Pull-tab lid is a recent introduction. Cans imported into the USA from countries other than Canada are often provided with lip and keys. Mediterranean packs are in the quarter club cans with net weight four and three-eigth ounce, half pound oval cans with net weight eight ounce as well as one pound tall or oval can with net weight fifteen ounce are being used in South Africa. Fish is packed in single layer, double layer or by cross packing methods depending upon the size of the fish to be canned.

In order to improve the quality of canned sardine CIFT has developed a process of packing sardine in its own juice, thereby saving labor and cost of oil. This process also improves the quality of the product, particularly, the flavour, which is very close to the natural one, especially because the natural flavour bearing constituents are not lost by pre-cooking.

With respect to oil packs, the presence of water in fill reduces the consumer acceptability in importing countries. The higher percentage of oil resulted in comparatively lower shelf life and there occurred increased incidence of Peroxide Value and Free Fatty Acids in the oil. Sardine packed in a 1:1 mixture of ground nut oil and sardine oil increases the acceptability. Agar agar can very well be used as a canning medium.

Canning of Oil Sardine-Natural Pack

A simple and economic process for canning of oil sardine (*Sardinella longiceps*) in its own juice having very good organoleptic characteristics has been developed. The landed sardine, while still in rigor, was transported in ice to the processing center. Fish after dressing, scaling were thoroughly washed in potable water till free of dirt, blood etc. and then dipped in brine containing potassium aluminum sulphate (potash alum) and citric acid, drained well, packed in quarter dingly cans in quantities sufficient to give a net weight of 106 g, exhausted in steam in an exhaust box, seamed and heat processed in retort.

Refined salt, food grade potash alum and citric acid were used. Fish used for processing were of uniform quality as regards size and chemical composition, the average figures being,; length-12-14 cm, average weight- 35 g, average moisture- 68 per cent, average fat- 8.16 per cent, average total nitrogen- 3.028/100g, average non-protein nitrogen- 339.9 mg/100g and average alpha amino nitrogen- 52.52 mg/100g.

Cans prepared out of sardine have to be heat processed to such an extent that the bones also become soft. In order to provide the desirable characteristics as regards the texture, particularly the softness of the bone to the required level, a period of 45 minutes heat processing at 1 kg/sq cm (120°C) steam is optimum.

When sardine is canned after treatment with brine alone, the self juice formed inside the can becomes turbid and discoloured. Besides on sufficient heat processing to render the bones soft, the flesh becomes very soft and disintegrated. Alum helps in providing the flesh a firm texture while leaving the self juice clear owing perhaps to its sedimentation action due to which the colloidal and dispersed solid particles (in the self juice) get sedimented during the process of brining itself and thus do not pass into the cans. However, treatment with alum alone along with brine imparts a slight bitter taste to the fish. In order to overcome this, a mixture of alum and citric acid is used. Citric acid, by lowering the pH of the fish muscle, imparts slight firmness to the flesh and a combination of citric acid and alum imparts a most desirable texture to the flesh with good taste and appearance.

Cleaning Cooked Tuna (The light meat used in canning is separated from the bones, skin, dark meat and other refuse)

Cutting Cooked Tuna for the Can

Filling the Containers with Shrimp and Checking the Fill-in Weight

Treatment with 15 per cent brine containing 1 per cent each of alum and citric acid for 20 minutes yield good products as the chances for salt induced rancidity of the body fat can be minimized.

Natural pack method effects a greater economy (pre-cooking to prevent the cook drip in the oil and avoidance of oil in the natural pack) in the production, side by side yielding better product retaining the natural taste of sardine.

**Horizontal Retort Used in Processing Tuna, Showing
the Metal Basket and Arrangement of Cans**

Canning of Tuna in Oil

Skipjack (*Katsuwonus pelamis*), yellowfin (*Neothunnus macropterus*) and bigeye tuna (*Parathunnus obesus mebachi*) weighing 2.5 to 4.0 kg, 30.0 to 43.0 kg and 70.0 to 82.0 kg respectively were kept in ice for 2-3 days before processing. The procedure adopted for canning the fish was as follows;

1. **Cleaning** – The fish after removal of viscera, heads and fins were washed thoroughly till free from slime and blood. They were then cut into chunks of suitable size for cooking in the autoclave.

2. **Cooking** – The species were arranged in cooking racks holding approximately 20-25 kg and cooked in steam at 0.84 kg/sq cm, selected at random until such time that the surface of the back bone attained a temperature of 93-94°C.

3. **Preparation** – Cooked fish were cooled either in a current of air or a cold room (3-5°C). The softened skin was scrapped off using a stainless steel knife and the flesh separated into halves from both sides of the back bone. The portions were again divided horizontally and the black meat removed completely. The loins were polished by rubbing off the loose fragments of flesh.

Oil-Dispensing Apparatus Used in Tuna Canning

4. **Processing** – The loins were cut to suit the size of the can. 170 gram material was packed in cans of size 301x206 followed by 50 ml hot refined ground nut oil providing for sufficient headspace. Cans were exhausted in an exhaust box, seamed in a double seamer, and sterilized in an autoclave in steam at 0.84 kg/sq cm for one hour, immediately cooled and stored.

In order that the water content in the filling oil be below 5 per cent, the fish should be cooked to a temperature of 90.5°C at the surface of the backbone, in which case the water content was 3 per cent. To be on the safer side, when bulk quantities are cooked as in commercial practice, it is advisable to prolong the cooking to such a period that the backbone attains a temperature of 93-94°C, when the water in the filling medium can be kept minimum. The same conditions were found to be applicable to fishes of all size grades.

The cooked pieces can be cooled either in a current of air or in a cold room. The latter has got a definite advantages over the former since the flesh becomes hard enough to permit easy scraping off the skin and cleaning the loins free from black meat with minimum flaking of flesh.

Dipping of cut pieces in 15 per cent brine containing 0.075 per cent sodium bicarbonate (added as soften agent to give a tender texture of the product) for 22 minutes, though imparts a proper saltish taste and good texture to the product, is invariably accompanied by the risk of increasing the volume of water in oil above that prescribed by IS specification, by the brine adhering to the pieces together with any water expelled from the flesh due to insufficient pre-cooking.

Alternatively, addition of solid salt to the meat packed in the cans can be advantageously adopted. Two per cent salt on the weight of packed meat containing 0.5 per cent sodium bicarbonate, has been found to serve the same purpose as the brining mentioned earlier. Concentration of bicarbonate above 0.5 per cent in salt definitely improves the texture, but is accompanied bleaching of the natural pinkish colour of the flesh and imparing of the flavour.

Fish to be pre-cooked for such a period that the surface of the backbone attains a temperature of 93-94°C so as to keep down the water content in the filling oil to the minimum and then cooled in a cold room to facilitate easy handling during the subsequent stages of processing. Addition of solid salt instead of brining dispenses with the possibility of introducing any water into the filling oil in addition to that from fish muscle. Incorporation of 0.5 per cent sodium bicarbonate on the basis of salt gives a good texture of the product.

Canning of Lactarius

Fresh lactarius after evisceration, gutting and cleaning was cold blanched in brine (15 per cent and saturated) alone containing different concentrations (calcium chloride-0.5 per cent, acetic acid-0.5-1.0 per cent, citric acid-0.5-1.0 per cent and alum-1 per cent) of citric acid, calcium chloride, alum as well as some of their combinations and then canned.

Salmon Canning Line; Closing Machine, Can Washer, Retorts

Lactarius cold blanched for 20 minutes in saturated brine yields sufficient soft meat, when heat processed for one hour. However, the bone is rendered soft only on heat processing for one and half hours, which in turn, renders the meat excessively soft and causes to lose the discrete shape in further handling.

Although all treatments in general are effective in maintaining proper texture of the meat, while rendering the bone soft, some treatment impart certain undesirable effects on the quality of the finished product, like, colour change of the meat, loss of luster of the skin. Potash alum though is known to impart firm texture of the meat, imparts slight bitter taste. Cold blanching in saturated brine containing 1 per cent citric acid yields a satisfactory product. It retains good flavour, texture as also colour of meat and the luster of the skin, while the bone is sufficiently soft.

Canning of Smoked Eel

Absolutely fresh medium sized eels (*Muraenesox talbonoides* and *M. cinereus*), procured from landing place and their surface rubbed with coarse salt to remove the slime. The fish was then cut longitudinally into two halves and the bones removed. Fillets of 18-19 cm length and 2.5-3.0 cm width were made and washed with potable water. The fillets were then dipped in 10 per cent brine for 15 minutes, drained and dried for 30 minutes in a tunnel drier (45-50°C) to remove surface water. They were

Salmon Canning Line; Clincher, Can Washer and Closing Machine

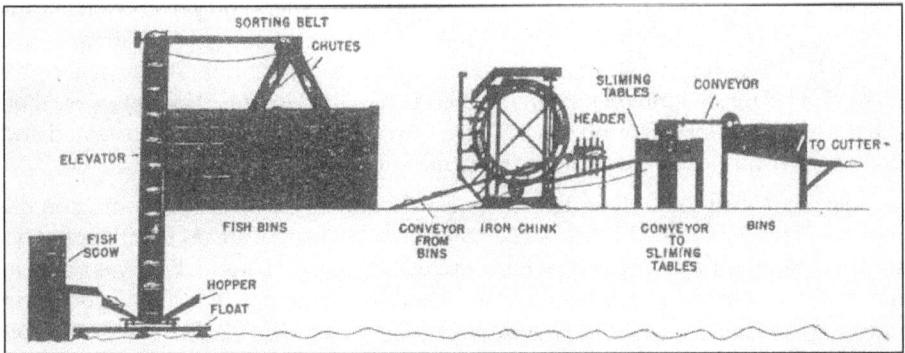

Salmon Canning; Unloading and Butchering

Salmon Canning Line; Fish Bin, Fish Cutter and Can Filler

then smoked for 8 hours. The smoked fillets were further dried for different periods to get varying moisture levels and cut into pieces of suitable size. In cases where moisture was less than 50 per cent, the skin was peeled off at this stage and pre-cooked in cans at 0.49 kg steam pressure for 15 minutes followed by draining and filling in cans (size 301x307). When the moisture content was above 50 per cent, the fillets were pre-cooked in perforated trays, skin removed and then filled in cans. The cans were then filled with hot refined ground nut oil (85°C), exhausted for 10 minutes, seamed and processed at 1.05 kg steam pressure for 45 minutes.

The moisture contents increased generally during canning. Depending on the moisture level of smoked fillets, 42 per cent to 54 per cent of steam volatile phenols were expelled by canning process. Even though considerable reduction was noted in non-volatile phenols during canning, the trend was irregular. The loss of cook drip nitrogen from smoked fillets was proportional to the moisture contents prior to pre-cooking. The same trend of change was noted in processing drip and drip nitrogen. This shows the loss of drip and drip nitrogen can be minimized by reducing the moisture content of smoked fillets to an optimum minimum level without affecting other qualities.

If the moisture content of smoked fillets was very high (66.6 per cent) the final product was found to be very soft crumbling and excessively juicy. As the moisture level was reduced to 51.94 per cent, the product become firm, and sufficiently juicy with appealing appearance, odour and flavour. At very low moisture levels (32.07 per cent), the final product become tough and lost the juiciness. So the optimum moisture level of smoked fillets at which sufficient juiciness, excellent odour and flavour were retained in the canned product was about 52 per cent.

Smoked fillets at moisture level around 50 per cent on canning remained in good condition for 10 months after which gradual deterioration in quality took place.

Canning of Smoked Sciaenids

Procedure for turning out a whole some smoked and canned fish from Sciaenids consists of dressing and cleaning of freshly caught sciaenids, kept in crushed ice and the dressed fish, divided into four batches and cold blanched in 15 per cent brine, 15 per cent brine containing 0.5 per cent potash alum and 0.2 per cent citric acid for15 and 30 minutes. The fishes were then drained, suspended in a smoke kiln in tail up position and exposed to smoke from saw dust for a period of 60 to 180 minutes at 40 to 50°C. The smoked fishes were then packed in S.R. lacquered cans (301x206 size) and pre-cooked at 0.35 kg/sq cm steam pressure keeping the cans in inverted position for periods varying from 25 to 65 minutes. The cans were then filled with hot refined ground nut oil, exhausted, sealed, sterilized at 0.7 kg/sq. cm steam pressure for periods varying from 30 to 60 minutes and cooled immediately in potable water. After surface drying the cans were kept at room temperature and analyzed for physical and organoleptic characteristics.

Blanching in 15 per cent brine for 30 minutes gives a fairly good product except for its soft texture. But incorporation of 0.5 per cent potash alum and 0.25 citric acid in the blanching brine improved the texture considerably. The processing time was chosen in such a way that the bones became soft and easily chewable, but the muscle was not over cooked. It was found that a minimum of 60 minutes of processing at 0.7 kg/sq cm steam pressure was required to achieve these quality criteria. It was also observed that a pre-cooking time of 50 minutes at 0.35 kg/sq cm steam pressure was required to bring down the water content in the filling medium to a desired level of less than 51 per cent, above which the canned product had an unappealing appearance with development of rancidity in the filling oil.

Smoking for 60 minutes was insufficient to impart a satisfactory flavour to the product, while 120 minutes smoking gave a product with good smoky flavour and the colour became appealing. A longer smoking time imparted a deep brownish colour and intense smoky flavour to the final product.

Blackening of Canned Prawns

Among the various technical problems associated with seafood canning industry, product blackening and sulphur staining of can interior are considered to be the most important. Canned crustaceans, particularly, prawns though packed in special lacquered cans are not free from these phenomenon. Sulphur staining is of two types, one caused by tin sulphide and the other by iron sulphide. Apart from mere staining, deposition of iron sulphide at break points of the can interior is also often observed. In the case of canned shrimp products, sporadic outbreaks of iron sulphide discolouration have been reported. In the case of canned tuna it is known that the formation of iron sulphide deposits is not favoured by the mere presence of hydrogen sulphide and exposed iron in the vapor phase, but is greatly influenced by the presence of either volatile acid or bases released at the time of heat processing and cooling of the cans.

Types of Blackening

Analysis of commercial wet pack canned prawns revealed that the blackening of the canned meat took place by direct inter-action between sulphur and copper or iron or by secondary reaction through iron sulphide deposits. In the former case there existed a linear relationship between the intensities of blackening and the copper content of the material. Slightly blackened products usually showed above 15 ppm of copper in the meat on dry weight basis (DWB), while moderately and heavily blackened samples contained 28 to 64 ppm of copper (DWB). Meat containing 200 ppm of iron (DWB) or above showed a deep brownish discolouration, which was distinct from blackening caused by copper. Iron sulphide blackening as against discolouration was characterized by deposition of black spots initially along the vertical seam joint spreading slowly along the internal seam curvature and imparting black discolouration to the meat in the can.

Source of Contaminants

The main source of copper and iron was the prawn tissue itself, which after canning was found to contain 3 to 12 ppm of copper and 12 to 69 ppm of iron (DWB). Water, salt, ice, copper and iron base utensils with which the meat came into contact during the various stages of canning and repeatedly used blanching brine in continuous blanching (the use of same blanching brine for a number of batch wise blanching operations) might also individually or collectively contribute to the mineral contamination. The ranges of heavy metal associated with water, ice, brine, citric acid and meat at various stages of handling was from 0.049 to 260 ppm with respect of iron and 0.0017 to 75 ppm with respect to copper. The accumulation of heavy metals in the blanched meat during continuous blanching was 20.37 ppm iron and 2.96 ppm copper for the first blanching which increased to 117.28 ppm iron and 3.87 ppm copper at 18th blanching. Inert type of utensils such as, stainless steel or plastic vessels or others, which form white metallic sulphides, such as, aluminum may be used for handling the material for canning.

Iron being the most abundant mineral in the world, the chances of contamination are more. However, blackening caused by iron sulphide is less on its deposition is not favoured under standard condition of canning and usually the iron content of meat does not exceed its critical limit. But canners very often drift from the standard technique of processing befitting their need and facility only to face some difficulties. Though the most modern "inert-lacquer" is used for coating the interior of the tin-plated iron-base container, it does not completely prevent the transmission of iron from the body of the can to the contents during storage. Experimental samples containing more than 200 ppm of iron in the meat (DWB) show some discolouration of meat which is usually uncommon in commercial cans. However, the effect of these higher levels of iron could be controlled by the incorporation of certain chelating agents such as salts of EDTA.

Besides the heavy metals, volatile sulphides are also necessary for the formation of black copper and iron sulphides. The main source of these sulphides are the sulphur containing amino acids of the tissue and the sulphides sometimes found in water. The release of sulphides from the tissue depends on (a) quality of the meat, the poorer the quality the more is the release of volatile sulphides. During ice storage of the meat, the volatile sulphides show an initial increase followed by gradual fall probably due to loss of solubles by leaching; (b) retorting conditions, the higher the retorting time and temperature, the more is the release of volatile sulphides due to break down of tissue; (c) concentration of brine in the can, the lower the concentration of salt (less than 2 per cent), the more is the release of sulphides; (d) pH of the contents, the higher the pH of the packed contents, the more is the release of volatile sulphides. It is known that cystine is the main precursor of sulphides in the muscle, its breakdown (at 100°C) occurring only in alkaline medium.

Control of acidity in canned production is very important in order to maintain its storage life and also to keep it free from undesirable influence of contaminations. Analysis of commercial cans indicate that deposition of iron sulphides takes place even at pH levels 6.4 to 6.6, which are usually taken as ideal, but not in cans having titratable acidity of 0.06 per cent level or above. On the other hand, different levels of acid added to cans under identical conditions of processing, also show the same pH values. The correlation between the change of pH and acidity of brine for commercial cans was worked out to be 0.156 which is not significant at 5 per cent level indicating its non-linear relationship. Emphasis has therefore to be made on the titratable acidity of the fill-brine in the processed can, rather than pH.

Generally, acidity of brine in can is controlled by addition of citric acid in the blanching or filling brine or both. Addition of (up to 0.5 per cent) citric acid in the blanching brine and finally packing the blanched meat in simple brine is a common commercial practice. In such cases, it has been found that final acidity in the cans vary significantly depending on the number of blanching carried out in the same liquor without replenishment of the acid Generally after each blanching about a gallon of liquid is removed with simultaneous addition of same quantity of freshwater and salt, but no care is taken for the maintenance of acidity. So with continuous blanching, the acidity of the blanching brine and consequently the titratable acidity of the fill-brine decrease gradually. As the number of blanching using the same brine increases, there is reduction in the final acid content in the can up to the tenth blanching, where after it tends to be almost constant. This indicates that in order to control the acidity it is desirable to add required quantity of acid to the filling brine depending upon the period of pre-processing, ice-storage of prawn, size of can etc.

In the standard methods adopted by the trade, levels of acid to be used in the blanching and filling brines are fixed by convention irrespective of the conditions of freshness of raw material. It has been found that this invariably results in fluctuations in the titratable acidity. The quantity of acid to be added to the filling brine was found to be more in the case of raw material preserved for longer periods in ice. This may probably be due to the leaching out of water soluble fraction and consequent shrinkage caused to the blanched meat during the later stages of ice-storage.

A four-day iced material requires almost three times the acid of that required by fresh prawn to maintain the same level of titratable acidity in the finished product.

For maintaining desired level of acidity to prevent iron sulphide deposition in the cans, the correct ratio of meat and brine is required to be maintained. But canners under standard conditions of canning usually add the same filling brine with or without acid in all cases irrespective of can size. Iron sulphide deposition is likely to be more in the 401x411 cans.

In order to maintain the same level of acidity in 401x411 cans, it is necessary to add 1.5 times more acid than that required by the other sizes (301x109; 301x206 and 301x307).

It is not the species of prawns have influence on the quantity of acid to be added for maintaining the acidity, but the size of the prawns exhibits significant effect, smaller size grades requiring more acid infilling brine compared to bigger size grades. It is therefore necessary to blanch the smaller size grades in lower percentage of acid (0.1 per cent) and then pack in 0.2 to 0.25 per cent acid (in fill brine) rather than the reverse conditions, which are commercially practiced.

It is recommended that material used for canning should be fresh and should have minimum level of heavy metal contamination. Only potable water free from copper and iron, as far as practicable but not containing more than 0.1 and 0.3 ppm respectively should be used for all requirements in the various stages of canning. Direct use of iron and copper utensils, particularly, the latter should be avoided.

Cans to be processed at a lower temperature for a longer period (22 minutes at 115°C) rather than at high temperature for a shorter period (12 minutes at 121°C). Higher temperatures and prolonged retorting should be avoided.

Titratable acidity (0.06 to 0.15 per cent) of brine in the processed cans should be maintained by the incorporation of specified quantity of citric acid in the filling brine depending on the period of pre-processing ice storage of the prawn.

Mere adjustment of titrable acidity does not prevent cut-end blackening in backwater prawns canned in brine. In these cases, addition of 50 mg per cent of disodium EDTA in the filling brine (3 per cent sodium chloride + 0.1 per cent citric acid) completely prevents blackening and improves the colour. Commercial scale application of the additive has established cent per cent reproducibility of the results. Storage up to an year, of the product showed no adverse effects attributable to the additive. The additive ia a permitted one and the cost is negligible.

Micro-Flora Involved in Spoilage of Canned Prawns

The pattern and cause of bacterial contamination of canned prawns are the same as those in all heat processed canned foods, namely, under-processing which leaves heat-resistant, thermophilic spore-formers in the cans or leakage through seams, which causes the entry into the cans of a wide variety of organisms of all types from air, water etc. The problem is of special interest to India, particularly from the point of view of its export oriented prawn canning industry.

Fifty pure strains were isolated and, collected from different factories showing visible spoilage by bulging of which 22 were gram positive spore formers belonging to the *Bacillus* type. Other types isolated included 7 gram positive non-spore forming rods, 8 gram positive *Cocci*, 7 gram negative *Cocci* and 6 gram negative rods.

From cans which did not show bulging, but which showed the presence of bacteria in them (as evidenced by positive growth in thioglycollate broth) 1436 pure cultures and 34 mixed cultures were isolated.

More than 30 per cent of the spore formers showed growth at 56°C and about 80 per cent of them produced hydrogen sulphide. Tolerance to sodium chloride varied from 5 to 10 per cent.

About 80 per cent of the strains isolated from cans were gram positive spore formers of the *Bacillus* type. No anaerobe could be isolated. The gram positive *Cocci* and gram negative rods are next in the frequency of occurrence to gram positive spore formers. Gram positive non-spore forming rods, gram negative *Cocci* and *Coccoids* are rarely observed. In the case of mixed cultures, generally the strain was found to be a gram negative rod. The other strain was either gram positive spore former or gram positive *Cocci*.

Canned prawns are processed at high temperature in steam under pressure. This procedure has the effect of eliminating all bacteria except those having spores of exceptionally high resistance to heat. The presence of only spore forming organisms growing at 37°C/or 55°C is generally indicative of under processing. Occurrence of spore forming as well as non-spore forming organisms in cultures might be due to the cans by passing the retort room without receiving adequate heat processing. The presence of mixed flora of rods and *Cocci* on microscopic examination and the isolation of these organisms on sub-culture is indicative of leakage. Spoilage due to leakage may be caused by excessive contamination of the cooling water or damage to the cans through rough handling.

The results indicate that in the bacteriologically defective cans (which average about 0.3 per cent of the total productions) spore formers predominate though present in some cases with a mixed flora. Analysis of cooling water from canning factories, where defects were observed, also showed the presence of spore formers. These point out that the cause of contamination of the cans could have been leakage through seams during the cooling process.

Only a few species of *Bacillus* type, namely, *Bacillus pantothenticus, B. firmus, B. brevis, B. pumilus* were frequently met with in the cans and those which were not spoiled at the time of examination, is an indication of their role in the spoilage of canned prawns.

Bacteriological Investigations of Prawn Canneries

A detailed survey of prawn canneries around Cochin, Kerala, revealed the micro-organisms present in the factory environment and their role in causing contaminations of the canned products. About 26 per cent of the total of 1030 strains isolated were found to be gram positive spore-formers of the *Bacillus* type; the cooling water being the major source.

In actual practice, the bacterial contamination of the product can happens in many ways. The various types of organisms harbor in the slime and guts of the fish, find slow entry into the flesh after death. During subsequent processing, extraneous contamination is a possibility unless due care is taken. This extraneous contamination is mainly by bacteria associated with the surface of the utensils and equipments with which the material comes into contact during transportation and processing. Further, the water used at various stages of processing is also a potential source of contamination of the products.

Twenty nine samples of raw material, 110 swab samples, 160 water samples, 15 ice samples and 112 cans of ten leading canning factories revealed about 40 per cent (410 numbers) of the strains out of 1030 were motile. Biochemical characteristics like catalase production, reduction of nitrate to nitrite, production of indole, hydrogen sulphide and acetylmethyl carbinol, liquefaction of gelatin, hydrolysis of starch, utilization of citrate as sole source of carbon, growth at 56°C and fermentation of sugars, namely, glucose, lactose, sucrose, mannitol and maltose, revealed that about 26 per cent of the strains isolated from the factory environment were gram positive spore formers of the *Bacillus* type. Majority of the spore formers were isolated from the water samples, especially cooling water. 33.7 per cent of the strains isolated from cooling water samples were gram positive spore formers. The cooling water samples are usually chlorinated so that almost all the vegetative cells are killed and this may lead to the predominance of gram positive spore formers. The percentage of gram positive spore formers increased after each cooling process and subsequent addition of ice. Analysis of ice samples has shown more than 12 per cent of the strain isolated were gram positive spore formers.

The swab samples from utensils, table surfaces and interior of the cans before filling mainly contained gram negative rods (about 31 per cent of the total strains isolated from the swab samples), 22.4 per cent of the strains were gram positive *Cocci*, gram positive spore formers constituting only 18.0 per cent. Of the gram positive spore formers majority was from the interior of the cans. Improper washing of the cans before filling might be the reason for predominance of gram positive spore formers.

The raw material samples contained less gram positive spore formers. Majority of the strains isolated from raw material samples were gram negative rods (27.7 per cent) and gram negative *Cocci* (25.0 per cent). In the case of filling brine, majority of the strains isolated was gram positive non-spore forming rods and gram positive *Cocci*. The findings clearly show that the chances of contamination of the processed cans by gram positive spore formers fron the raw material and filling brine are rare.

Among the gram positive spore formers isolated from factory environment, the predominating types were, *Bacillus subtilis* (19.5 per cent), *B. pantothenticus* (18.4 per cent), *B. megaterium* (15.7 per cent), *B. brevis* (12.4 per cent) and *B. pumilus* (7.5 per cent). Similar type of organisms were found to predominate in bacteriologically defective cans. The chances of contamination by gram positive spore formers entering the cans before processing are rare, since none of them can withstand more than 10 minutes at 121°C, the usual processing temperature. Also the raw material and filling brine samples contain less gram positive spore former. The cooling water as a potential source of gram positive spore formers may enter the cans during cooling through defective seams, since gram positive spore formers were found to be predominating types in defective cans. So the most probable source of contamination of the can seems to be cooling water.

Maintenance of Bacteriological Quality in Canned Prawns

The most important criterion for judging the suitability of canned foods for storage and human consumption is its freedom from micro-organisms capable of producing spoilage, toxins and diseases in man. Occasionally at least, every canner experiences a case of a rejectable product being turned out, the defect being lack of commercial sterility. Hence it is essential that the processors are aware of the probable sources of micro-biological defects in canned products.

1. Empty cans should be stored in such a manner that no contamination takes place from dirt or any other extraneous material.

2. At the time of use, the cans may be checked for constructional defects, defect in lacquering etc. Defective cans should not be used for packing the product. Defects if any may be brought to the notice of the manufacturer for rectification.

3. Before filling with blanched prawn, the can should be washed properly with water.

4. The meat should be cooled immediately after blanching by circulating cold air and the cooled meat should be canned immediately.

5. Temperature and pressure of retorting (processing) should be recorded on a log book and during retorting the vent pipes of the retort should be kept slightly open. Coming up time and release time should be more or less constant. Care should be taken to give proper time and temperature for sterilization.

6. The processed cans should be immediately cooled in chlorinated running water in which free chlorine level shall not be less than 3 ppm.

7. For adjustment of seam, seaming machine should be tested as frequently as possible. At least 200 cans may be sealed empty for testing both by applying pressure and by actual seam measurements.

8. Utensils coming in contact with raw and blanched meat should be washed initially with the detergent like 'Teepol' solution of 0.5 per cent concentration, followed by a disinfectant like chlorinated water containing

100 ppm of available chlorine. Ice used with the material should be prepared from potable water. Overhead and storage tanks should be kept well covered and should be cleaned at least twice a week.

Bacteriological Quality of Cooked Frozen Prawns

Strict bacteriological quality standards for cooked frozen prawns are laid down in the importing countries (USA and Australia). In order to sustain and expand the market it is essential to give due care in processing so that the product conform to the standards specified.

Investigations have shown that contamination of the material with faecal organisms result from unhygienic handling practice and use of bad quality waters and ice and un-cleaned utensils. Workers hygiene is the main factor which influences contamination with coagulase positive staphylococci. Material gets generally contaminated with staphylococci, when the worker's hands are not washed especially before start of the work, since the organisms are inhabitants of wounds, ulcers, mucus etc.

To prevent contamination from the above sources and to prepare good quality products certain minimum precautions have to be taken in the processing plants as given below.

1. The workers should be instructed to maintain proper cleanliness always. Before starting the work, they should clean their hands using a detergent (soap) and then should disinfect the hands by using chlorinated water (chlorine at 200 ppm level). Workers suffering from communicable diseases, ulcers, wounds, skin eruption etc should not be allowed to handle the prawn at any stage.

2. The various utensils used for handling the material will be sources of bacterial contamination unless they are cleaned and disinfected before use. All materials should be washed first with a detergent (0.5 per cent Teepol solution) and then for disinfection, they should be kept immersed in water chlorinated at 100 ppm level for 4-5 minutes. They should be finally washed with freshwater. When prawns are handled in the factory along with frog legs at the same time, separate utensils should be used for each of the items.

3. Ice used should be prepared from chlorinated water and it should be handled carefully and hygienically, since material coming in contact with contaminated ice will peak up bacteria from it.

4. After cooking the raw material it is cooled in ice-cold water. If the water used is of poor bacteriological quality, as is found some times, it will be a source of heavy contamination of the material. The cooked material should be cooled in water chlorinated at the level of 10 ppm, so that contamination from this source can be completely checked.

5. Before packing the peeled and washed material into the freezing trays, it should be dipped in water chlorinated at 20 ppm level for 10 minutes. This will effect in considerable reduction of bacterial load in the material.

6. As a general rule, the time interval between cooking, peeling and admitting the material into the freezer should be kept minimum, that is, should be processed within the shortest possible time so that bacterial multiplication during the storage can be kept minimum.

7. The water used for glazing and re-glazing should be chlorinated at the level of 10 ppm and 50 ppm respectively, and the re-glazing water should be changed after dipping at the most about 16 blocks in the same water. When more blocks are dipped in the same water, it becomes more polluted and the bacterial loads becomes high.

8. The packing of the frozen material should also be done carefully under strict hygienic conditions. If gloves are used, they should be washed thoroughly and properly disinfected.

Products processed even from the freshest raw material will be sub-standard bacteriological quality if utmost care is not exercised during handling and processing of the material.

Factory sanitation

Maintenance of a high standard of cleanliness is quite essential in any food processing plant. Food material processed from even the freshest raw material can turn out to be of unacceptable quality or even dangerous to human life if it is handled and processed under unhygienic conditions. This is especially so in the case of fish products as fish is known to be one of the most perishable of food materials.

Sources of Bacterial Contamination

In fish processing plants bacterial contamination of the material occurs from the surfaces of tables and various utensils with which the material comes into contact, from the water used for various purposes like, washing material, glazing and re-glazing of the frozen product etc, and also from ice, used for preservation of the material. It is further observed that in almost all cases of contamination of the material by faecal organisms, like, *Escherichia coli*, Enterococci etc, the source is mainly water and/or ice. In order to maintain a satisfactory hygienic standard, it is essential that all the utensils and tables etc that come in contact with the material are kept always clean so that the bacterial load on the surface is kept below 1000 organisms per square inch. The water and ice must not contain bacterial loads of more than 100 organisms/ml. Insufficient cleaning of overhead storage tank, storing water in open tubs in processing halls and using it carelessly and careless handling of glazing water are the main reasons for contamination of water. Ice gets contaminated when water of low bacterial quality is used for preparing ice, when it comes in contact with unclean surfaces, for example, where the ice cans are dipped in thawing tanks containing impure water, when stored on or dragged along dirty floors etc. The water used in the processing plant should be potable. It should be chlorinated to contain 5 ppm of residual chlorine. The ice used should be that prepared from potable water. Utmost care should be exercised in handling it under hygienic conditions. If by some means the ice gets contaminated it should be washed with chlorinated water (chlorine at the level of 5-10 ppm) before using with the fish.

Adequate Cleaning Program

An adequate cleaning program is the first requisite in maintaining good sanitary condition in the fish processing plant. The cleaning work carried out should be effective in (1) the removal of visible product wastes, foreign matter and slime (the presence of slime can be felt by dragging the finger over the surface; if the surface is slippery it contains slime), (2) the destruction of spoilage bacteria and bacteria that reflect upon general sanitation, (3) the removal of undesirable chemicals such as those used as detergents or germicides. In removing slime dirt or deposits, a detergent is necessary. High pressure water is effective in some cases. Tenacious deposits are removed sometimes only by use of scrappers. For destroying the bacteria a germicide is to be used. The detergent and germicide are to be used only at correct levels required.

Cleaning Schedule for Fish Processing Plants

1. Rub the surface with brush so as to remove all solid organic matter.
2. Apply a suitable detergent (Teepol 0.5 per cent) to remove slime followed by washing with freshwater (chlorinated at 10 ppm level).
3. Apply a suitable disinfectant (sodium hypochlorite or bleach liquor solution) containing 100 ppm of available chlorine for 4 minutes (rubbing with a brush or coir gives better results). For disinfecting floor surfaces, gutters etc, the chlorine dose should be 500-800 ppm.
4. Finally wash thoroughly with freshwater (chlorinated at 10 ppm level).

Cleaning Schedule After Each Shift of Work

The above schedule of cleaning be followed for cleaning various utensils table surface etc. The utensils, table floor surfaces gutters etc should be cleaned as per the schedule after each shift of work. Before the start of the next shift they should be washed well again with water chlorinated at 5 ppm level.

Cleaning of Boat Decks, Fish Holds, Wooden Boxes etc.

It has been observed that the quality of freshly landed fish depends to a great extent on the care with which it is handled and stored in the fishing vessel. Apart from the use of ice, the hygienic condition of the fish hold and fish containers is the important factor which determines the quality of landed fish. It is therefore essential that the boat deck, fish hold, the containers for fish etc should be thoroughly cleaned daily before going to fishing. The cleaning schedule recommended for boat decks, fish hold and the fish containers is as below.

1. Rub the surface with brush so as to remove all solid organic matter.
2. Apply a suitable detergent (Teepol 0.5 per cent solution) to remove slime, followed by washing with freshwater.
3. Apply a suitable disinfectant (sodium hypochlorite or bleach liquor solution) containing 1000 ppm of available chlorine for 4-5 minutes.
4. Finally wash thoroughly with freshwater.

By cleaning the surfaces as per the schedule the load of spoilage organisms can be reduced and kept within limit and also complete reduction in the pathogenic organisms can be achieved.

Deodourisation of Fish Containers, Fish Carrier Vans and Refrigerated Wagons

Removal of fishy odours from the fish carrier boxes, vehicles and rail wagons has been an/all important problem faced in the transport of fish. It has been observed that by proper cleaning, the fishy odour can be completely removed. The schedule of cleaning worked out is as below:

1. Preliminary washing and scrubbing with 0.5 per cent soap solution to remove all adhering slime.

2. Spraying with 100 ppm available chlorine solution to remove the fishy smell followed by washing with water.

3. Spraying with 50-60 ppm sodium thiosulphate solution to remove the residual chlorine smell.

4. Final washing with freshwater

A chart showing the quantity of bleach liquor of different concentrations required for chlorinating known volume of water at the desired chlorine level has been prepared as detailed below:

Chlorine Level (ppm)	Bleaching Powder of 30 per cent Available Chlorine to be Used in gram	Bleach Liquor Containing 10 per cent Available Chlorine to be Used in ml
1	0.03334	0.1
5	0.1667	0.5
10	0.3334	1.0
50	1.667	5.0
100	3.334	10.0
1000	33.34	100.0

Teepol solution (0.5 per cent solution of Teepol) is to be used as detergent in the cleaning schedules, recommended. 5 ml of Teepol added to one litre of water will give one litre of the solution of the required concentration and on this basis the quantity of Teepol for known volume of water can easily be calculated.

Chapter 11
Curing of Fish

Fish curing comprises all the methods of preservation except refrigeration and canning. It includes drying, smoking, salting and pickling of fish.

Preservation of fish by the method of drying either by sun, fire or smoking is perhaps the oldest preservation technique known to mankind. In dates back to pre-historic days of stone age, the evidence of fish drying is available in the Magdelenian period. Preservation by salting is of later origin and can be related to the period of bronze age. The ancient Egyptions extensively practiced salting of fish. In Mesopotemia fish was salted, smoked, dried as early as 3500 BC. Salted and dried fish became a major commercial commodity for international barter trade. This method came down to Indus Valley civilization from Arabian countries

Prior to the advent of freezing and canning, fish curing was a major industry the world over. Subsequently its importance has considerably declined. It has now stabilized itself and it still occupies an important position even in developed countries. Out of 23.8 million tons of fish used for processing in 1970s, 34 per cent was processed by curing alone.

In India the status of fish curing is still more significant. By 1970 onwards, curing alone stabilized itself to 20 per cent of total landings.

Traditional Curing of Fish in India

The methods of preservation of fish employed in India were;

1. Desiccation with or without salt (curing)
2. The use of antiseptic preservatives, such as, brine, vinegar etc (pickling) and certain miscellaneous processes, namely, making spice paste.

The object of desiccation is to withdraw just so much moisture as will prevent putrefaction and to do so by such available methods as shall suit the contemplated market. Bacteria, the causative agents for putrefaction, can thrive only in presence of moisture. If this be sufficiently withdrawn they become incapable of action and reproduction and hence putrefaction ceases. The moisture to be got rid of is the intracellular moisture, which amounts to about 75-80 per cent of the weight of raw fish. The moisture is contained in the millions of tiny cells, which lie from the surface of the fish to the full depth of the tissues. To obtain a wholesome product, therefore, the desiccating process must be carried through enough rapidity and thoroughness to dry the inmost cells to a degree sufficient to prevent bacterial action before taint can set in. This can be achieved (a) heat, (b) dry air, (c) salt and other extractives and (d) salt and pressure. In case of (a) and (b) moisture alone is removed from the cells by hot or dry air; in case of (c) salt, by the process called osmosis, displaces the fluid contents of the cells including not merely water, but also the nutritive elements like nitrogen, phosphorus. Pressure acts by driving out the contents from the cells together with what has already been extracted from the cells by the salt.

Drying was preferred to wet-salting by the fishermen in India owing to the process being simple and handy and owing to the presence of powerful sun always at hand to dry without cost. In India, as salt was expensive, wet-salting was rarely practiced. Even when salt was used it was applied only sparingly, putrefaction being prevented by the subsequent heavy drying. In other countries salt was used to remove most of the moisture and drying in the sun or in artificial drier was only for producing finished product.

But simple sun-drying is difficult in the tropical sea coast, where the air though hot, is very moist and where the sun quickly dries the exposed surface into a hard crust, which hinders the moisture from within the tissues from evaporating. Except in case of thin or flattened fishes, like, Bombay duck, ribbon fish, prawns, silver bellies, eels etc, simple sun drying does not give satisfactory product. Consequently in most curing operations, the use of salt can not be avoided in order to reduce the amount moisture in the fish prior to exposing it to the heat of the sun.

Bombay-Sind area

(a) Sun drying – Sun-dried fish exceed in quantity fish cured in any other way. The chief varieties sun dried are Bombay duck, ribbon fish, prawns and eels.

(b) Curing with salt – The following varieties were salted and dried; mackerel, pomfret, silver-bar fish, sardine, seer fish, jew fish, sharks and rays. The quantity of fish cured by wet salting was comparatively small and almost all this wet cured fish was exported to the Madras Presidency.

(c) Tamarind pickling – Packing of fish with salt and tamarind in barrels, was not also practiced to any great extent. It was carried out by Moplahs who come from Madras Presidency and visit certain fish curing yards annually for this purpose.

In addition to the curing conducted in the fish-curing yards with duty-free salt, there was a small amount of private curing conducted by the fishermen with bazaar salt. In the Konkan coast, Moplahs, who catch sharks and deal in shark fins, worked independently of the yards. In Sind all the curing was done in this way.

The types of fishes dealt with by curing in the yards may be classified as (a) shoaling fishes like sardine, and mackerel, which were caught in enormous quantities during certain seasons, the curing being resorted to liquidate glut and (b) the less abundant, but more valuable "table fishes" like seer, pomfret, Indian salmon etc which were cured to get better prices in the inland markets

For smoked fish there was no demand at all and no smoking of fish was done.

Fish pastes were also not made in the Bombay Presidency. There was however, good scope for developing an industry in the manufacture of fully flavoured prawn and Bombay duck pastes.

Prawns were mostly sun dried. When they have dried hard (2-3 days exposure usually), they were trampled to remove the shells.In the Sind coast prawns were boiled first in weak brine for about half hour and spread out for drying. When they have dried hard, they were"thrashed" out with sticks to remove the shells. After cleaning, the prawn were stored according to size.

Madras Presidency

Curing with Salt in the East Coast

Very large fish, such as, sharks and skates were gutted, slit from head to tail (are often also cut into chunks) and washed in sea water. Deep incisions were made in several places along the length of the fish and finely powdered salt was rubbed into the flesh. For stacking the fish round porous earthenware troughs several feet in diameter was used. At the bottom of the vessel a little salt was sprinkled and the split fish were laid skin side down. More fish were laid in layers upon the first with a sprinkling of salt between two consecutive layers. The pile of fish was weighted with heavy stones and the trough was usually closed by inverting a second trough on it. The fish remained in salt for one night only; next morning they were washed in sea water, mixed with the brine which had exuded from them and dried in the sun. Salt used was in the proportion 1:6 by weight, that is, 1 part salt to 6 parts of fish.

Medium sized fish, such as pomfret, seer, bhetki etc were treated in a similar fashion, but the flesh was not cut into pieces. The operations were conducted with a little more care.

In the case of small fishes, splitting and scoring of the flesh were dispensed with; the fish being only gutted and washed. Salt was dissolved in sea water and in the resultant strong brine, the fish were given a preliminary "rousing'. The brined fish were stacked in an earthenware trough and any quantity of brine left over was poured on the top of the pile. The fish were weighted with stones and allowed to remain in brine for one night. The next morning they were washed and dried in the usual manner.

At Muthupet on the east coast large fish were gutted, entirely split through the back and then several large longitudinal incisions made in the flesh. They were then washed in old brine obtained from the previous batch of cured fish. This curing was done in closed sheds, in the soil of which several pits lined with brick and cement have been made. The quantity of salt used was 1:5 or 1:6 by weight. The fish were stacked in layers in the pit and salt is rubbed into the flesh of each layer. The pit is finally covered over with a mat and the fish left over till next morning when they were removed and restacked in a similar manner in a second pit. Fresh salt is not used but some old brine obtained in a previous curing operation is allowed to remain in the pit and fish was stacked again along with this brine. On the third day fish were removed and stacked again in a precisely similar manner in a third pit. On the fourth day they were taken out and washed in the brine left behind in pits 1 and 2, cleaned and sun dried.

The procedure is the same for medium sized fish. Salt was used in the ratio of 1:8 and the fish were merely gutted. In the case of small fish, they were stacked in pile in thin layers with salt (1:8 or 1:9) sprinkled between the layers and allowed to remain in the pit undisturbed for 2 days. At the end of this period they were taken out, washed in the resultant brine and dried in diffused sun light.

At Adirampatnam, the sheds were provided with a hard mud floor, which slopes on all sides and drains into a circular masonary pit in the center. On the floor over a palmyra-leaf matting, the fish under cure were piled in rectangular heaps about 1-2 feet high. The self-brine drains into the pit in the center. On the second day, the fish were restacked, the top fish on the previous day becoming the bottom layer in the new pile. The restacking process continued for 4-5 days after which the fish were let out of the yard. Before being put on the market, the fish were often washed with strong brine and sun dried. The smaller fish were cured by salting in a pit for one day with 1:7 salt and sun drying for 2 days.

The method adopted by curers in Tuticorin was still more primitive, though the principle was the same as that in "kenching" followed by the curers in the west. Pits were dug in the soil in open air and lined only with palmyra mats. After arranging the fish, each pit was closed with another mat and over this a heap of sand was placed. The sand was often trampled over to make the pile more compact. Curing was over in about 36 hours. The fish were neither washed nor dried. Salt was used liberally, 1:4 by weight being the usual ratio. Smaller fish were cured in earthen pots with 1:8 salt. In this case salt and fish were arranged in alternative layers and the fish allowed to remain in contact for about 12-18 hours after which they were washed in sea water and sun dried.

Curing with Salt in the West Coast

The west coast yards were generally more clean and the operation were conducted in a more satisfactory manner than in the yards of east coast. The commonest method of curing was salting followed by sun drying except in the very small "miscellaneous" fishes which were only sun dried.

Shark, dog fish, skate, cat fish, ribbon fish, seer, horse mackerel, mackerel and sardine were the fish commonly cured. Shark fins were treated separately, the curing being mainly for the export market. The swim bladders, where ever present, were removed carefully and were dried in the sun.

The essential curing process was the same for all fishes. Only there were minor variations of detail in procedure (such as gutting and scoring) adopted to suit a particular kind of fish. For example, sharks were cut into chunks before salting, cat fishes were gutted, split and the flesh deeply scored, seer were treated in a fashion similar to cat fishes, but were not beheaded and sardine and mackerel were simply gutted and split. The fish were then cleaned and thoroughly washed in sea water. The clean fish were brought to the fish curing yard, where salt was issued on the weight of the fish, a larger proportion being usually allowed in the wet weather. Curing was done in cement tanks or in wooden tubs inside the closed sheds. The cement tanks were made with rounded ends at the bottom to facilitate cleaning. The fish were first "roused" with salt and then neatly arranged in curing vats. When full, the vat was covered with a mat or gunny. After 12-18 hours in salt, the fish were removed, well washed in sea water and spread on leaf mats in the sun for drying. The fish were dried hard and the finished article were stored either in bundles, gunnies or in simple heaps. The cured fish was largely exported to Sri Lanka.

"Ratnagiri" Curing

This was a special system of curing adopted by the Bombay fishermen at Malpe (South Canara). These fishermen came down to Malpe in the cold weather. The fish cured was mainly seer (*Cybium* spp). Gutting, splitting and cleaning were done by the crew in the boats, so that the fish was ready for salt immediately on landing. Salt was issued in the proportion of 1:3 on the weight of cleaned fish. The salt was divided into three portions; one portion (half the total quantity) was used on the first day and 0ne-fourth on each of the next two days. The first portion of salt was applied un-powdered to the fish and well rubbed into the incisions. The salted fish were piled in rectangular heaps on the floor of the curing shed over leaf mats, with some salt sprinkled between successive layers. The pile was usually 3 feet in height. On the second day the fish were repacked, the top layers of the first day became the bottom layers and a fourth of the quantity of salt was sprinkled between the layers. The stacking was done again on the third day when the final installment of salt was used. The pile was then left for the next 8 days un-disturbed. On the 11th day fish were sent out of the yard without any drying whatsoever.

Pickling

Pickling was practiced only to a limited extent on the Timil Nadu coast. At Cochin, Ceylonese fishermen conduct some true pickling in barrels. The system was called the "Colombo cure". The fish usually the mackerel were gutted, rubbed with powdered salt and packed into barrels in alternate layer of fish and salt. A quantity of Malabar tamarind was mixed with the salt. The cask was packed to the brim with the fish and weights were placed on the top. After 2-3 days, considerable shrinkage was observed and the cask were filled with brine. The self-brine was first drawn off through a bung hole near the bottom and the cask was again packed to the bream with fish from the same pickling. It was generally observed that 12 casks could now be packed into eight. Finally the cask was headed up and completely filled with old brine. The product which was called "Jadi" was almost wholly exported to Sri Lanka.

Jadi can also be made with seer, sardine etc. though the product keeps good for only about 6 months as against 2 years with mackerel.

Fish Paste

Tamarind fish was usually prepared out of seer fish, which were first cut into slices, cleaned and well salted.

The moisture-free slices were treated one by one in a special spiced paste and arranged in a clean dry stoneware jar. The paste was usually made by first grinding chillies, mustard, cumin seed, tamarind and turmeric with vinegar to which green ginger and garlic cut into small bits were added next and the paste boiled in ghee or coconut oil and allowed to cool. If any paste remained behind after treating the slices, it was poured on the top of the pile in the jar. The pickle was ready for use in a fortnight's time. Mackerels, prawns etc were also treated in the same way.

Prawn Curing

There are two processes in vogue for prawn curing; (i) boiling and drying and (ii) sun drying. The former method, largely resorted to during the wet weather, consisted in boiling the prawns in wide flat copper vessels with small quantity of water. During boiling the colour of the prawn changes to reddish brown. The blanched prawns were then dried in the sun with the shells on. From the dried prawns, the shells were removed by putting them in jute sacks and beating hard against a block of wood. The kernels were sorted out and once again dried before being packed in gunnies for export. In the second process fresh prawns were simply sun-dried and the shells were not removed from the dry product.

Smoked Prawns

The Ganjam on the Chilka Lake prawns were preserved by smoking. Smoking was not practiced anywhere in the Madras Presidency. In a trench dug out on the ground, a smoky fire was kept burning. Over the trench a bamboo frame-work was erected and prawns were spread on this frame-work and over the fire. The drying was rapid and the process seems to be one of drying than real smoking.

Semi-dried Prawns

Fresh prawns were dipped for a minute or two in boiling water or in 6 per cent boiling salt solution until some of the prawns begin to float. The blanched prawns were shelled and then salted by immersion in saturated brine for about 20 minutes (according to size). After draining off the brine, the prawns were semi-dried either in open air or in a drier. The drying was only partial and was stopped when the prawns were so firm that fairly strong pressure between the thumb and the finger was needed to make an impression on them.

Semi-dried prawns when stored in a sealed tin in an atmosphere of carbon-dioxide keep it in sound conditions for several months. If the salted prawns were immersed in water for about 30 minutes, the excess salt is leached away and the prawn imbibes water and looks and tastes like a fresh prawn.

Bengal Presidency

Sun Drying

The common variety of sun-dried fish was called "sutki" and was manufactured in Sylhet (now Bangladesh). The drying was conducted in open yards called "khola". Large fishes were beheaded, gutted and the entire viscera removed. Very large varieties were cut into slices. Smaller varieties were dried with the heads on and the entrails were removed by squeezing. In certain places the entrails were trodden out when the fish was half dry. The smallest fish were dried entire. The fish were seldom washed after gutting. The fish were laid flat side by side on mats made of nal-reeds and exposed to the sun. Sometimes there were bamboo platforms, 3-4 feet high to spread the fish. In some parts fish were dried on the hot sand on the banks of Brahmaputra. The drying process was continuous and extends for a week or 10 days depending on the sun shine. The fish were not removed at night, but were protected from the depredations of birds and other animals by stretching a netting over them.

The dried fish were preserved in earthen jars, called matkas, which were well smeared with fish oil on the inside. These jars were buried in shallow trenches dug in the ground, which were lined with straw and mats. After putting the fish the trenches were covered with earth. Sutki was only made from fish which have a firm flesh.

Curing with Salt

Only hilsa fish was cured in this manner, The fish was slit and eviscerated. The cavity was filled with salt and closed up. Salt was also rubbed over the fish. The prepared fish were placed in a pit with more salt and the pit kept closed. Sometimes weight were placed on the fish to facilitate expression of fluids.

The roes of hilsa were also preserved by drying in the sun, by smoking over an oven or by immersion in brine.Sometimes the roes were pickled in mustard oil.

Curing in Oil

The fish was dried as for sutki until crumbled into a powder between the fingers. They were then collected in baskets and immersed in water for few hours. The baskets were then hung up for draining. The moist fish are then packed into large earthen jars which were specially prepared and the interior was well soaked with fish oil. The fish were packed closely inside with a sprinkling of fish oil between the layers. The mouth of the jar was closed with an earthenware plate and the edges luted with kneaded clay. The jar was buried in the ground for several months to allow the fish to develop the taste.

Fish Pastes

In Jalpaiguri, dried fish was powdered, mixed with the crushed stems of caladiym (kachu) and made into balls. In Nowgong, a fish paste, called "hidal-khunda" was manufactured in the following manner;

Certain varieties of small fishes were placed in a hole dug in the ground and kept covered for nearly a month. The mass is then sun-dried, powdered and packed into bamboo tubes. Sometimes a little ground pepper and some alkaline ash (called khar in Assam) was mixed with the fish at this state. The Maghs in Cox's Bazaar prepared a fish paste of prawns, called "nga-pee'. The prawns were pounded into a paste with 3:1 salt and the fluid drained off. The paste was sun-dried and again pounded. This process was repeated till the product is stiff and clay-like. The paste was baked and sold as 'nga-pee". In Chittagong Bazaar nge-pee was also called "ba-lee-chong".

The nga-pee was kneaded with water, strained through a piece of cloth for the removal of pieces of shell etc and the liquor was added to the curry.

Prawn Curing

Prawns were dried in the sun or artificial drier and beaten to remove the shells.

Bacteria and enzymes can not be active without water. Similarly in high salt content, bacterial action as well as enzymatic activity are retarded. Hence the main principle involved in sun drying is the speedy removal of water content of fish to such a level that the microbial and enzymatic activity is retarded. Salting also substantially help to reduce the moisture content of the fish and the balance of water is made so highly saltish that bacteria and enzyme cannot function. Salt by itself if not antiseptic or bactericidal, but has bacteriostatic properties at high concentration. Similarly under such high salt levels the enzyme also gets substantially deactivated. This is the basic principles of preservation in drying and salting.

In case of smoking, in addition to partial salting and drying as above, it also gets the support from some of the preservative components of smoke, like, phenols, formaldehyde, acetic acid etc.

Different curing methods currently practiced in India are, a) sun drying, b) dry curing, c) mona curing, d) smoke curing, e) pit curing, f) Colombo curing and g) wet curing.

Sun Drying

Generally small and thin types of fishes, like, white bait, silverbellies, ribbon fishes, soles, jew fishes, Bombay duck etc are cured by sun drying. The idea is that the fish must get adequately dehydrated in the shortest time possible, before the tissue start deteriorating. No salt is used at all. The fish is first washed and straight away spread out in the open sun, either on mats, hard ground or even on sandy beach. Drying by hanging on ropes is very effective as in the case of Bombay duck. In beach dried products there will be often too much sand. This process is extremely simple and cheap. Roughly about three days drying will be required depending upon ambient conditions. From the nutritional point of view the sun dried products has an advantage in that at no stage nutrients are lost by salt leaching as in other curing methods.

However, too much of sand is a serious handicap. If properly dried and stored, the sun dried product has shelf life of about 2 to 3 months and gives an approximate yield of 25 per cent.

Dry Curing

In this case, the fish are generally split open, viscera etc removed, washed and are then intimately roused in specific proportion of good salt. The salt proportion varies according to the size of the fish, the season and fat content of the fish. After salting, the fish is neatly stacked in tanks or other receptacles with copious supply extra salt on top. By osmotic activity salt penetrates into the fish tissue and in turn it expresses out water from the fish which accumulates in the tank as self brine and engulf the fish. Usually heavy weights are placed on the top of the fish stack to hasten the extraction of water. During this stage the fish muscle changes from translucent to opaque stage, looses stickness and becomes more fibrous in texture. This is called the struck through stage. The fish is allowed in this condition for one or more days, when the proper ripening or maturation takes place. After this, the fish is taken out, rinsed in water, and then dried in the sun for 2 to 3 days. In the rainy season if the drying conditions are not favourable, it is allowed in the tank till drying conditions are favourable. Drying is done on mats or by hanging on ropes. After drying it is packed and dispatched to market. Yield is 48 per cent and shelf life 90 days approximately.

Mona Cure

This method is followed mostly in Maharashtra region in the case of mackerel, lactarius, otolithus etc. The method is almost identical to dry curing, except that the fish is never split open. The viscera is removed by putting it out through the buccal cavity. It is then washed and salted in the ratio of 1:4 and is stuffed in belly cavity and then stacked. After two days the salted fish is dried as in the case of dry cure. The product looks elegant and whole. Yield is 70 per cent and shelf life 50 days (approximately).

Wet Curing

Some times called Ratnagiri method is almost akin to kench cure of western countries. In this case, the fish is split and gutted and then salted in 1:4 or 1;3 ratio and stacked in cement tanks or platform. Some weight is applied on top and the self brine is allowed to drain out into the gutter. There is no drying at all in this method, hence called wet cure. After three days the fish is marketed. The moisture content is as high as 50 per cent. The salt content is also high (24-30 per cent MFB) and this give the fish the required protection. The shelf life of the product is very low, only about 2-3 weeks. In some areas of Kerala, fish after salting is straight away stacked in bamboo basket where the self brine flows out by itself. After the cure is completed the basket as such is sent to the markets for disposal. Yield is 70.4 per cent and shelf life 2 to 3 weeks (approximately).

Pit Cure

Pit cure is confined to certain areas of Tamilnadu state only. In this case fish is split and gutted and salted in 1:3 or 1:4 ratio and is then stacked in 2 feet deep pit in the sand lined with palmyra leaf mats. When the pit is three fourth fill it is covered over on the top by another mat and then sand. The maturation is strictly under anaerobic condition for about 36 to 48 hours. The fish is then dug out of the pit and straight away marketed without any drying. The product is very un-wholesome. The moisture content and salt content are very high as in the case of wet cured products. Yield is 69.4 per cent and shelf life about 20 days.

Colombo Cure

This is perhaps the only type of commercial pickling in India, carried out mostly in South Kanara and Malabar region. It is almost solely intended for Colombo market. The fish usually mackerel is gutted and split. Mixed with 1:3 salt and a small piece of Malabar Tamarind (1:10 ratio) is then very neatly and tightly packed in huge wooden drums. Salt brine is fully retained in the drum. After 3 days, the fish shrinks. The empty space is fully packed with more number of similarly treated fish from another drum. The packing is very tight. The top lid is then fixed in a water tight condition, The bung hole of the drum is opened and extra quantity of self brine or saturated brine is poured into the drum till it is completely full to the brim. This replaces the air pockets inside the drum. Now the bung hole is finally closed and the drum as such is exported to Sri Lanka. In this case also there is no drying at all. But the fish is in a medium of concentrated brine fortified by a small quantity of tamarind, which gives a particular flavour and added protection. The fish is also out of contact with air and spoilage due to rancidity is also appreciably reduced. Yield is 75.2 per cent and shelf life is more than six months.

Smoke Cure

Though smoke curing yields very tasty products, its life is very short unless stored in refrigerated conditions. Smoke curing is not widely practiced in India. There are very large number of smoking procedures giving a variety of products. But the general principles are that big fishes are fully split, gutted, scored and washed. They are then immersed in brine for specified period. Fish is then taken out, thoroughly drained and surface dried in the sun. The fish are then removed to smoke kiln, and hung there. Smoke generated by burning saw dust, wood shavings, coconut husk etc. and the temperature is raised according to the particular type (about 16 hours, a temperature of 50 to 60°C). The fish are then removed, packed and stored.

Yield is about 40.8 per cent and the shelf life about 40 days under ambient temperature conditions depending on the type of smoking.

Though there are highly sophisticated smoke kilns, like Torry Kiln, the cost is very high for Indian conditions. Cheaper smoke kilns can be easily fabricated in India, which will serve quite well for commercial purposes.

Between Catching and Curing Mackerel

The quality of shelf life of cured fish are known to be ultimately related to the initial freshness of the fish at the time of curing, inadequacy of which particularly accounts for the poorer quality and shorter shelf life often observed in Indian cured products. Unsold stock of fish, waiting for considerable length of time for their disposal with quality deterioration usually find their way into curing industry. Such raw material may have remained unpreserved for indefinite periods and consequently would be highly variable in their degree of staleness and decomposition. Unless specific steps are taken for preservation, the freshly landed fish deteriorate beyond the standards of human consumption within a short period, which varies according to prevailing climatic conditions and fish species.

Though the importance of climatic conditions is well realized, the temperature and humidity variations, for practical purpose may be taken to represent tropical conditions.

In a temperature range of 17 to 36°C and humidity range of 55 per cent and 100 per cent, the spoilage at room temperature conditions and effect on raw materials have been observed basing on the physical conditions of fish. Based on the colour, odour, general appearance, and the condition of the gills, eyes, abdomen and muscle tissue, it can be seen that the deterioration of fish during the first 5 hours was very gradual and almost imperceptible. After this stage, the deterioration appears to have gained acceleration till a critical stage is reached at the 8th hour. Then onwards there was abrupt deterioration, which have been substantiated by the chemical factors, like, TVN, TMA, NPN and total bacterial count.

The organoleptic qualities of the cured fish products prepared from materials from different sub-samples indicate a close relationship between the period of spoilage, the raw material had undergone and the quality of the finished product. Accordingly, finished product obtained from raw material of up to 5 hours spoilage were adjudged to be good, base on odour, flavour texture and colour characteristics. The products of 7 hours and 8 hours spoilage were found to be just satisfactory. The finished product from more than 8 hours of spoilage was found to be very poor in quality with perciptable off-odour. Their texture also became very soft tendering to be pasty and were unanimously unacceptable to the taste panel body.

The spoilage indices of the finished product also indicate a direct relationship with the period of storage. This is true with respect to the initial values of these samples as well as when they are kept in storage for shelf life studies. Finished products from fresh raw material had a TMA value of just 1.3 mg per cent and TVN value of 37.5 mg per cent. These values steadily advanced almost proportional to the period of exposure to room temperature conditions and indicate that after the 8 hours stage, the finished products in the initial stage itself was in a very poor condition and they further deteriorated when they were kept in the storage.

The fish kept in ice storage for three days remained in good condition, after which they deteriorated rapidly, which fully corroborated by chemical and bacteriological findings. Compared to samples at room temperature conditions, some of these chemical factors recorded lesser values, which may partly be due to the leaching out of these water soluble constituents during iced condition.

The finished products from raw material of up to 3 days icing were acceptable and satisfactory. But the tasting qualities were found to be poor in the case of all the iced samples, due to the leaching out of water soluble flavour and taste components. The general trend of variation of the important spoilage factors were almost of the same pattern as in the case of samples prepared from raw material under room temperature conditions.

It may therefore, be concluded that under room temperature conditions, the maximum permissible delay between the catching of the fish and its curing is to be limited to 8 hours.

Although 8 hours held raw material gave reasonably satisfactory and acceptable finished products, their shelf life and tasting quality decreased proportionately as the period of spoilage prior to curing advanced.

With 1:1 icing, the curing of mackerel may be delayed up to 3 days without unduly undermining the quality of the finished product. Only very little difference could be detected in the finished product prepared from raw material up to 3 days in ice after which symptoms of spoilage were more pronounced.

Curing of Mackerel

As mackerel, like sardines, is a seasonal fishery and the bulk of the landings takes place in a narrow strip of time, preservation of excess catch often becomes necessary. Earlier a substantial portion of the mackerel catch used to be salt-cured by dry or wet process or pickled according to Colombo curing method. A portion of such products used to be exported to Sri lanka.

With regard to preservation of mackerel, curing used to be the most important process applied and a good amount of technological research has been carried out in this field.

In pit curing of mackerel, initially a ratio of 1:5 salt to fish and salting period of 2 days improved the organoleptic properties of the fish in imparting a characteristic flavour and softening flesh. Subsequently, working on the effect of varying proportion of salt to dressed fish on the quality of sun dried mackerel proposed that a ratio of 1:7 or 1:8 would be adequate if the salted fish is dried at the end of 18-24 hours salting without stacking. When dried to about 40 per cent moisture level, the sodium chloride content would be more or less sufficient to saturate its moisture content and would not be in solid phase, thus imparting a better appearance of the product. It was reported that at a salt to fish ratio of 1;5 maximum water loss took place in 24-26 hours for gutted fish and in 18 hours for split open fish. It was found that this proportion of salt was sufficient to saturate the moisture content of fish and, as such, any excess amount of salt will be in the solid phase and so useless.

The use of curing mixture containing potassium sorbate, sodium benzoate, sodium acid phosphate and common salt was recommended for curing mackerel.

Fishes like mackerel could be successfully pickled and preserved for a long period by giving a pre-dip treatment in propionic acid bath followed by usual heavy salting. For the control of reddening and mold growth, usually met with in stored dry curing fish, it was suggested that dipping eviscerated and split open fish in 4 per cent propionic acid for 10 minutes followed by salting (1:4 salt to fish) for 48 hours and subsequent sun drying. Storage life could be extended to 62 weeks as against the normal 19 weeks by this method.

Pickling mackerel in saturated brine fortified with 0.5 per cent and 0.25 per cent propionic acid has been found useful for keeping the fish in good condition for periods of one year and five months respectively. It was suggested that by smearing a mixture of 3 per cent sodium propionate, 0.5 per cent butylated hydroxyl anisol and 0.5 per cent sodium sulphate in dry powdered salt over cured fish at 10 per cent level, the product can be kept for 9-12 months free from any visible signs of spoilage; browning or rancidity. In the case of wet cured fish, sodium propionate at 2 per cent level is sufficient.

An improvement over the traditional Colombo curing of mackerel by incorporating sodium benzoate as a chemical preservative in combination with salt and "gorukapuli" is normally used. Smoke curing of mackerel after treating the fish with brine containing propionic acid and turmeric extract separately followed by smoking turned out an attractive product.

Freshness of the fish as well as the quality of salt used for curing are factors which can influence the quality and storage life of the processed product. The maximum permissible time lag between catching of mackerel and its curing should be 8 hours when the fish is kept at room temperature and 3 days if iced.

Marinating Fish

Marinades are semi-preserves, which involves the use of acid, such as acetic acid and salt. If acid and salt mixture are added to fresh fish, the bacterial and enzymatic action can be retarded. The product will have an extended but limited shelf life. All food poisoning bacteria and most of the spoilage bacteria are prevented from multiplying at a pH 4.5 or below. If the product is stored at 2 to 4°C, it can remain several months provided pH do not exceed more than 4.5.

Some bacteria and enzymes will remain active in marinades throughout the storage even in salt and acid, this residual action is desirable.

Marinated fish have long been popular in Northern Europe, and increasing quantities are now being marketed in U.K and North America.

The procedure to produce marinated fish (Dhoma) are as follows

Use absolute fresh fish, or chilled fish in ice within one day of capture. Washed the raw material in freshwater so as to remove blood, slime, dirt etc. The material was then kept in perforated wire mesh basket so as to drain the excess water for 15 to 30 minutes. The drained material was then salted in 1:8 proportion (salt: fish) and allowed to stand for 16 to 20 hours at 0°C. When the fish was hardened, it can be dressed removing head, scales, visceral parts. The salted fish can then be filleted and the vertebral column can be removed using a semi-circular knife. The dressed fish is then washed in 1 per cent salt solution, so as to remove blood, slime etc. The washing may be repeated at least twice in 1 per cent salt solution (for one litre salt solution 2 kg material). The washed material may be allowed to drain for 30 minutes and weighed. The yield of the material up to this step is almost 43 per cent. The drained material was then treated with acetic acid (7 per cent) containing 14 per cent salt (high grade over 98 per cent pure sodium chloride) in closed glass jars with fish to liquid ratio 2:1. The material was stirred occasionally and stored at plus 2°C. The fish absorb acetic acid and salt from the liquid until an equilibrium is attained. The material can remain in good condition for at least 6 months at low temperature, provided enough care is taken to cover the fish completely with liquid.

After a minimum storage of 3 to 4 weeks at low temperature, the fish are inspected. If any brown or dark coloured pieces are there, they are removed. Good pieces are trimmed, if required and packed in suitable lots in glass jars with a pickle liquid containing 1.5 per cent acetic acid or citric acid along with 2 per cent salt until the jar is full. If desired clove (whole) one or two garlic pod can also be added. The lid is screwed on. This pack has a shelf life of about one month at plus 2°C.

The following type of spoilage may occur in certain cases if the acidity of the pack is not carefully controlled. The pH should be between 3.0 and 4.5 and the meat pieces are to be held below the surface of the liquid. The product requires low temperature for its storage and distribution though at ambient temperature (30°C) the product remain without any adverse changes upto 7 days.

Spoilage

1. Bacterial spoilage affecting the clarity of the covering liquid.
2. Mould growth on the surface of the meat.
3. Softening and disintegration of fish due to enzymatic action.

Chapter 12

Radiation in Fish Preservation

Radiation of Fish for Long Shelf Life

The potential application of gamma irradiation for shelf life extension and quality improvement of seafoods have been extensively reviewed. Radurization (radiation pasteurization) offers the duel advantages of minimizing the spoilage rates and staggering the shelf life of the fish.

The fish were beheaded, eviscerated and washed thoroughly. Each fish variety was separately packed in the polythene bags (300 gauge) and heat sealed. The net weight of each bag ranged between 250-500 gram in case of small and medium sized fish varieties and about 1 kg in case of seer fish.

With a view to enable packaging of fish as rapidly as possible for large scale operation, a part of the fish after washing and evisceration were packed in polyethylene pouches without any additional sealing. For easy transportation the unsealed end of the pouches was tied with thread.

The sealed as well as unsealed bags containing fish were exposed to a radiation doze 100 Krad in a Cobalt 60 package irradiator (100000 curies) at 2°C.

All the radurized samples along with the respective unirradiated batches were stored at 0-2°C. The samples were assessed for keeping quality at regular intervals during storage.

Two key factors which are responsible for the loss in quality of fishery products are the initial bacterial load and the temperature of the storage. For radurization process initial bacterial load up to 1000000-10000000 could be tolerated, but storage temperature should be critically maintained at 0-3°C. These exact requirements will therefore help in transforming the prevailing practices of handling, processing and storage to prerequisites which are necessary to obtain maximum benefit from the radurization process.

The process parameters, namely, a doze of 100 krad and storage temperature in the range of 0-3°C are stipulated to meet the requirements in preventing the outgrowth of *C. botulinum*. As processing of trash fish should be commensurate with its availability in large quantities, higher radiation doze may reduce the throughput of the irradiated products in the plant. Also higher doze may decrease the quality attributes relating to textures, flavour and colour. In terms of higher throughput requirement, dozes above 100 Krad do not offer any specific advantage.

Currently most of the trash fish varieties (trash fish formed 30-35 per cent of the total catch) are converted to fish meal, cattle feed and organic manure as such, they are not being carefully handled. The bacterial contamination and the accompanying biochemical putrefactive reactions would greatly reduce the quality of the raw material. This in turn would reflect on the quality of the secondary product. The utilization of trash fish variety in the preparation of reconstituted fillets from deboned meat, fish protein concentrate. Protein hydrolysate and other food formulations is receiving greater attention and will also help in enhancing the commercial value of the trash fish varieties to a greater extent.

From the point of quality control, the radurized trash fish can be stored out species wise or graded into categories depending upon the type of the products to be prepared. Thus products will qualify for grade I during storage when organoleptic score lies between 7-10. This quality of irradiated fish can be marked either as such or as products for human consumption or diverted for isolation of FPC grade 2 products would have organoleptic score between 7 to 5. This grade of trash fish would still be suitable for preparation of protein hydrolysates, ensilage and peptones. Subsequently during the terminal stage when the organoleptic score decreases to less than 5 and irradiated or unirradiated samples are not fit for human consumption, the fish together with offal and other waste materials, collected as a part of the processing could be converted into manure.

The irradiated trash fish, as a raw material, thus provides almost 20 days for product development in contrast to un-irradiated fish which rapidly loses its quality attributes. In the conventional ice storage therefore it becomes manadatory to complete product development within a short span of 2-3 days.

Increased Storage Life of Fish by Gamma Irradiation

Ionization radiation is an important tool for extending storage life of food including fish which is by nature very susceptible to spoilage. The hot and humid climate, particularly prevailing in most part of India is conducive to rapid spoilage of fresh fish. Due to unscientific method of handling even fish transported under ice may not reach the destination in a marketable form. The problems of handling, storage and transportation are particularly important in India, where consuming centers are far away from fish landing areas.

Bactericidal effect of gamma radiation have been utilized extensively for extension of storage life of fish due to suppression of rapid spoilers. But radio-pasteurization process does not reduce all food-borne micro-organisms in a fixed proportion due to the differences in their radiation sensitivity, repair mechanism and doze modifying capacity. Usually radiation doze levels of 0.1 M rad to 1.0 M rad are applied for extension of storage life of seafood. Fish radurised in dozes of 100-200 K rad, can be consumed safely provided the temperature of storage is strictly maintained below 3.8°C.

For internal transportation and distribution of marine and freshwater lean fish, radio pasteurization has a scope with significant results.

Fish were eviscerated, degilled, washed and sealed in a polythene pouches and irradiated in Cobalt 60 Food Package Irradiator at specified doze levels. Irradiated and control samples were packed separately in insulated ice boxes with fish to ice ratio of 1:3 to 1:4 and transported by rail to Calcutta. They were received at the destination on the 3rd or 4th day of dispatch and held in ice (0-2°C) till the examinations and observations were over.

Irradiated samples of Bombay duck (100 K rad) were acceptable up to 22 days, while the control samples were rejected after 5-6 days. In the case of Rahu irradiated samples (100 K rad and 200 K rad) were organoleptically acceptable up to 3 to 4 weeks at 0-2°C depending upon the doze, while the control sample was rejected after 12-14 days.

Radiation Pasteurization of White Pomfret (*Pampus argenteus*)

It has been observed that the acceptability of the frozen fish decreases progressively on account of textural alterations and development of characteristic yellow discolouration on the skin surface. Reckoning the limitations of the conventional methods of ice and frozen storage, the advantages of radiation pasteurization process have been examined for augmenting the shelf life of white pomfret in "as is" form.

Earlier studies have shown that white pomfret provides a system susceptible to radiation-induced oxidative changes, skin tissue being more sensitive than the muscle. Pre-packing under vacuum, however, effectively suppressed the oxidative changes in fish both on irradiation and during subsequent storage at 0-2°C. Organoleptic, chemical and bacteriological analysis indicated that the un-irradiated fish fillets packed under aerobic or un-aerobic conditions and stored at 0-2°C, spoiled within 8 and 12 days respectively. Doze-dependent extension in shelf life was, however, exhibited by the irradiated fish. An-aerobically packed fish fillets exposed to radiation dozes of 0.1, 0.3 and 0.5 M rad showed extended shelf-life of 35, 50, and 60 days respectively without undergoing any loss in quality attributes of colour, odour or texture. The fish samples packed under aerobic condition were, however, prone to development of rancid odours on irradiation (0.5 M rad) and yellow discolouration (0.3 and 0.5 M rad) during storage and also exhibited lesser extension in shelf life when compared to an-aerobically packed fish.

Yellow Discolouration in Irradiated White Pomfret

Among the major chemical constituents of white pomfret, the triglycerides isolated from skin and muscle tissues, phospholipids from skin and sarcoplasmic proteins of the muscle contributed appreciably towards 2-thiobarbituric acid reactive substances

(TBRS) on irradiation contribution by the other constituents of fish was comparatively less.

Formation of insoluble yellow coloured aggregates were also observed by interaction of fibriller proteins of white pomfret with malanaldehyde. It may be concluded, therefore, that the yellow patches encountered on the skin surface of irradiated white pomfret fillets may be a manifestation of the two major reactions; 1) formation of TBRS by radiation and 2) reactivity of TBRS with the amino nitrogenous compounds in the fish described below

| Tissue constituents Neutral lipids, FFA Sarcoplasmic proteins Fibrillar proteins | Gamma $\rightarrow$ TBRS Radiation | + | Food components Amino acids Proteins and Amino compounds | $\rightarrow$ | Yellow Pigment complex |

In *vitro* reaction under aseptic condition of the skin tissue with pure malanaldehyde produced yellow colouration of the former. Comparison of the spectral characteristics of yellow colour formed by this in *vitro* reaction with those of the colours formed in *situ* in irradiated skin samples suggest that the three pigments were similar in nature.

Control of Radiation Induced Oxidative Changes

White pmfret was examined for its response to irradiation treatment. Skin and muscle tissues of the fish, analyzed separately showed greater concentrations of triglycerides, free fatty acids and phospholipids in the former than the later. Assessment of fish tissue samples in terms of peroxide, carbonyl, free fatty acids (FFA) and 2- thiobarbituric acid (TBA) values indicated that the skin was more susceptible to radiation-induced oxidative changes than the muscle.

Difference in response of skin and muscle of white pomfret to irradiation were clearly distinguished by the type of sensory changes occurring in the tissues and by the higher susceptibility of the skin to radiation-induced oxidative changes than the muscle.

It has been postulated that rancid odours arise from the lipid components, while degradation of amino acids and soluble proteins accounted for development of burnt feather type odours. Further oxidative changes in lipids also lead to yellow or brown discolouration in certain fish species, subjected to irradiation. Due to high concentration of lipids (16 per cent) in the skin of white pomfret, this fish is susceptible to radiation-induced oxidative changes as well as for the development of rancid odours and yellow discolouration.

When irradiated under vacuum, the skin samples did not develop rancid odours or yellow discolouration, concomitant with suppression of the indices of oxidative changes

Development of rancid odours (0.5 M rad and higher dozes of radiation) and yellow discolouration (1.5 M rad) were noticed in the irradiated skin as against burnt type of odours in muscle. Pre-packing under vacuum effectively suppressed the oxidative changes and associated undesirable alterations in colour and odours of the skin tissue on irradiation as well as during storage at 0-2°C.

Chapter 13

Antibiotics in Fish Preservation

Antibiotics in the Preservation of Prawns

A great variety of species of both freshwater and marine fishes have been tested using ice containing either chlorotetracycline (CTC) or oxytetracycline (OTC) or dipping them in very dilute solutions of these antibiotics. Information available on treatment of shell fish with antibiotics is limited, with sucked oysters, cooked lobster meat, sucked clams and raw shrimp and crab.

Treating *Metapenaeus affinis* and *M. dobsoni* with 5 ppm antibiotic ice (5ppm CTC ice) for determining viable plate count, for trimethylamine nitrogen (TMAN), volatile acid number (VAN) and organoleptic studies, it was found that the effect of CTC in the keeping quality of prawn was pronounced only after 8 days of storage. Organoleptically little difference could be detected between prawns kept in ordinary ice and those in CTC ice until 8 days of storage. The spoilage became apparent after 11 days of storage in ordinary ice, whereas the CTC ice stored samples indicated symptoms of spoilage after 21 days. The cooked muscle of CTC-iced sample after 13 days of strrage was as good as that of 8 days stored control sample. The latter became completely unacceptable on or before 11 days of ice storage. This would mean a shelf life extension of at least 5 days over the control.

The bacterial counts of prawns in ice storage generally fall during the first few days and then increase. But such increase was slower in the CTC-treated sample. Until 8 days of storage the bacterial count did not differ much. During latter period of storage, the counts of untreated sample showed marked increase, while that of CTC-iced sample increased slowly. The chemical indices (TMAN and VAN) supported the general trend noted in the organoleptic and bacteriological assessment of quality. The TMAN values of untreated sample were always above those of treated samples.

Though not much difference was noticed until 8 days, VAN values of the untreated samples were greater than those of CTC-iced samples, during later stages of storage.

The CTC-uptake of the muscle of CTC-iced sample gradually increased with storage time. In the case of whole prawn, a CTC level of 4.064 micro-gram per gram muscle was obtained after 21 days of storage, while 7.80 micro-gram per gram CTC was absorbed by the muscle of headless prawns during the same period of storage.

CTC when incorporated in ice at 5 ppm level can extend the shelf life of prawn by at least 6 days. In most species of fish especially, cod, haddock, plaice with 5 ppm CTC-ice, a shelf life extension of 8 to 10 days over the untreated samples were recorded. The alkaline nature of the shrimp and the presence of calcium anf magnesium ions in its meat would contribute to the comparative instability and consequent poor activity of the ice. This was the reason why CTC was found less effective for preservation of prawns.

The use of higher levels of CTC in ice is not advisable, since it would result in higher amount of absorbed CTC in the muscle, which is not completely destroyed during cooking.

Since the effect of the antibiotic treatment becomes pronounced only after 8 days of storage, the use of CTC-ice would be restricted to those fishing trips, where vessels remain off shore for more than 8 days.

For preparation of CTC-ice "Acrohize" (American Cyanamid Co) was used, the appropriate amount of which was dissolved in tap water and frozen in aluminum trays of 7 litres capacity. Water used for ice preparation was always acidic (pH 6.4-6.5). CTC concentration of each lot of ice was determined so as to ensure that appropriate CTC activity was present in the ice used.

Chapter 14

Storage of Fish and Fish Products

Protective Covers for Frozen Fish

In order to help retard the changes that fishery products undergo during frozen storage, some form of protective coating or covering is used. In the early days of freezing and frozen storage, very little was known about ways for preventing desiccation and other undesirable changes. In fact there there was no suitable wrapping materials available. Probably the first "package" that came into general usage was the ice glaze, that is formed by dipping the frozen fish in cold water. Because of the wide spread practice of using the glaze in the earlier days of the industry, very little attention was given to other methods of packaging. Not until the introduction of quick frozen foods in prepared form, any serious attempt was made to improve upon protective coverings for frozen fishery products. During the past 40 years, in particular, great strides have been made in the development of these coverings, and although the perfect package is yet to be found, there are available a number of types of materials and packages that offer excellent possibilities for minimizing adverse changes in fishery products during extended period of frozen storage.

Glazes: Definition and Properties

Glazes may be defined as any continuous thin film or coating that adheres closely to the product. The glaze is usually applied on the frozen product by either dipping the product or by spraying it with a solution of the glazing agent and allowing the glaze to solidify. Some desirable properties of glaze for frozen fish are; (1) non-cracking, (2) strength to withstand rigors of handling and shipping, (3) low water vapor pressure to minimize evaporation of glaze, (4) light weight and low bulk, and (5) attractive appearance.

Ice Glaze

Although many types of glazes for fishery products have been introduced, the ice glaze remains the only one of commercial importance. The ice glaze is formed when frozen fish are given a short dip in cold water or are spread with water, which freezes into a thin coating of ice.

The chief advantage of the ice glaze are its susceptibility to cracking, its brittleness and its high vapor pressure, which necessitates re-glazing if the product is stored for any length of time. Its advantages far outweigh its disadvantages however. The ice glaze is inexpensive, easy to apply, readily adaptable to a production line, and provides a satisfactory protective covering for a variety of fishery products. When properly applied, and maintained, the ice glazes are effective in preventing loss of moisture from the product and also in preventing ready access of air to the fish, thus retarding the onset of rancidity.

In ice glazing fish, best results are obtained if (i) the fish is thoroughly frozen and is sufficiently cold, preferably at 0°F or lower, to freeze rapidly the film of water on its surface; (ii) the water used for glazing is pre-cooled to a temperature (34 to36°F) slightly above the freezing point, and (iii) the room in which the ice glazing is carried out is maintained at a temperature slightly above freezing to avoid warming the fish. The following is a description of the commercial methods used to apply ice glazes to whole and dressed fish, fish frozen in blocks, fish steaks and shell fish

Whole and Dressed Fish

The ice glaze is the principal means of covering fish held in bulk storage. In some plants, conveyers are used to bring the fish from the freezer to the glazing tank and then to frozen storage. In other plants, fish carts are used to transport the frozen fish to and from the glazing room. In this room the fish are transferred from the carts to a large metal baskets, which is raised and lowered into the dipping tank with an electric hoist. The ice glaze is formed by dipping the frozen fish momentarily in the cold water and removing them to allow the film of water to freeze. This process may be repeated several times in order to secure a thicker ice glaze. In at least in one plant in Alaska, the glaze is applied by covering the frozen fish through a series of sprays. Such a system is a recent innovation in Alaska and lends itself well to mechanization, which reduces labor costs.

The thick body section of the fish, which possesses a large capacity for refrigeration than to the thinner sections, such as, tails and fins, is capable of acquiring a thick ice glaze, whereas the tail fins and snouts may take only a thin ice glaze. During the cold storage of the fish, it is these areas where the ice glaze is thin that must be checked and re-glazed frequently in order to keep the ice glaze intact

Fish Frozen in Blocks

Small fish are often frozen in blocks in pans, and the blocks of fish are then dipped in cold water to acquire an ice glaze.

Fish Steaks

Steaks cut from frozen fish with a band saw are either dropped into the dip tank and removed on a conveyor belt or are placed on a wire mesh conveyor belt and sent through a water sprayer to apply the glaze.

Shell Fish

Crabs may be frozen and then glazed whole, as legs in the shell, or as meat alone. On the west coast, the meat is usually pan frozen in a block, the block of frozen meat is covered with water to fill the voids and to provide the ice glaze, and is then returned to the frozen storage room.

Shrimp may be frozen individually or in a block in a pan and then ice glazed. Pan-frozen shrimp are also placed in a cart on, covered with water and refrozen. Alternatively, the shrimp are packed into cartons. They are then frozen. The cartons are opened, and cold water is added by immersing or spraying. They are then returned to the freezer.

A new glazing technique, for shrimp is to quick freeze the shrimp in a pan and then pack the frozen shrimp in the carton over a bottom layer of slush ice with a second layer of slush ice on top. The slush ice freezes on contact and completely surrounds the shrimp. This new technique is said to provide protection equal to regular ice glazing, yet the refreezing and double handling of cartons are eliminated, and the amount of ice weight is substantially reduced.

Re-glazing

In order that the glaze on unpackaged fish can be properly maintained, it is necessary to make frequent examinations in the storage room to be certain that the glaze has not disappeared. Re-glazing should be done as soon as the condition of the glaze indicates that it is necessary. Quite often, the re-glazing is done by sparying cold water on the fish in the frozen storage room.

Evaporation of the glaze can be retarded by simply by placing a tarpaulin over bulk stored fish or by storing the fish in corrugated fiber boxes or in wooden boxes lined with waxed paper. With fish frozen in blocks placing them in waxed cartons after they have been glazed will enable them to be held for longer periods without the necessity of having to renew the glaze.

No hard and fast rule can be established as to how often the re-glazing should be done, as a number of factors are involved. Glazed fish stored without any additional protective covering may need to be re-glazed as often as every two weeks. It is best to examine the fish at frequent intervals and not rely on some arbitrary time for re-glazing.

Additives to Ice Glazes

One of the chief disadvantages of the ice glaze is its brittleness and its susceptibility to cracking. In order to overcome these disadvantages, various additives to the glazing water have been investigated, and patents have been taken out on additives, such as, salts, sugars and alcohols. The additives however, have not been widely used in most commercial operations, for various reasons.

Bedford (1939) has a patent to produce a substantimately transparent non-cracking glaze from aqueous solutions containing approximately 0.2 to 0.3 per cent alcohols, particularly edible polyhydric alcohols, aldehydric alcohols and ketonic alcohols.

Anti-oxidants may be incorporated directly into the dip solution, if desired. Experiments have shown that whole fish steaks and fillets can be glazed conveniently from a solution containing 0.5 to 1.0 per cent ascorbic acid. The dipping solution should contain 1 to 2 ounces of ascorbic acid per gallon of water and be kept about at 33 to 34°F. Fillets and steaks may be given a single dip of 10 to 20 seconds; whole fish may be dipped 2 or 3 times in order to give them a heavier glaze. The glaze formed is slightly opaque and shows considerable resistance to cracking.

Glazing Fish for Locker Storage

A convenient method of glazing fish for locker storage has been demonstrated. Un-frozen fish are wrapped in vegetable parchment. The package is then soaked in cold water for a few seconds, immediately wrapped in moisture-vapor proof material and frozen. Because of the excess water retained by the parchment wrapper a heavy glaze is formed over the entire fish upon being frozen. The advantages of this method are (i) the elimination of prior freezing and extra handling involved in glazing after freezing and (ii) the obtainment of a close fitting package by having the fish in an unfrozen condition at the outer moisture-vapor proof is applied.

Other Types of Glazes

Though most of various types of glazes were tried only on experimental scale, none have replaced ice glazing on a commercial scale for fishery products.

Pectinate Films

Pectinate has been suggested as a gel coating in place of ice glazing on frozen foods, including fish fillets. A 3 per cent solution of sodium hydrogen pectinate in water, preferably in the pH range of 4 to 6 is applied on the product by dip or spray coating. Present pectinate films are not effective barriers to moisture, but may find use as carriers of anti-oxidants.

Gelatin-Base Coatings

Hall and Griffith (1933) have taken a patent for gelatin-base protective coating that consists of 12 parts, by weight of pure gelatine, 25 parts of cold water, 1.2 parts of potash alum and 2.4 parts of glycerine. The solution is applied by dipping or spraying at temperature between 135 and 140°F.

Another glazing formula consists of 7 pounds of gelatine, 10 ounces of glycerine and 60 ounces of water.

A "freezer warp" based on purified, vacuum-dried derivative of Irish moss that is mixed with sorbitol and water can be applied by the dip or spray method and is reported to set in 30 seconds.

Combination of Chemical and Irish Moss Extractive

Irish moss extractive solution were effective in prolonging the storage life. Gallic acid (0.1 per cent) and ascorbic acid (0.2 per cent) in Irish moss extractive solution and nordi-hydroguaiaretic acid (0.2 per cent) in cotton seed oil delayed rancidity in the fillets for 7 or 8 months.

Waxes

Thermoplastic waxes appear promising for coating frozen foods. They consists of blends of various waxes combined in number of ways with non-waxy materials to produce strong, tough, moisture-vapor proof films. The foodstuff grades of thermoplastic waxes are odourless, tasteless and non-toxic and may be used with frozen foods. In application, frozen foods are dipped in the melted liquid thermoplastic wax. The thermoplastic wax hardens immediately into a tough and flexible wrapper that withstands ordinary rough handling at low temperature. The thermoplastic wax adheres closely to all surfaces of the food and is easily removed by stripping

One thermoplastic wax has a dipping temperature range of 140 to 158°F, with 145°F being the ideal temperature. The manufacturer recommends it for use with non-oily fish and with shell fish, such as shrimp and crab. For oily fish the manufacturer recommends wrapping the frozen fish in very light-weight pliofilm, saran or waxed paper before dipping to prevent the wax coating from cracking and separating away from the fish.

This thermoplastic wax appears to give effective protection to products that can be coated directly. For products that must be given an initial film wrap, the advantage of the form-fitting coating may be lost to varying degrees; additional handling will be involved, and the cost of the protective coating will increase.

In comparison to the ice glaze, the wax glaze is more expensive and requires more handling and a longer time to apply but is tougher, stronger and longer lasting.

Edible Oils

With blue fish, sea trout and butter fish, Lenon (1932) reported substituting the ice glaze with a thin coating of edible oil (cotton seed oil, corn oil or peanut oil) by either dipping the fish into the oil or spraying them. Since the oil film does not evaporate, one coating of oil reduced the drying effect of air during the 12-week cold storage period. All the oils tested were superior to the common ice glaze in preventing moisture evaporation; however the use of edible oils for coating fish has never found practical application.

Films and Wraps

Films and wraps make another class of protective covering for frozen fish. A good wrapping material for frozen fish should be strong enough to resist tearing and puncturing from handling during the packing operation, pliable enough to make a tight wrap, easily sealable, grease proof and durable at low temperature; should impart no odours or flavours and should have a low rate of moisture-vapor and oxygen transmission.

There are a large number of films, foils, resin-coated papers, lacquered papers, waxed papers, plastic coated papers and various combinations of laminated paper on the market. Each has its own set of characteristics. The choice of the packing material is governed by such factors as the protection requirement of the frozen product, the cost of packaging material, the cost of packaging operation, and its merchandising features.

Cellophane

Cellophane is transparent, grease proof, odourless, tasteless and flexible. It is available in about 150 different grades or types and in a large variety of colours and of printed patterns and designs. Cellophane lends itself readily for use with automatic wrapping machines.

The type of cellophane used for frozen fishery products is coated to be moisture-vapor proof, grease proof, readily heat sealable, durable at low temperature and to have low oxygen permeability. The common gauzes are 300 (0.0009 inch) and no 450 (0.0013 inch). It is used as sheets for direct wrapping or as bags for whole fish.fillets and steaks. It is also used as liners and over-wraps for frozen fish cartons.

Polyethylene

Polyethylene is widely accepted as a packing film for frozen foods. It is tough, durable, flexible even at low temperatures and chemically inert It has a low level of taste and odour transfer, water absorbency, and moisture-vapor transmission. The disadvantages of polyethylene are that it transmits oxygen and solvent vapors and it may be penetrated and softened by many types of fats and oils. The gauzes commonly used are 0.002 inch for the plain polyethylene and 0.001 inch for the laminated type.The film is heat sealable but is apt to stick to the heated surfaces of the sealer if the temperature is not carefully regulated.

Aluminum Foil

Aluminum foil is non-toxic, tasteless, odourless, pliable at low temperature and grease proof. It has a low degree of gas permeability, even in the lighter gauzes, and minimized oxidation and rancidity of the packaged products. The water-vapor transmission rate is negligible in foil thicknesses as low as 0.0035 inch. Aluminum foil is available with one shiny surface and one mat or satin finished surface and may be decorated by embossing, gloss coating or laminating, and colouring or printing. Aluminum foil is also widely used in combination with other packaging materials, such as, paper, plastic or cellulose films and paper board. The aluminum foil with a thermoplastic coating is heat sealable.

Aluminum foil is one of the best wrapping materials for frozen foods. For locker use, the best thickness of plain foil is 0.0015 inch, which is known as, "locker foil", for home freezer use a 0.001 inch gauze may be used. In addition to ita excellent protective properties, aluminum foil is fully moldable and when properly applied hugs the product closely at every point, thus eliminating air pockets. By folding the edges of the foil, twice or more, until it lies flat and tight, the package requires no taping, tying or heat sealing.

Vinylidene Chloride (Saran, Cryovac)

Vinylidene chloride film has the lowest water permeability of any of the plastic films. It is very strong, tough, flexible at low temperatures, completely transparent, chemically inert, heat sealable, an effective barrier to oxygen and moisture-vapor, as is resistant to common solvents, oils and chemicals. This material has been modified to make a film (Cryovac) that is particularly useful for packaging irregularly shaped frozen foods. In the form of bags, it can be made to shrink skin-tight over the food by the momentary application of heat after the air has been drawn from the bag.

Rubber Hydrochloride Film (Pliofilm)

Rubber hydrochloride film is manufactured in a large number of combinations of types and amounts of plasticizers. It is extremely tough and makes a strong seal. Its stability is affected by direct sunlight, but is not affected by changes in relative humidity. It may be laminated to itself to paper, or to other bases to improve its desirability. A special rubber hydrochloride film, which is plasticized by heating and stretching can be shrunk around an object by applying moderate heat. A pilofilm particularly adapted for frozen foods is said to remain flexible at minus 20°F, is heat sealable, can be sealed closely against the frozen product, and has a negligible rate of moisture-vapor transmission. Gauze no. 120 (0.0012 inch) and no 140 (0.0014 inch) are the most commonly used.

Coated Papers

A variety of coated papers that are suitable for frozen-food packaging are on the market. Among the most important are wax, vinyl and saran and polyethylene.

Waxed Paper

Waxed paper is tasteless, odourless and non-toxic. Paraffin wax is the principal material used in most waxed-paper coatings. Polyethylene and micro-crystalline waxes are used in varying quantities as additives to improve the qualities of the regular wax. The resulting coating produces greater moisture impermeability, better flexibility, at low temperatures and higher seal strength. These properties have led to the extensive use of specially prepared waxed paper in the packaging of frozen foods and other food products. It is available in various thicknesses. Opaque waxed paper lends itself to the effective use of colour for brand identification of a packaged product.

Vinyl-and Saran-Coated Papers

Papers coated with these plastics are quite resistant to water and water vapor and to the common gases, such as oxygen. The materials are used quite extensively for cartons that hold very wet foods and foods that contain considerable quantities of fatty materials.

Polyethylene-Coated Papers

Polyethylene-coated paper (usually a 40-pound white paper with a ½ to 1-mil coating) is a popular locker paper for frozen foods. It gives better moisture protection than surface-waxed and wax-laminated papers especially at creases and folds in the wrap. It has very good low-temperature-flexibility and is heat sealable. It is rated as only a moderate barrier to oxygen.

Polyethylene-Coated Cellophane

Cellophane with a 1- to 2- mil coating of polyethylene offers considerable promise for the packaging of seafoods. It combines the advantages of cellophane transparency, grease-proofness, scuff resistance, gasproofness and good printing surface with the additional advantages of moisture-proofness low-temperature flexibility, improved film strength and heat sealability.

Antioxidants in Papers for Food Wrappers

Antioxidants have been incorporated into paper wrappers to improve the keeping quality of fats and fatty foods. Antioxidants can easily be applied to parchment glassine and paper board. The un-waxed papers are treated with an aqueous emulsion of an antioxidant during the drying stage or directly after the final drying operation of the paper manufacture. With waxed paper, the antioxidant may be dissolved directly into the molten wax before its application to the paper.

Various antioxidants may be used. Food-inhibitor grade butylated hydroxyl anisole (BHA) was found to be effective for most paper application. Good results can be obtained by treating un-waxed paper with 0.05 to 0.10 per cent BHA. Its effectiveness may sometimes be increased by addition of 0.02 to 0.05 per cent citric acid to complex such contaminating metals as iron and copper. Food-inhibitor-grade n-propyl gallate (0.02 to 0.03 per cent) may also be used to advantage in some applications.

The keeping quality of fats dispersed in cardboard was greatly improved by impregnating the cardboard with citric acid, pyrogallol or galacetonin. Citric acid and gum guaiac mixed with sizing on parchment, improved the keeping quality of lard wrapped with this material.

Modified Atmosphere Packing (MAP) of Seafood

Replacing the air in a pack of fish with a different mixture of gases, typically a combination of carbon dioxide, nitrogen and oxygen is termed as modified atmosphere packing (MAP). This technology when adopted in conjunction of chilled storage of seafood or seafood products can double or treble the shelf life of the product. The proportion of each gas component is fixed at the time of introduction of the mixture, and no further control is exercised during storage. A gas flush or evacuation and backfill is used to replace the air in a package with gas mixture specific to the product. Carbon dioxide, oxygen and nitrogen individually or in combination are the gases most frequently used in MAP. Nitrogen is used as an inert filler to counteract the pressure when vacuum is drawn and maintains package integrity to prevent the product from being crushed and/or sticking together. Oxygen is added to meats to maintain colour and to white fish to reduce drip, but may be omitted in fatty fish to prevent the development of oxidative rancidity.

The elimination of ice in refrigerated modified atmosphere containers or packages result in economic and sanitation advantages. Longer shelf life expands the market potential and reduces waste during distribution and retail display. The consumer benefits with dry, high quality, ready-to-cook fresh product. But caution has to be exercised in the implementation of retail application, since the antimicrobial effect of carbon dioxide changes the traditional spoilage pattern of seafood in MAP and botulism may occur from the consumption of improperly handled MAP seafood. The purpose of seafood MAP is to slow down or prevent the sequence of bacterial spoilage.

The simpler and cheaper machines pack the product in a flexible bag or in a tray inside a beg to form a pillow pack. At the most expensive end of the available range are sophisticated machines that continuously form packs with rigid bases from rolls of thermo-formable plastic film. Some of these machines have a duel purpose and can be converted to vacuum packing when required. Gases can be obtained either ready mixed or separately for use in machines that mix the gases before packing. Where gases are mixed in the machine, the gas composition in the packs should be measured at the start of a run, and monitored throughout the day, particularly when faults are suspected or adjustments to the gas mix are made. Only the highest quality fish should be used for MAP packs, in order to get the most of benefit from any extension of storage life. Fish should be handled hygienically and kept chilled from the time of capture until they are packed. Whole fish and fillets should be kept in ice, while awaiting processing and smoked products should be held in a chill room at 0°C. Ideally an air blast chiller should be provided in the processing line, either before or after the packing machine, since the fish may warm up significantly during the packing operation. Layering of products within a pack should be avoided; a single fillet or portion is more fully exposed to the action of gases. Wet fish products that are likely to exude drip can be laid on a pad of absorbent paper inside the pack. Packs with faulty seals can be detected by pressing them with the hands; faulty packs will collapse. Packs should be clearly labeled according to the existing regulations, and should be marked with a sell-by or consume-by date.

Packaging fresh seafood in carbon dioxide can delay spoilage and provide consumers with a high quality product. Generalizations can not be drawn about shelf life extensions and acceptability since spoilage criteria differ among studies. Carbon dioxide concentrations of 20 to 100 per cent have been recommended. The ideal concentration is dependent on the fish biology, initial microbial population, gas-to-fish ratio and packaging method, but carbon dioxide concentrations of 40 to 60 per cent are used most often. MAP is not effective alone, but will be so as an adjunct to refrigeration below 3.3°C. The low temperature inhibits *C. botulinum* and the gas mixture extends the lag phase and slows generation times of spoilage bacteria. The packaging methods maintains the atmosphere and protects the product from contamination. Additional treatments, such as, potassium sorbate and low dozes of irradiation may enhance the effects of MAP.

Insulated Container for Long Distance Transport of Fish

One of the most effective and commercially tried means to retard spoilage of fish is to reduce its temperature. Crushed ice is usually employed for this purpose and fresh fish keeps well for limited period of time in contact with it. But ice melts rapidly necessitating re-icing at frequent intervals during transport to distance places to maintain the fish at sufficiently low temperature. The use of more ice will increase the cost of fish at the consumers end. Well insulated containers which can retard rapid melting of ice, therefore, essential. CIFT has developed an insulated container which can be used for transport of iced and frozen fish to long distances. The details of the container and the process adopted are as follows;

Container

Wooden boxes, made of 3-ply plywood sheet, nailed to 2 cm thick wooden frames (ordinary tea chests) can be used as the container. Such boxes are fairly robust and stand the transport well. A box of the size 48cmX48cmX62cm will hold approximately 100 kg ice and fish. 20 mm or 25mm thick thermocole slabs cut to the inside size of the box and sealed in 200 gauze polythene bags to prevent the thermocole slabs from coming in contact with the ice melt water, are used as insulant in the box. Six thermocole slabs are required to insulate the complete box, four in all the four sides, one at the bottom side and one inside the lid.

Transport of Iced Fish

1. Fish and ice should be kept in the ratio of 1:1.
2. Put a layer of crushed ice at the bottom of the box.
3. Place one layer of fish and then another layer of ice and so on, with a layer of ice at the top finally.
4. Fix up the lid to the box properly, wrap the entire box in gunny and sew.

By this method the fish maintains its freshness for 55-60 hours. Cost of one box of the size 48cmX48cmX62cm insulated in the above manner may be 3000/- rupees. Even the box get damaged in transit due to rough handling, the thermocole sheets are fairly shafe and can be re-used. If proper care is taken in handling it can easily last for two to three seasons.

Transport of Frozen Fish

Since re-icing enroute is not generally practicable due to various reasons and maximum period of storage of fish in insulated container is limited to 60 hours and since prolonged storage of fish in ice adversely affects its quality, it become necessary to evolve a suitable method of transportation of fish to long distances. The best alternative has been found to be packing frozen fish instead of iced fish in the insulated container.

1. Freeze the fish, either individually or as blocks to a temperature of minus 20°C (small fish like sardine, mackerel etc should be frozen as glazed blocks. Big fish like seer, tuna etc can be frozen individually or as small pieces).
2. Pack the frozen fish in thermocole insulated plywood boxes of convenient size.

3. If individually frozen fish is packed, fill up the gaps between fish with crushed ice.

4. Fix up the lid and transport in rail wagons or trucks.

By this method fish can be transported to very distant places as may require 3-4 days journey. Fish will reach the destination in a partially thawed condition and can be marketed as fresh fish.

An insulated container of the size of 48cmX48cmX62cm can hold about 60-65 kg of fish frozen as glazed blocks (sardine). The method was tried from Verabal to Bombay and Delhi and from Kakinada to Delhi and Calcutta.

Ice Box

The benefit of using ice in fishing are; (a) all fish landed could be in good condition, and command high first sale prices; (b) boats could stay at sea for longer time without the catch deteriorating; (c) boats can land their fish at any time of the day without catch deterioration. Fish can be kept until the next day if the landing is too late for the first market of the day and (d) during multi-day fishing trips, the fish from early catches can be kept fresh rather than salted.

Design of Ice Box

Following factors need considerations, while designing fish ice box for fishing boats; (a) the capacity requirements and type of fish being caught by the boat; (b) the space available on board and stability of the boat; (c) the high ambient temperatures and the need to conserve ice; (d) the need to occasionally, but not regularly, remove the box from the boat for maintenance of the boat and or box and (e) the additional capital, running and maintenance costs for the fishing operation.

The box can be made of fiber reinforced plastic (FRP) laminated directly on to 70 mm thick polyurethane foam. This sandwich construction gives a wall thickness of roughly 75 mm with external dimensions as 130cm length X 70cm width X 60cm depth. The box is cuboid with internal dimensions 115cm length X55cm width X45cm depth. Both internal and external surfaces of the box are smoothed with a gel coat to give an easily cleaned surface.

Access to the box through a square hole 50cm X50cm located in one end of the top surface. A simple drop in lid, insulated and constructed in the same way as the box fit into this hole.

In order that ice alone and ice/fish mixture can be kept separated, the boxes are fitted with removable boards which divide the inside longitudinally. The four wooden boards are approximately 10cm thick by 10cm deep and slot into channels internally moulded into the ends of the box.

A single brass 1cm diameter plug is fitted through the box at the bottom of one side of the box at the lid end. This is constructed so that it is at the lowest point when the box is on the boat and it has a screw-in stopper so that drainage can be regulated.

Carrying handles are provided at each end to assist in lifting the box on and off the boat.

The box is designed so as to able to carry sufficient ice and fish for normal operations. It has a nominal capacity of 175-200 kg ice/fish mixture. The box is designed to be stored on board the boat below the thwarts. It is not envisaged that the box would be removed from the boat on regular basis.

In the best cases profits up by over 20 per cent with the pay back periods on the box of about one year. It is estimated that the boxes will last at least 5 years and probably 10 years without major repairs.

Stowage on Quality of Fish

Quality of boxed cod and haddock was compared with that of penned fish in terms of organoleptic grading, trimethylamine values, Electronic Fish Tester readings and fish and fillet yield. Grading were of three categories, Grade-I top quality fish; Grade-II slightly lower than Grade I and Grade-III, the fish which were not suitable for human consumption. The boxed haddock landed in October were 90 per cent grade-I and 10 per cent grade-II, while the fish from the same lot held in pens were only 52 per cent grade-I and 48 per cent grade-II. The boxed cod landed during May and June was all grade-I, while the fish held in the pens was 55.17 per cent was grade-I. In warmer weather (July) the quality of both the lots of fish suffered, however, the boxed fish were 46 and 60 per cent were grade-I, while one lot of penned fish had none in grade-I and the other had only 13 per cent grade-I

The Intelectron Fish Tester V gave a relative measurement of muscle damage due to physical mishandling and tissue break down due to bacterial and enzymatic activity. The fish stored in boxes had higher readings than those held in the pens. The fish held in pens were subjected to pressure from the fish stored above them. In many cases the appearance of the fish showed the effect of this pressure and they were squeezed and partially flattened. The boxed fish were packed no more than 25 cm deep, thus the pressure on the lower fish becomes negligible. The average fish tester reading for boxed fish was 68.1, while penned fish scored an average of 46.95. TMA values however, did not differ greatly. There was less physical damage due to squeezing with boxed fish and consequently a greater yield of landed round fish and fillets.

Storage of Tuna Onboard after Catching

Tunas are bled as soon as they are hauled aboard and chilled as per the standards laid down for export. With regard to the aspects of production of quality sashimi and loins of tuna and their processing for export, the icing of tuna on board, it is emphasized the need to replace the ice used once in every four hours. The harvested fish may be placed in ice slurry or in refrigerated sea water tanks on board. In Australia, ice or chilled water in refrigerated sea water tanks was used for preservation of tuna on board. Fishes may be hung into the slurry for some time and at the right time they may be transferred for packaging. Slurry damage skin of fishes sometimes. Preservation of frozen loins had to be made in minus 65°F. Loins may be sliced out of chilled tuna. In order to meet the ice requirements on board, it is recommended to install an ice maker of half ton per day capacity on board.

Storage of Cured Fish Products

Reddening caused by halophillic bacteria and growth of molds, popularly known as "dun" constitute the main type of spoilage in cured fish under tropical condition as in India. Fish cured by wet or dry curing method under normal conditions keep only for a period of 2-3 weeks. By this time, the above type of spoilage set in and make the product unacceptable.

With the experience gained in the field and as a result of experiments carried with propionic acid and its sodium salt, a simple and effective method of treatment has been evolved with the chemical sodium propionate for preservation and storage of cured fish products.

When the finished product is ready for packing, sprinkle it with an intimate mixture of 3 per cent sodium propionate in powdery refined salt.

In case of wet cured products, the proportion of the preservative can be reduced to 2 per cent level.

After application of the preservative mixture, pack the products as usual in good containers. Containers, lined inside with polythene paper are more effective to prevent further exposure to contamination and also to prevent excessive dehydration of the product.

It is roughly assessed that 10 kg of the cured product will require about 1 kg of preservative mixture. This proportion need not be followed strictly, however, it is essential that the mixture is uniformly dusted on the entire surface of the cured fish.

Prior to cooking, the treated fish may be rinsed in a little running water to wash away the adhering preservative and it then be soaked in freshwater in the usual way.

It has been observed that dry-cured product preserved by the mixture of sodium propionate in refined salt as mentioned above will keep for a period of 9 to 12 months free from fungal attack by the "red", whereas the normal storage life of dry cured product is only 8 weeks. Wet cured product treated by the preservative mixture has been found to keep well for 3 months as against 2 weeks in the case of untreated sample.

Advantages of cured products

1. The method is safe, simple and easy to apply even by the fishermen.
2. The storage life of the treated products is extended considerably.
3. As the preservative is in the dry form and acts superficially, the chance of imparting undesirable flavour to the product is very much limited.
4. The cost of treatment is reasonable, provided the chemical is available at duty free rate.

Quality and Shelf Life of Dried Sharks Produced in India

Samples of cured sharks were collected from various production centers, like, Quilon, Cochin, Calicut, Tuticorin, Bombay and Verabal for determination of moisture, acid insolubles, ash, sodium chloride. Total Volatile Nitrogen (TVN) and Trimethyl

The factory covers 1,500 m² in total floor area with 40 workers.

Surimi (stabilized meat of coldfish) is ground in a mortar.

Ground meat is fed into a shaping machines.

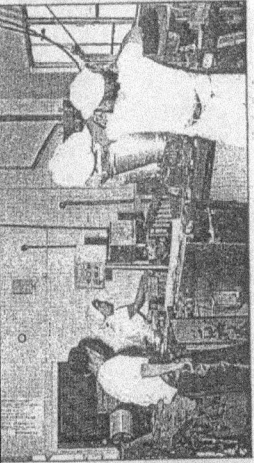

Shaping is done on a belt conveyor.

Head and viscera are removed.

It is cleaned in water. This industry needs much water to rinse the fish meat.

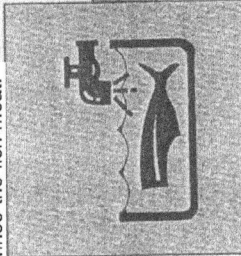

Meat is chopped, shieved and ground.

It is ground in a mortar with seasonings.

It is made up in a semi-circle shape on a piece of wood.

Processing Line of 'Kamaboko

Salt and Seasonings are Added to Chopped Meat

amine Nitrogen (TMA-N). Variability of moisture was less for most of the samples, when compared to the overall variation. The average moisture content of samples from Bombay and Verabal were less than those from other places. TMA and TVN values show high variability in different samples. Shelf life of commercial samples varied between 2 to more than 12 weeks for moisture content of above 50 per cent and below 35 per cent respectively. Only 33.53 per cent of the samples had shelf life above 4 weeks. Curing of shark flesh for 12 hours or more yield samples with 30-32 per cent salt. When moisture in these samples are below 45 per cent, they keep well for more than 3 months. It has been shown that nearly 50 per cent of the urea is leached out during the curing process. The lowered urea level with high salt content may delay the development of ammoniacal odours. Salting shark flesh in 1:3 ratio and allowed to cure in self brine for more than 12 hours followed by drying give a product with satisfactory shelf life.

Hand moulding of large sized Kamboko.

It is ground thoroughly in a stone mortar. Then it is churned to improve the texture.

Freeze Drying

Prawns belonging to the species, *Metapenaeus dobsoni, M. affinis* and *Parapenaeopsis stylifera* freshly procured from trawlers were washed, cleaned, peeled and deveined, spread in a single layer on the tray of freeze drier and frozen in plate freezer. Freeze drying and packing of the dried material were conducted to observe the broad biochemical changes that fresh prawn muscle undergoes during the process of freeze drying and further storage.

Practically no denaturation takes place in the muscle of prawns due to freeze drying with respect to water extractable proteins. However, salt extractable proteins are rendered insoluble to the extent of 21 per cent during this process. Water extractable nitrogen, salt extractable nitrogen and non-protein nitrogen are not affected to any appreciable extent due to prolonged storage of the product, whereas free alpha-amino nitrogen contents exhibit a remarkable increase after 3 to 4 months of storage. The freeze dried products remain in good edible condition even after 32 months of storage.

Chapter 15

Fish Processing and Fish Products

The Alternative Uses of Fish

Fish processing is one function besides handling, transport and storage involved in to products flow from the production sectors to the consumption sectors.

Fish products is the commodities, which can be divided into primary or basic fish products (in the round or dressed with minimum processing), low-cost minimum storage products and convenience products.

Each method of processing depends on the one hand upon the supply of suitable fish and hence is related to the structure of the catching side of the industry and on the other hand upon a suitable system of distribution and sale.

In the developed human food fisheries, the evolution from traditional products to fresh and frozen fish has been achieved because of a combination of factors. Of these, the development of a rapid transport system and urbanization have been crucial, but it is doubtful whether these alone would have enabled such developments to take place. Technological advances in vessels and gears, establishment of ancillary repair and ship building industries, growth of a suitably flexible marketing and processing industry (which is again only possible with the development of fast methods of communication), have all contributed to the growth of fresh and frozen fish industries. Other types of processing, such as, canning or fish-meal manufacture, require certain additional technical facilities beside supply of fish.

In determining the best ways of utilizing the fish catch, the available choices will often be limited by economic or biological factors. Traditional methods of preserving the catch, although in some ways the most wasteful and nutritionally least desirable, nevertheless still commend themselves as being the most generally applicable. There is room for a considerable amount of field work by trained people in improving techniques which have often remain unchanged for centuries. It does not require high scientific skills, but it needs men who can slowly raise the technological level.

In so far as new products need to be developed, it appears that those most likely to be successful will be those based on traditional methods of preservation used in the area in question. There is no doubt considerable scope for the application of modern food technology here.

There would be value in;

1. Preparing a manual for field workers on general methods of preparing of various types of traditional product, with advice on how primitive methods can be improved.

2. Carrying some detailed studies of costs of establishing and running various types of operation, for example freezing, improving traditional methods in some chosen developing countries. This would assist in making decisions on how fisheries might be developed.

3. Collecting the available information on factors affecting consumer preferences in developing countries. This would be of assistance to those attempting to produce new types of products.

General Principle for Production of Good Quality Products

Quality Control in Food Service

When the definition of quality control is applied to food service, it becomes the standard to which all steps of the operation must, of necessity, conform in order to ensure that changes in characteristics of a food item do not take place. This role promotes quality control to a broad, encompassing and highly significant activity in the development work. Many factors are responsible for poor quality food. Most of them can be traced to poor sanitation, faulty handling, malfunctioning equipment, incorrect preparation, and carelessness. The following are the prime factors responsible for significant quality changes.

1. Spoilage due to microbiological, biochrmical, physical, or chemical factors.
2. Adverse or incompatible water condition.
3. Poor sanitation and ineffective ware washing.
4. Improper and incorrect pre-cooking, cooking and post-cooking methods.
5. Incorrect temperature.
6. Incorrect timing.
7. Wrong formulation, stemming from incorrect weight of the food or its components.
8. Poor machine maintenance program.
9. Presence of vermin and pesticide.
10. Poor packaging.

Any of these factors, either singly or in combination, will contribute to poor quality and effect changes that will be evident in the food's flavour, texture, appearance and consistency.

Controlling Microorganisms in Raw Materials

Major losses in raw agricultural and fishery products throughout the world result from microbial, insect or rodent attacks. The increasing limitation being placed by regulatory agencies on the use of pesticides, antimicrobial agents and rodenticides, make it more necessary than ever that physical methods be used for protecting raw materials from contamination prior to their receipt at food processing plants.

Corlett (1974) proposed a scheme for the use of microbiological testing in acceptance of raw materials and ingredients. While buying raw products or ingredients, the general list of specifications such as, organoleptic, physical and chemical characteristics including a microbiological specification should be considered. Raw materials and ingredients must be tested on a routine basis to ensure that they meet the suppliers specification.

Sanitation

Contamination may take place through the following channels.

1. During production on farm or at sea 2. During harvesting, slaughtering, or catching, 3. In handling prior to processing, 4. During processing, 5. During distribution and selling, also in retail, 6. In cooking and feeding in home, 7. Storage.

Special attention should be given on (a) surrounding of packaging plant, (b) water supply, (c) construction of buildings and equipment from the sanitation point of view, (d) personal hygiene, (e) plant clean up at the end of processing and (f) sanitation during processing.

The performance of a plant sanitary program is reflected in the end-of-line bacteria counts. Counts exceeding the bacteriological standards indicate inadequate plant sanitation. Cooked product ready to serve as food without further preparation, public health authorities are very particular about the number and types of bacteria present in the product. Absence of *S. aureus*, a food poisoning bacteria is required.

Physical Parameters Involved in Fish Processing Technology

Many a technique in fish processing technology, whether it applied in freezing, dehydration or canning, involves always a type of heat transfer which is dependent to a certain extent of external physical parameters, like temperature, humidity, pressure, air flow etc and also on the thermodynamic properties of fish muscle in the temperature ranges encountered. Similarly, information on other physical values, like dielectric constant and dielectric loss in the design of quick thawers and in quality assessment of frozen or iced fish refractive index and viscosity in the measurement of the saturation and polymerization of fish oils and shear strength in the judgement of textural qualities of cooked fish are also equally important.

The impact of physics in fish processing technology can be seen in the development of sophisticated techniques and instrumentation needed for research like radio chromatography in biochemistry, Intelectron Fish Tester V based dielectric behavior of fish tissue, physical techniques employing nuclear magnetic resonance and molecular absorption properties of water molecules (or hydrogen atom) as a measure of moisture in fishery products, electron microscopy for identification of bacteria and studying the ultra structure of fish muscle etc.

Physical Properties of Fish

1. Specific resistivity (50 c/s) (Jason) in minus 30°C – 26X 1000000 ohm/cm, in 0°C –800 ohm/cm.

2. Dielectric constant 40 Mc/S) (Jason) – in minus 30°C –3 and in 0°C –65.

3. Thermal conductivity (Jason and Long)–plus 1°C–1.30 Cal/sq cm/Sec/0°C/cm X 1000; in minus 10°C 3.96 and in minus 20°C 4.40 Cal/sq cm/Sec/0°C/cm X 1000.

4. Specific heat (Riedel) –in minus 40 and plus 20°C – 0.44 and 0.89 Kcal/kg/0°C respectively.

5. Enthalpy for shell fish (83.6 per cent water, Riedel) – Between minus 40 and plus 20°C – 94.38 K cal/kg.

6. Thermal diffusivity (Jason) – in minus 30 and 0°C – 10.4sq m/h and 1.5 sq m/h respectively.

7. Initial freezing point depression – 1.1°C.

8. Per cent of water frozen (Mahadevan) – in minus 7 and in minus 12°C – 86 per cent and 91 per cent respectively.

9. Bound water (Riedel) Freezing at – 70°C –0.385 g water/g total solid.

10. Density – in 21.1 and 1°C –60.31 and 65.79 lbs/cubic feet respectively; Bulk density (whole fresh fish) (Waterman) – 62.5 lbs/cubic feet; Bulk density (whole gutted frozen cod) – 55 lbs/cubic feet (compact packing).

Biophysics in Freezing

Three important parameters, namely, apparent specific heat, thermal conductivity and ice/water fractions in frozen fish tissue, are required for the development of a suitable refrigeration system for freezing, thawing, cold storage or transport of frozen/chilled fish. Fish muscle contains protein, fat and organic and inorganic salts in solution. As the temperature of fish is progressively decreased below its initial freezing point, minus 1.1°C (corresponding to 0.28 M Nacl solution), inorganic salt solution throws out pure crystalline ice, itself getting more concentrated in the process. The freezing point of salt solution thus decreases and progressive freezing takes place.

Thus the enthalpy (H) of the fish muscle of a given composition is a function of temperature (theta) of fish. For thawing frozen fish most of the latent heat must be supplied in the temperature zone minus 5 to 1°C, so called "thermal arrest", where (dH/d theta) becomes very large. This thermal arrest zone is of interest, as biophysical investigations have revealed that at critical freezing rate, when the muscle cools from 0 to minus 5°C in about 60 minutes, each cell contains just one ice crystal (Love, 1962) and this is also dependent on the initial temperature of the material. This initial temperature effect is attributed partly to changes in thermal conductivity values of frozen and unfrozen fish tissues and partly to differential cooling rates. X-ray diffraction examination of the ice component of frozen fish muscle has revealed the common hexagonal form but with preferred crystallographic orientation of ice crystals (Aitken, 1966). During freezing, protein becomes increasingly dehydrated as more tightly bound water is freed and to avoid denaturation, low temperatures in freezing are to be avoided. Love (1963), in one of his classical investigation, has found that freezing of bound water at minus 78°C (and subsequent storage at minus 14°C) is irreversible. Riedel (1957) defines bound water to be water unfrozen at very low temperatures. It is actually chemically combined in the muscle at the rate of one molecule of water for every two molecules of amino acid, that is, 0.385 kg water/kg of dry substance.

Since the thermal conductivity of frozen fish at minus 30°C is about three and half times that of chilled fish at 1°C (Jason and Long, 1955) for same temperature gradient, it is easier to freeze than to thaw. However, the electrical conductivity at low frequency of fully thawed fish is 250 times that of frozen fish at minus 28°C, a property utilized in the design of electrical resistance thawing (Sanders,1963). Enthalpy or apparent specific heat of fish in the freezing zone is considered to be the algebraic sum of specific heats of salt solution and ice, latent heats of solution of salt, melting of pure and eutectic ice and specific heats of protein gel and unfrozen water. Enthalpy measurements useful in the design of refrigeration equipment are worked out for cod, haddock, white fish etc by Riedel (1956) and Jason and Long (1955). To cut ice requirement for chilling and storage of fresh fish, knowledge of specific heat of fish above 0°C is necessary. So also the thermal conductivity of five above 0°C is important in arriving at optimum thickness of fish layer for rapid chilling by ice. The bottom of a 10 cm thick layer of prawns iced on the top will be at 10°C even after an interval of 3 hours due to poor conductivity of chilled fish (Rao, 1968).

Kinetics of biological reaction rates and frequency factors for such reactions are studied to predict the physical and chemical changes the frozen fish undergoes during fluctuating or steady low temperature storage. Recent investigations try to link these changes to the unusual dielectric behavior of frozen fish. He has observed frozen cod muscle to exhibit characteristic resistance behavior of intrinsic semi-conductors, like crystalline proteins. Energy of activation of protein denaturation is close to that of charge carriers and there is a possibility that protons are charge carriers and essential for the chemical process in the frozen state. Knowledge of variation of density with temperature is required in the calculations of freezing or thawing times and bulk densities for stowage rates. The determination of density is purely theoretical (Long, 1956) and as anticipated it falls as freezing proceeds, as large portion of moisture in fish is converted into ice.

Texture of processed fish, an important quality factor linked with the structural quality of the material is often defined in terms of hardness, fluidity, elasticity, adhesiveness, chewiness etc. The manner in which these properties manifest themselves though difficult to analyze can be expressed in terms of energy or power required to chew, which can be measured in an equipment designed to determine its shear strength, elasticity and Poission's ratio to effect that change (Charm, 1963).

Temperature and Humidity

Temperature and humidity are other important physical parameters in fish preservation. It is observed that the rate of spoilage of fsh is related to the temperature and at 2.5°C it would be twice as fast as minus 1.1°C (Hess, 1934). The storage life of a fish product at minus 18°C is reduced by six weeks if it is kept for three days at minus 9.5°C, prior to storage at minus 18°C (Deyer *et al.*, 1957). Three types of portable instruments for temperature measurement of frozen and iced fish under field conditions are available in the temperature range of minus 30°C to plus 30°C.

The humidity in cold storages is required to be near saturation, that is, the partial pressure of water vapor in air should be equal to the vapor pressure of ice on the surface of fish at the storage temperature (minus 30°C and below) to avoid dehydration (freezer burn) of frozen fish. Psychometric charts show that at these low temperatures even a wet bulb depression of 0.5°C is enough to reduce the humidity from saturation by 5 to 6 per cent and the drying potential could be easily imagined, The capacitance resistance hygrometer having anodized aluminum oxide as the dielectric material can be used with certain amount of accuracy at relative humidity above 90 per cent and temperature up to minus 15°C (Jason, 1957). During initial stages of drying too, the drying rate of fish muscle is controlled by environmental conditions, of which humidity is an important parameter.

Drying and Salting

Preservation of fish by salting and drying has been practiced since long in the developing countries. Yet approach to these practices had been mostly practical and empirical, scientific investigations into the basic mechanisms involved in the process being scanty (Jason, 1938; Del Valle and Nickerson, 1967). During initial stage of constant rate period, the drying rate of the fish muscle is controlled by external conditions and is equal to that from a saturated surface of same shape. The duration of this period is expressed by a relation containing the rate of evaporation per unit area, effective diffusion constant, thickness of fish sample and free water content The subsequent falling rate has two distinct phases. The drying behavior in the first phase is in accord with the solution of a diffusion equation based on Fick's law, the effective diffusion coefficient being independent of shrinkage of fish. The diffusion coefficient in the second phase is about $1/5^{th}$ of that of the first phase. The transition from the first to the second falling rate period appears to be associated with uncovering of the unimolecular layer of water which covers the protein molecule. The process of evaporation and diffusion are characterized by a scheme of energy levels involving the heat of adsorption of unimolecular layer of water, the heat of liquefaction of water and activation energies corresponding to each of the two phases of falling rate periods. The effect of presence of fat is a reduction in the effective values of diffusion coefficients in the first and second phases of falling rate period.

Studying the dynamic aspects of salting of fish, Del Valle and Nickerson (1967) observed that salt concentration in fish muscle and tissue water increases with brine concentration. The equilibrium distribution coefficient based on muscle volume and water content increases at first, passes through a maximum and then decreases with brine concentration. The distribution coefficient is roughly equal to unity and is independent on salt concentration in the brine. The migration of salt into fish muscle has been studied deeply by observing other factors like equivalent conductance and sodium and chloride ion transference number in muscle and water. The diffusion coefficient and equivalent conductance of salt in fish muscle increase with temperature and time and are in agreement with predicted values. The activation of energies for salt diffusion in fish muscle and in water of infinite dilution are found to be in the region of hydrogen bonding energies.

Other physical methods like measurements of optical density of fish muscle extract by colourimeter and refractive index and opacity of muscle fluid for assessing the quality of ice stored prawns and fish, determination of the variation of electrical resistance with loss of moisture as a measure of bound water content in fish, non-destructive type of testing freshness of edible fish by Fish Tester V, based on the fact that the AC resistance of fish contains capacitative and resistive components which are conditioned by properties of cellular skins, measuring cold storage changes in frozen fish muscle by the cell fragility apparatus, determination of quality of smoke constituents deposited during smoke curing by smoke density integrator etc illustrate the significant role of physics both in fish processing research and quality assessment of fishery products.

Processing of Indian Mackerel

Though in the past, more than 50 per cent of the mackerel landed used to be salt cured, in recent years, is consumption in fresh condition has been increased due to the increased demand from the interior part of the country, ready availability of ice in the landing centers and the employment of improved know-how in handling, preservation and transportation.

Curing

Since mackerel, like sardines is a seasonal fishery and the bulk of the landings takes place in a narrow strip of time, preservation of excess catch often becomes necessary. Since early times a substantial portion of the mackerel catch used to be salt cured by dry or wet process or pickled according to Colombo curing method. A portion of such product used to be exported to Sri Lanka.

As far as preservation of mackerel is concerned, curing used to be the most important process applied and a good amount of technological research has been carried out in the field.

In preliminary studies (1958) on pit curing of mackerel, it was proposed a ratio of 1:5 of salt to fish and salting period of two days; the pit curing improved the organoleptic properties of the fish imparting characteristic flavour and softening flesh. Subsequently (1961) it was proposed that a ratio of 1:7 or 1:8 would be adequate if the salted fish is dried at the end of 18-24 hours salting without stacking. When dried to about 40 per cent moisture level, the sodium chloride content would be more or less sufficient to saturate its moisture content and would not be in solid phase, thus imparting a better appearance of the product.

At a salt to fish ratio of 1:5 maximum water loss took place in 24-26 hours for gutted fish and 18 hours for split open fish. This proportion of salt is sufficient to saturate the moisture content of fish and as such any excess amount of salt will be in solid phase and so useless.

CTC (50 ppm) followed by salting prevented browning to a large extent. Dip in 10 per cent sodium propionate for 5 minutes checked the fungal attack. Sodium carbonate, sodium bicarbonate and sodium hexametaphosphate improved the texture at moisture level above 35 per cent. The use of a mixture containing potassium sorbate, sodium benzoate, sodium acid phosphate and common salt for curing mackerel have been recommended in 1967.

It was reported (1961) that fishes like mackerel could be successfully pickled and preserved for a long period by giving a pre-dip treatment in propionic acid bath followed by usual heavy salting. For the control of reddening and mold growth usually met with in stored dry-cured fish, it was suggested (1962) dipping eviscerated and split open fish in 4 per cent propionic acid for 10 minutes followed by salting (1:4, salt to fish) for 48 hours and subsequent sun drying. Storage life could be extended to 62 weeks as against the normal 19 weeks by this method.

Pickling mackerel in saturated brine fortified with 0.5 per cent and 0.25 per cent propionic acid has been found useful for keeping the fish in good condition for periods of one year and five months respectively. By smearing a mixture of 3 per cent sodium propionate, 0.5 per cent butylated hydroxyl anisole and 0.5 per cent sodium sulphate in dry powdered salt over cured fish at 10 per cent level, the product can be kept for 9-12 months free from any visible signs of spoilage, browning and rancidity. In case of wet cured fish sodium propionate at 2 per cent level is sufficient

Mackerel subjected to smoking after treating the fish with brine containing propionic acid and turmeric extract separately turned out an attractive product

Freshness of the fish as well as the quality of salt used for curing are factors which can influence the quality and storage life of the processed product. The maximum permissible time lag between catching of mackerel and its curing should be 8 hours when the fish is kept in room temperature and 3 days if iced. In the former case, the shelf-life and organoleptic characteristics showed proportionate downward trend as the period of storage prior to curing advanced. The rate of penetration of salt has no relationship to the impurities content even at a level of 0.75 per cent.

Spoilage of mackerel takes place quickly. A generous use of ice during transport and handling at processing plant is essential. Even after irradiation, mackerel undergoes rapid deterioration which is brought by the enzymes present in them. To oviate this the enzymes have to be inactivated by hot water or steam blanching before irradiation if the desired shelf life is to be achieved.

Processing of Prawns and Shrimps

Among different processing methods, the most important are freezing, drying and canning of prawns. Besides other diversified prawn products are dry prawn pulp, breaded prawns, prawn flakes, shrimp extract and prawn pickle. Prawn by-

Processing of Shrimps

Processed and Packed Product

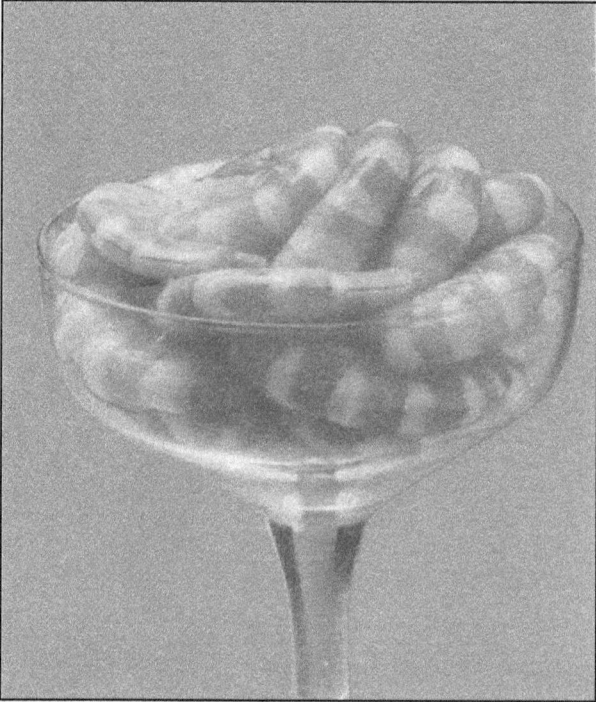

Ready to Eat Product

products, which include prawn manure, prawn meal, chitin and chitosan are also very important from industrial point of view.

Freezing of Prawns

Frozen prawns have made a tremendous headway during last few decades and considerable improvement in the quality of the products have been achieved by employing newer and more efficient techniques of handling and processing. In commercial practice, several types of packs of frozen prawns are turned out for export. Basically, these are based upon the method of preparation of the raw material for freezing. The different types of packs of frozen prawns are;

(i) Headless shell on, (ii) Peeled and deveined (P&D), (iii) Fan-tail/Butterfly, (iv) Peeled un-deveined (PUD), (v) Whole/head on, (vi) Cooked and peeled (CP), (vii) Cooked, peeled and deveined (CPD) and (viii) Peeled deveined cooked (PDC).

The flow chart of freezing the prawns are:

Raw material → Washing → Dressing → Washing → Draining → Size grading → Weighing → Packing → Freezing at minus 40°C in plate freezer → Glazing → Packing in duplex cartons → Packing in master cartons and strapping → Frozen storage at –20°C.

Frozen prawn product has become the vital item of the seafood freezing industry in India because of the export value. The final quality of frozen prawn depends upon the freshness of raw material, handling of fresh prawn and the processing techniques to be followed. Some of the common defects occurring in frozen products are dehydration (Freezer burn) and discolouration of shell and meat. Under-weight conditions are generally encountered in frozen peeled and deveined prawns as they hold large quantities of free water in the fresh condition and the water holding capacity is reduced during thawing after freezing. Weight losses in frozen prawns generally varied from 7 to 12 per cent in P&D, 5 to 7 per cent in HL and 7 per cent in cooked and peeled prawns during storage for a few days. Black spot formation or melanosis is chiefly because of the contact of prawns with atmospheric oxygen. This occurs by a reaction involving free amino acids, phenolase enzymes and oxygen, which converts phenols to melanin pigments and these pigments generally occur on the internal shell surfaces or in advanced stage on the underlying shrimp meat.

Loss of quality during early period of storage is mainly caused by autolysis and with longer storage, spoilage occurs mainly through bacterial action. Since prawns live only a few minutes after removal from their natural habitat, microbial spoilage starts immediately through bacteria on the surface and in the digestive system and through micro-organisms which happen to contaminate the prawn on the deck during handling and from ice, used during their storage. Removal of heads reduces the bacterial count some what, because the head carries approximately 75 per cent of the total bacterial load. Frequent washing of the raw material reduce the microbial contamination to the minimum.

Thaw drip loss is also an important problem in the frozen prawn industry. Generally, thaw drip loss occurs to the extent of 5 per cent in headless shell on, 10 per cent in peeled and deveined and 3 per cent in cooked and peeled style of prawns.

This can be alleviated by a dip treatment of the prepared raw material in an aqueous solution of 12 per cent sodium tripolyphosphate and 8.6 per cent sodium dihydrogen phosphate or 2 per cent citric acid. Thus it maintains correct drained weight and improves the yield during prolonged frozen storage and also protects the frozen products from denaturation of proteins.

Frozen prawns must be prepared from clean, whole some and fresh prawns and there should not be any visible sign of spoilage. Colour, meat texture and odour of raw material must be having characteristics of freshly caught prawns.Standard size gradings must be followed. The frozen block, if having any white patch due to dehydration (freezer burn) or black spot must be noted carefully. Thaw drip loss of prawns must be observed. Any discolouration of shell and meat is to be noted quantitatively. Number of deteriorated pieces are to be counted. Overall odour of the thawed material and flavour of the meat need to be assessed. A pleasant flavour indicates the product is of prime quality. Presence of any extraneous materials, like, legs, bits of veins, loose shells, foreign materials etc should be noted. Total bacterial plate count, counts of *E. coli* and *Enterococci* should be determined before thawing. High bacterial count indicates the unhygienic handling of raw material.

Canning of Prawns

A method of preservation where heat treatment is given to food material in air-tight sealed containers, so that all the pathogens will be killed and other micro-organisms may or may not be killed, but made inactive. Also the heat treatment inactivates the enzymes, responsible for causing spoilage and thus contributes to the preservative action. The main advantages in canned products are that they can be stored for years at ambient temperature and are ready-to-eat products. But the cost is the only and major criteria for its production.

The flow diagram for preparation of canned prawns is;

Raw material → Washing → Peeling → Deveining → Washing → Size grading → Blanching in boiling 10 per cent brine for 4 to 8 minutes → Draining → Fan drying → Weighing for different can size → packing → Adding salt etc → Filling till net weight for the can size is reached → Vacuum sealing → Processing at 10 pounds pressure per square inch for 18 minutes → Cooling → Coding → Packing in cartons → Storing at room temperature.

Some common defects have been observed in canned prawns. The poor appearance and odd flavour of the product are mainly due to the stale raw material and prolonged storage of raw material in ice. Overfilling in no case is desirable. Generally low vacuum is caused due to delay in seaming, defective seams and microbial activity. Reduction in filling media causes non-uniform heat penetration. Blanching is also an important step in the preparation of canned product. An over-blanched material generally takes up moisture and results in over-weight and under-blanched material loses moisture and thus becomes under weight. Besides, neither over-cooking nor under-cooking is desirable. Under-cooking results in microbial survivality, while over cooking destroys the food quality. Sudden cooling is recommended in potable water. Blackening of can interior and can contents has been a major defect in canned prawns. This is due to formation of iron sulphides from the hydrogen sulphide released from sulphur containing amino acids of the proteins and the metals which contaminate the material during processing.

The taste and smell of the canned prawns should be pleasant. The consistency of meat should be firm but not tough. The microbiological control has a very important role in maintaining the high quality of finished product in the canning industry.Swelling of cans and spoilage of raw material are due to microbial action. Not only the count of microbes, but also species composition should be determined. Special care must be taken for estimation of *Escherichia coli* (indicator organism), *Staphylococcus aureus* (toxicogenic organism). Vacuum inside the can must not be less than 10 cm of Hg. Head space shall be 5 to 7.5 mm and drained weight shall not be less than 65 per cent of water capacity of the can.

Drying of Prawns

Drying in the sun is one of the oldest and most widely practiced methods of preservation of prawns. Here the moisture content of the material is lowered to such an extent that the micro-organisms, responsible for bringing out the spoilage are destroyed and the enzymes are denatured and inactivated. Thus the material can be preserved for longer period.

The flow diagrams for the preparation of dried and semi-dried prawns are;

Dried Prawns

Raw material → Washing → Sun drying or mechanical drying → packing in polythene bag or gunny bag → Storing at room temperature.

Semi-Dried Prawns

Raw material → Washing → Blanching by dipping in 4-6 per cent boiling brine for 2 to 3 minutes → Cooling → Peeling → Immersing in saturated brine for 15 minutes → Drying at 50°C till moisture content attains 40 per cent → Packing in polythene bags → Storing at room temperature.

The moisture content of the dried material is very important. This content must be 8 to 10 per cent in case of dried prawns, while 40 per cent in semi-dried prawns. Even though sun drying is widely practiced, artificial dehydration has got some advantage over the conventional methods. In the artificial dehydration method, the drying is quicker and more efficient and turns out a much more hygienic product. But disadvantage is the added cost involved in the new method. There are different types of dryers used for the purpose, namely, hot air drier, tunnel drier, rotary drier, fluidized bed drier, solar drier etc.In this method, temperature, relative humidity and wind speed are regulated to a desirable level for obtaining better product.

Raised platform must be provided for sun drying with crow-proof and preferably fly-proof enclosures. Good storage rooms with moisture-proof floorings and rodent-proof walls and ceilings must be constructed for storage until it is marketed. Better hygienic packing materials, like polyethylene bags, gunny bags etc must be employed. Moreover, retail packing should be done in polyethylene bags.

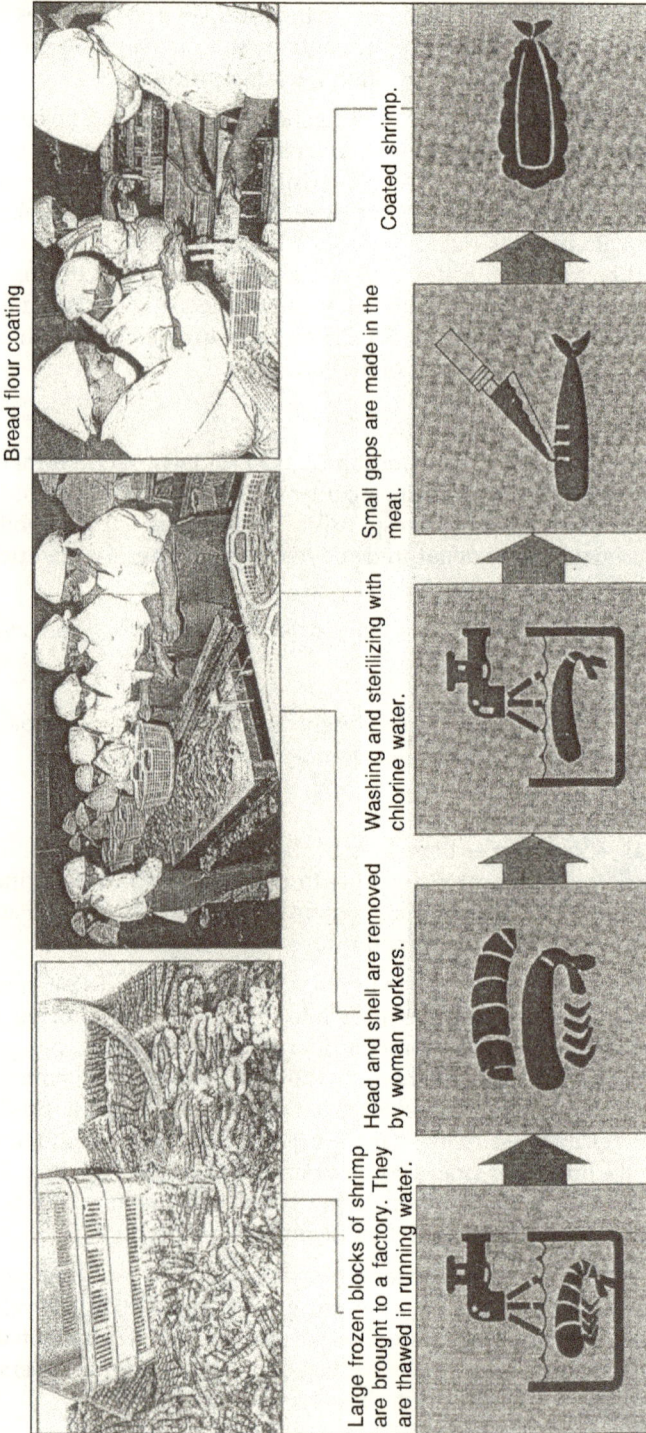

Bread flour coating

Large frozen blocks of shrimp are brought to a factory. They are thawed in running water.

Head and shell are removed by woman workers.

Washing and sterilizing with chlorine water.

Small gaps are made in the meat.

Coated shrimp.

Processing Line of Value Added Coated Shrimp

Thawing in Running Water

**First, coated with wheat powder, dipped in egg yolk,
they are then again coated with bread flour.**

Other Prawn Products

Dry prawn pulp is one among the special cured products. The prawns are washed in sea water. Some amount of common salt may be added at times. The raw materials are then cooked in brine water for sufficient time to obtain desirable characteristic pink colour and curling. The cooked prawns are sun dried over a mat and packed inside a gunny bag and beaten with a wooden mallet to separate the shells. Thus the edible pulp is separated from shell particles and packed in polyethylene bags.

Prawns (P&D) are the raw material for break prawns. Here they are blanched in 7 to 8 per cent boiling brine for 12 to 15 minutes, cooled and kept aside. Then batter is prepared using maida, water egg and salt. The blanched prawns are dipped individually in the batter and coated with bread powder by rolling the battered prawn in the powder. This is to be fried in edible oil and served. Sometimes the products can also be frozen and preserved for a long time and fried in oil as and when required.

Prawn flakes can be prepared from minced prawn meat. To the minced meat, starch and other ingredients in suitable proportion are to be mixed well and then steamed followed by cutting into required size. This needs to be dried in any artificial drier till moisture content of the product attains at about 5 per cent. It becomes a good side dish when fried in edible oil resulting into swelling and becoming crisp in nature. This can be packed in polyethylene bags and stored in room temperature for long period.

Shrimp extract is a partially hydrolyzed prawn protein concentrate powder. The product is very much popular in USA, Canada and UK. It has a very good taste and at the same time, it contributes to the daily protein requirement of the consumer. Generally shrimp powder of *Acetes* spp is used for this preparation. The shrimp powder after mixing with water is to be hydrolyzed partially and the mass is concentrated using agar agar. The product is generally filled in cans. Vacuum sealed and processed in a retort at 115.3°C. The shelf life of the product is over a year at room temperature.

Small varieties of prawn (P&D) are most suitable for preparation of prawn pickle. In the preparation, ingredients are mixed to the blanched raw material in a fixed proportion. The method of mixing the ingredients consists of heating the oil at 180 to 200°C in a vessel to which mustard seeds are fried, chilies and lemon pieces are next added and stirred well. Salt, ginger pieces, red chilly, white pepper, turmeric powder, peeled garlic etc are to be added in the sequence and heated for about 2 minutes. Vinegar is added then and mixed well. This is to be cooled and then sodium benzoate is added for preservation. The product with its pH around 4.5 is most desirable. The pickle is packed in bottles and their mouths sealed with acid proof caps. The shelf life of the pickle is for about one year.

Prawn By-products

In a prawn processing industry, the important waste materials are its head, shell and tail. All these can be readily converted into prawn meal. This is a highly concentrated nutritious feed supplement containing proteins, minerals, vitamins and some unknown growth factors. Generally they are used along with cattle and poultry feed. The product is prepared just by drying the waste materials in the sun and then grinding to powders. The product is packed in gunny bags.

Chitin is a polysaccharide which is found in the hard shells of prawns. Chitosan is derived from chitin by a process of deacetylation. The protein is extracted from the shell with hot dilute sodium hydroxide. The insoluble material is treated with hydrochloric acid and the product is filtered and separated solids are air-dried for 48 hours and ground to yield chitin. Chitin and chitosan are used by industries for sizing rayon, synthetic fibers, paper wool, cellophane as adhesive, stabilizing and thickening agents. It is also used as a chromatographic base, ion exchange resin etc.

Chitosan from Prawn Shell

Prawn shells form a good raw material for preparation of chitosan, a valuable industrial product.

Process

1. Boil the prawn shell in about 3 per cent solution of sodium hydroxide (quantity just enough to immerse the shell for about 30 minutes Drain off the solution and repeat the treatment with the residue. Treatment with sodium hydroxide removes the protein content of the shells. Wash twice with water.

2. Treat the protein free residue with bleach liquor containing 0.3 to 0.5 per cent available chlorine at room temperature for one hour. Wash two times with water.

3. Treat the bleached residue with 10 per cent hydrochloric acid (commercial) at room temperature for two hours so as to remove the calcium and phosphorous contents in the residue completely. Wash the residue free of acid, which is almost pure chitin.

4. Treat the chitin at 100°C with 1:1 solution of sodium hydroxide for 90 minutes.Wash with water to free it of the alkali and then dry to get the final product. The yield of chitosan is about 3 per cent of the weight of the fresh prawn shell.

Chitosan finds use as sizing material for cotton, wool, rayon and other synthetic fibers. It may also be used in the preparation of cosmetics and pharmaceuticals, and also as a water clarifying agent. Chitosan dissolves easily in dilute solutions of organic acids to give viscous solutions. It can be precipitated by neutralizing the solution with alkali. Chitosan under specified conditions is beneficial in giving a permanent organdie type finish to cotton fabrics and also in preventing shredding of jute.

Chitosan solutions have good film forming properties and are therefore, potentially useful in gels and coatings. The new promising technologies integrating edible coatings with natural anti-microbial substances (chitosan may be one) are being developed to kill or inactivate undesirable micro-organisms in more environment friendly ways without effecting the food quality.

Chitosan is anti-microbial against wide range of target organisms. But this activity is influenced by a number of factors including type of chitosan, degree of polymerization and other various physical and chemical properties. Chitosan with low molecular weight (MW) is more effective at inhibiting the growth of *S. aureus, Escherichia coli, Saccharomyces cerevisial, Candida albicans* and *Fusarium culmorun* than chitosan with high molecular weight. Yeasts and moulds are the most sensitive group, followed by gram-positive bacteria and finally gram-negative bacteria.

Chitosan despite certain difficulties and its limited efficacy as a preservative, is still an interesting compound with considerable potential for improving the quality and safety of our food.

Handling and Processing Tuna

Fresh and Chilled Sashimi Grade Tuna

Handling of Tuna on Board

Immediately after the harvest and taking fish on board, a stinger is pushed inside them behind the head in a way that it would get into the vertebral column and paralyze the entire nervous system. Immediately there after, before transfer to the fish hold (as per Japanese system), the gills and the viscera is removed. Thereafter the fish should be hung for ten minutes for draining the blood. After this, each of the tuna should be cleaned with water for 1.5 minutes. The cleaned fish should then be put in an ice box or chilled refrigerated sea water tanks. Before this transfer, it should be ensured that (a) the fish under transfer would not struggle, (b) there would be no dropping of blood and (c) the "Double Shine" (the process of totally incapacitating the brain and the spinal chord) was perfectly done.

In several countries, the practice followed was to transfer the harvested fish from the vessels to the landing point at the fishing harbor concerned using a crane. The landed fishes are then washed for 30 seconds in cold water at the infrastructure facility for the purpose set up as part of the processing plant, a few meters distant but within the crane's reach. After the washing is done, the cleaned fishes would be sent for grading at a point nearby at the plant. The grading would be done taking into the account of skin colour and quality of flesh. The skin would be scanned all over with hand as a part of the process so as to check any damage. The fishes are there after conveyed by crane into the processing plant proper.

The checking of quality as mentioned above is done by using a gadget known as "Sashi bou". This is needed to be inserted into each of the fishes and later pulled out. When this is done, some meat would also come out sticking all around "Sashi bou". This is then tested by a process that would also include testing by tasting. The Sashi bou test would indicate how long the fish be kept under preservation. After Sashi bou testing the fishes would be packed in dry ice or jelly ice. The testing system with "Sashi bou" for assessing quality is followed in Philippines and Indonesia. Two sizes of Sashi bou are used, one for larger fish and another for smaller ones.

Tuna for Export

1. Fresh chilled tuna, also termed as "Sashimi" grade tuna for export shall be caught by vessels equipped with facilities on board for handling the fish as required. The vessel should also be equipped with refrigerated sea water tanks or a slurry tank of adequate capacity to cool the fish to zero°C prior to storage in ice.

2. The fish shall at all times be maintained at a temperature of zero°C and handled with utmost care at all stages including transport.

3. Fish holds, including linings, pen boards, shelving etc shall be designed and constructed in such a way that they can be easily cleaned and disinfected between trips.

4. The vessel should have all essential tools/equipments so that the fish will be dealt with quickly for maintaining quality of the product. All surfaces of the fishing vessel that come in contact with fish shall be kept hygienically clean

5. Adequate arrangements shall be in place for.cleaning the deck.

6. Provision shall be made on the vessel for proper storage of oils and other substances of such a nature that may contaminate or taint the fish at any time.

7. Such approved vessels will be registered by MPEDA as vessels for harvesting tuna for export.

8. No other vessel will be permitted to catch and handle tuna on board as fresh, chilled tuna for export.

9. Fresh, chilled tuna exporters shall always buy their raw material only from such vessels registered with MPEDA and having such specifications as approved.

Standards for Fresh, Chilled Tuna Handling Centers

The handling center for tuna meant for export in fresh, chilled form should have the following specifications and standards.

1. The handling establishment shall be housed in a building of permanent nature affording sufficient protection from normal climatic hazards, such as wind, blown dust and rain.

2. Separate storage facilities for packing material, such as, thermocole boxes, plastic sheets etc. shall be provided.

3. The lay out of different sections, namely, raw material receiving section, cleaning and washing section, icing, packing section etc shall be such as to facilitate the smooth and orderly and unidirectional flow of work to prevent any possible cross contamination.

4. There should be adequate natural and artificial lighting.

5. There should be two chill rooms attached maintaining a temperature between 0 to 4°C and also an ice store.

6. There shall be a change room with facilities for changing dress, gum boots, head gears etc before entering in the work area.

7. Wash basins with non-hand operated taps and liquid soap shall be provided at the entrance to the work area.

8. Suitable measures shall be taken to prevent entry of insects, birds, rodents and other animals into the premises and handling area.

9. The floor of the food handling area shall be water proof and easy to clean.

10. The internal wall of the handling areas shall have the smooth surface and shall be durable, impermeable and easy to clean.

11. The ceiling or roof linings shall be easy to clean.

12. The packing area will be air-conditioned and to maintain a temperature of 15°C.

13. Sorting, grading and washing are to be done on top of stainless steel tables.

14. Facilities for chilled potable water and quality ice is essential.

15. Facilities shall be provided for easy disposal of solid waste and the wastewater with out cross contaminating the products handled.

16. Mechanical fish handling equipments, like fork lifts, pallet, insulated/refrigerated trucks are essential.

17. HACCP plan for fresh chilled tuna should be prepared and enforced meticulously.

18. Trained grader should be employed.

19. The center should have histamine testing facility.

20. For export to EU, sashimi tuna needs to be frozen at minus 20°C for 24 hours. This is to kill parasites if any. This must be applied to the raw product or the finished product. This is not required if epidemiological data are available indicating that the fishing grounds of origin do not present a health hazard with regard to the presence of parasites and the competent authority so authorized.

Product Specifications of Sashimi Grade Tuna

Generally the quality of sashimi produced from blue fin tuna is the highest, followed by the quality of sashimi produced from big eye tuna and yellow fin tuna. However, sashimi produced from top-quality big eye tuna is considered to be of better quality than sashimi produced from average quality blue fin tunas.

The quality of tuna for the sashimi market is evaluated on the basis of the objective criteria, such as, species, the period and region of capture, the method of conservation (fresh/refrigerated or frozen) and the fishing device used. It is then evaluated on the basis of organoleptic criteria, such as, the presence of fat, the appearance of the skin, protuberant, clear and moist eyes, intact stomach and fresh smell.

The best sashimi is produced by large tuna individuals caught during the season preceding the reproductive season. Usually sashimi grade tuna is graded as follows;

Grade 1 + — Tuna whose flesh is bright red, compact, clear and fat. This sashimi of outstanding quality is produced by tuna caught with a hand line or long line, refrigerated on board.

Grade 1 — Tuna whose muscle tissue is red, with compact, clear and fat flesh, caught by long liners, refrigerated on board.

Grade 2 — Tuna whose muscle tissue is red, with compact, fairly clear but lean flesh, which can be used for the production of steaks as well as for lower quality sashimi. It can be either refrigerated or frozen at sea.

Grade 3 — Tuna whose muscle tissue can be both red or brown and whose flesh is compact but opaque and lean. It is frozen at sea and used for the production of steaks.

Grade 4 — Tuna whose muscle tissue is grayish-brown and whose flesh is soft and opaque. This tuna is sent to the processing industry.

It may be noted that the difference in quality between tuna for sashimi or steak would depend on the price level obtained in the market.

Different sashimi cuts from different species have various market values, depending on fat contents, the higher the fat content, lighter the colour and the more valued, the sashimi will be. The best sashimi comes from Toro, the peripheral layer of lighter coloured tuna meat with a fat content of 25 per cent. Toro is further divided into Otoro, Pink which gives the prime sashimi and Chutoro, darker pink. Within Otoro there is a smaller part, called Sunazaki, whose texture is marked with thin lines of fat, with distinctive looks and flavour. The central block of red meat, with lower fat content (around 15 per cent) is called Akami and fetches lower prices than Toro.

Frozen/chilled sashimi tuna, which is for eating raw, must guarantee a bacterial number (survival number) of less than 100000/g tested material, test negative for coliform group bacteria and have a MPN (most probable number) of *Vibrio parahaemolyticus* of less than 100. The product must be labeled to indicate "for eating raw".

Processing of Tuna–Masmeen

" Masmin" is the most popular product made out of tuna in Lakshadweep and 90 per cent of the tuna caught in the islands are processed into this traditional product. This product resembles the Katswobhushi of Japan. It has a shelf life of about a year and has excellent taste and is popular in south west and south east coasts of India and Sri Lanka. The annual production of masmin in Lakshadweep, during 2009-10 was 1276 tons valued at 45.93 crores of rupees. In absence of no processing facilities and no avenues of marketing tuna in fresh condition, the fishermen in Lakshadweep Islands, adapt the traditional local system of processing and marketing of tuna. The tuna are converted into the local processed form known as "Masmeen" or "Masmin", a traditional cooked, smoked and dried tuna product. This system made its entry in the past into the Lakshadweep Islands from the Maldives. Masmeen is traditionally exported to Sri Lanka from India.

After landing, the fishermen cut the heads of tuna from their bodies. The bodies are cleaned as the first step in dealing with the catch. The vertebral column and the bones of the fish are removed from the body and the rest of the body is neatly filleted. All the waste materials including head and bones are buried in a pit excavated at the sea shore itself. The by-catches are sold directly to the people that gather at the landing centers.

The cleaned tuna fillets are boiled in a large sized pot or container filled with sea water. After boiling, the fillets are removed and smoked using coconut leaves and wood. Thereafter, they are sun dried to remove the residual moisture. Then they are packed in gunny bags and exported to the main land markets. Modern post-harvest facilities and value-addition to the by products of tuna are very limited. The end product, obtained after nearly three months of processing time, will be hard, but whole some in texture.

The quality of dried tuna depends on the degree of excellence of boiling, heating and drying. The traditional masmeen making centers have a raised platform with coconut fiber mats spread across. The wood of coconut trees is used as a fire production source for placement in the pits beneath the platform. The time taken for total sun-drying of the fishes thereafter varies from a few hours to 2-4 days depending on the availability of sun light. Fresh batches are spread along side, near the old ones, to facilitate continuous processing. In general, with the residual and semi-colloid water, the fish are boiled two to three times or even more to make a cream like product, which is locally called " tuna charkara ". It has a high nutritious value and is used as a special type of teeth soother. Masmin has good demand and is therefore expensive. Different products are developed from masmin, which include, mastuna, masfinger chips, mas powder and the tuna mas pickle. In the canning factory at Minicoy, tuna pickles are produced. A masmin and fish meal unit has been established in Agatti to produce processed masmin.

Processing of Cephalopods

Cephalopods have high commercial value as they constitute a good source of quality aqua-products. The flesh of cephalopods is firm and it blends to provide a variety of processed and preserved products owing to the absence of bones, ease of cleaning and leaving behind a marginal quantity of wastage. They have also a high food value. The meat content in cephalopods is much higher than fish, containing 40-70 per cent consumable fraction (Takshashi, 1965). Cuttle fish contain high protein, a low carbohydrate and lipid. Edible portion in a squid is about 76 per cent and that in cuttle fish is about 65 per cent with a higher value of protein (18-20 per cent) than many fishes and the protein is composed of 20 amino acids, eight of which belong to a group of essential amino acids. The proximate composition of cephalopod flesh shows moisture at 75-80 per cent, crude protein (TN X 6.25) at 16-21 per cent, crude fat (1.0-1.5 per cent) and ash (1-2 per cent). Due to high nutritive value and better taste, cephalopods are widely accepted as a choice of food in various parts of the world.

Cephalopods meat is prepared in many ways for food; such as, cutting them into slices and treating them with spices and frying or cooking into curries, cutlets and

Artisanal Tilapia Processing

Tilapia fillets

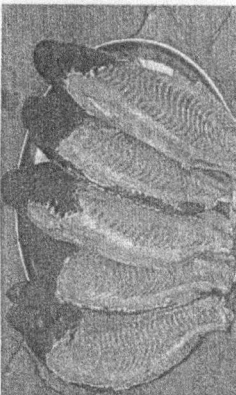

Tilapia products display

Skinning of Tilapia

Head-on Tilapia

Packaged Tilapia products

Tilapia harvest

Tilapia drumsticks

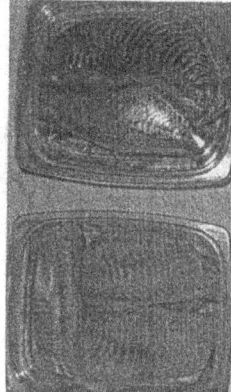

A range of processed Tilapia

Processed and Value-Added Tilapia

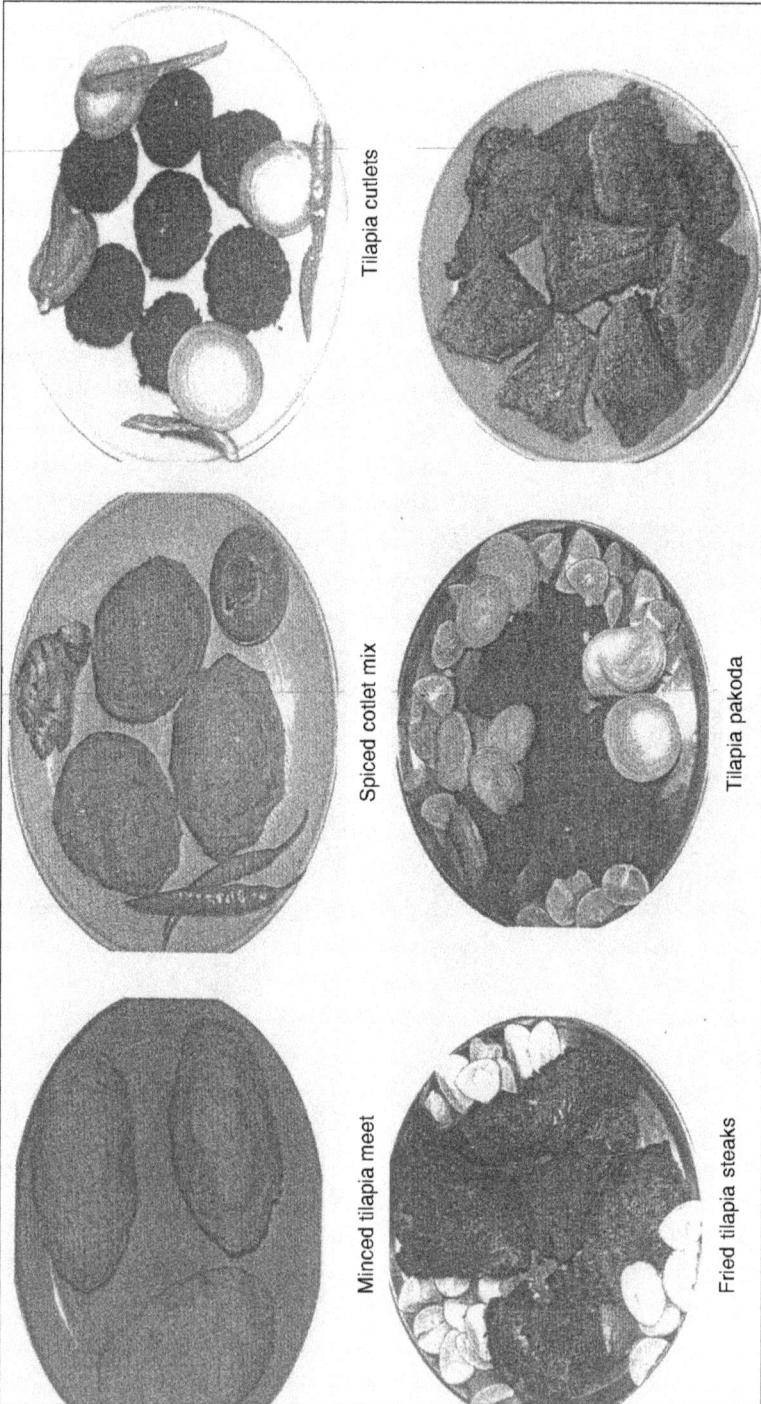

Tilapia cutlets

Dried fish tilapia

Spiced cotlet mix

Tilapia pakoda

Minced tilapia meet

Fried tilapia steaks

soups. In India, cephalopods were initially used in dried form. In the Philippines, the meat of cephalopods is first boiled in vinegar with crushed garlic and then fried in oil and spices. In east Asian countries, baby octopus is consumed live, considered as delicacy and is highly relished. In Britain, squids are incorporated into fish stews to enhance the taste. The Japanese have mastered a large variety of culinary preparations out of squids, cuttle fishes and octopods. In Korea they are mainly salted and fermented. Of late, products, such as cephalopod pickles etc have been developed and are gaining popularity.

Cephalopod products meant for export are frozen whole by IQF (Individually Quick Frozen) or peeled and frozen in the form of slabs of different commercial grades. Various high value products are made out of cephalopods. Some of these are whole squids, stuffed tubes, squid rings, frozen tentacles, fillets, roe, squid-rings battered and frozen. Some of the other products are seasoned squid, fermented squid, smoked squid and pickled squid meat. Cephalopod products such as, canned, dried, salted, smoked and fermented are also produced.

The non-edible parts of the squid, namely, viscera, skin, pen, beak and eyes comprise about 30 per cent of the total weight of the animal. These can be converted into meal, similar to fish or prawn meal. Skin as well as damaged low grade raw materials can be ensilaged and sold as high protein feed. Sometimes it is also used as manure. Viscera of the squid are rich in amino acids, minerals, vitamin B etc and therefore, it forms a good constituent for poultry feed.

The cuttle bone is commercially used in preparing abrasives. Cuttle bones are also used as grinding stone for the beaks of cage-birds. It is also a good source of food for poultry and cage birds as it is rich in calcium. Artists have used the ink of cuttle fish as a natural sepia pigment. Ambergrees (obtained from sperm whales, which is used as a fixative for perfumery) is found around the beaks of squids, consumed as food.

Live cephalopods are good experimental animals, in the field of research and education. They are also gaining popularity in aquarium market.

Processing of Major Carps

Carps have been the mainstay of fish farming in India due to their low production cost and good marketing potential in several states. Carps are cultured for 9-10 months and are harvested as and when there is demand for internal markets. As the availability of carps is seasonal, the fish processing industry can utilize their processing and production facilities to process these freshwater fishes during monsoon period to overcome the problem of raw material shortage during fishing ban and lean period.

Presence of bones is a major attribute of almost all species of bony fishes. However, the presence of intra-muscular bones poses a specific problem in carps. In general, there are 43 intra-muscular bones on each side of the fish, 26 above the lateral line and the remaining 17 below the lateral line. Among the 26 intra-muscular bones, which are above the lateral line, 20 have a typical Y-shape. These intra-muscular bones are found in the flesh at approximately one-third of the depth below the body surface. The recent developments have proven the efficient removal of these intra-muscular bones using sophisticated equipment which is available in the market. The problem can also be eliminated by cutting the bones into small pieces either manually or by using equipment, which consists of a row of revolving blades, which slit the intra-muscular bones into small pieces when the fillets are made. To maintain the integrity of the fillets, the cut should not be too deep. However, the yield from the filleting is only around 35-40 per cent

The negative image of muddy taste is mainly because of the presence of geosmin. The muddy flavour is commonly observed in bottom dwelling fishes. Among Indian major carps, *Catla catla*, *Labeo rohita* and *Cirrhina mrigala* exhibit surface, columnar and bottom feeding behaviour in their natural environment. As most of the modern day aqua-farming completely depends on artificial feeding, the problem of muddy taste can be overcome by proper management of culture ponds.

Chilled Products

Chilling is an effective method of maintaining freshness of fishery products. This normally involves keeping fishes in melting ice or slurry ice to maintain the fish temperature around 1 to 4°C, which delays the enzymatic action and microbial activity, thereby extending the shelf life of the products. Chilling in ordinary air packs will generate limited shelf life, but it can be increased by using advanced vacuum packaging, modified atmosphere and active packaging techniques.

Vacuum packaging can minimize the oxidation of fishery products, but it creates anaerobic environment which encourages the growth of anaerobic spoilers and the most dangerous *Clostridium botulinum* growth and toxin production. This can be minimized by using good quality raw material and maintaining storage temperature less than 3.3°C throughout, which prevents the growth and toxin production of the organism. In modified atmosphere packaging, the air inside the pacakage is replaced with favourable gases like carbon dioxide, nitrogen and oxygen depending on the fat content of the fishes. Major carps, the category of medium to fatty fish are normally packed with 60:40 (carbon dioxide: Nitrogen). However, introduction of oxygen has advantage in overcoming the growth and toxin production of *C. botulinum* Active packsging technique is either gas-flushing or more recently gas-scavenging or emitting systems added to emit (namely, nitrogen, carbon dioxide, ethanol) and or to remove (namely, oxygen, carbon dioxide, odour) gases during packaging or distribution. Major active packaging techniques are concerned with substances that absorb oxygen, ethylene, moisture, carbon dioxide, flavours/odours and those which release carbon dioxide, anti-microbial agents, anti-oxidants and flavours. Whole, cleaned, headless, steaks and fillets forms of carp can be used for preserving in vacuum, modified atmosphere packaging or active packaging.

Freezing

Freezing is an age old practice to retain the quality and freshness of fishery products for a long time. This involves the conversion of water present in fishery products to ice, that is, a phase change from liquid to solid. This retards the microbial and enzymatic action by reducing the water available for their action. Freezing can be either slow or quick freezing. Slow freezing affects the quality, whereas quick freezing preserves the quality. Quick freezing is normally accomplished using any of the following four methods; air blast freezing, indirect contact freezing (plate freezing), immersion freezing and cryogenic freezing. Normally products are frozen till they attain a core temperature of minus 18°C or lower and are stored in cold storage maintained at this temperature. The freezing and frozen storage of fish have been largely used to retain their sensory and nutritional properties before consumption. Frozen products form one of the largest portions of fishery product traded all over the world. Indian major carps either as whole fillets or as steaks for retail pack can be adopted for freezing.

Thermal Process

Thermal sterilization is one of the most efficient methods of food preservation, widely practiced world wide. The main objective of thermal processing is to achieve long term shelf stability. Thermal processing generally involves heating the food products packaged in hermatically sealed containers for a predetermined time at a pre-selected temperature to eliminate the pathogens of public health significance as well as those micro-organisms and enzymes that deteriorate the food during storage. Historically, heat processing started in glass containers. Over the years different containers, like,metal, rigid plastic containers and flexible retortable pouches are developed for thermal processing. Unlike other marine fish products, the thermal processing of Indian major carps need special attention as the texture of meat becomes very soft after heat processing. A mild frying before packaging for thermal processing will overcome this problem. Thermal processed Indian major carps are mainly preferred with curry medium. A wide variety of curry styles can be adopted depending on the intended consumer. Thermal processing of IMCs is also advantageous as all the bones, whether central back bone or intra-muscular bone will become soft, ready for consumption without the need for separation.

Fish Mince

Minced fish is prepared by concentrating only on edible muscle part, removing all the other parts like scale, skin, gut, bones. The minced fish technology offers a great possibility for the processing of bony fishes particularly, Indian major carps which possess intra-muscular bones. Fish mince act as base material for majority of the value added products. It is prepared by split opening the fish ventrally and passing through the meat bone separator. Larger bones are normally removed manually and fine pin bones are then separated by passing the mince through pin bone remover, so that the bone content in the final mince should not exceed 2 per cent as it affects the quality of mince. Normally fish mince is stored under frozen condition. Fish mince is used for the preparation of surimi, fish balls, fish fingers and other specialty products.

Fish Finger

Fish fingers are regular rectangular sized fish portions made from either frozen fish fillets or fish mince. Skinless and boneless fillets are frozen to get the correct shape of the finger. For ease of cutting operation, frozen slabs of 1.5 cm thickness are used. The frozen slabs are passed through a motor operated band saw to cut into suitable size. A typical British fish finger weighs around one ounce (28 grams) and in

Asian countries it varies from 20-25 grams each. They are battered and breaded before freezing or cooking. They are normally frozen stored and consumed as snack food after frying in vegetable oil.

Fish Balls

Fish balls are reconstructed convenient product which believed to have originally come from China. These are similar to the products like Kofta of India, Polpette of Italy. It is prepared from a mixture of fish, fat particles, water, carbohydrate,ginger, garlic, pepper and salt. During the processing, meat is mixed with the ingredients and carbohydrate source, which will bind the particles directly or indirectly. The mixture is then formed to desired shape and this shape is retained after freezing or cooking. Minced meat of IMC can be utilized for preparation of fish balls and they are normally battered and breaded before freezing or cooking.

Fish Cutlet

Cutlet is a spicy snack food popular in Asian, European and South American countries. Fish cutlets are prepared by mixing the cooked fish meat with the cooked potato and fried onion, chilly, ginger, garlic, pepper, turmeric, coriander, oil, salt and other ingredients based on consumer taste. After proper mixing, they are shaped desirably and battered and breaded. Cutlets can also be prepared from cooked meat, from skeletal frame remaining after filleting the fish. Frozen storage is the normal practice and the product is consumed after frying in oil.

Processing of Tilapia for Value Addition

Tilapia grows fast and a prolific breeder. The hardy tilapia, commonly known as aquatic chicken. Its meat is typically white, moist and resilient. The firm and flaky texture of tilapia has a sweet mild flavour. The fish ia a great choice for a low carbohydrate diet or for any healthy meal. It is an excellent substitute in any white fish recipe and has a higher moisture content and good body shape. Cooked meat is opaque in colour. Hundred grams of raw tilapia flesh contains 98 calories, 20 fat calories, total fat 2.4 gram, saturated fat 0, cholesterol 0, sodium 52 mg, potassium 0, protein 18.5 gram and iron 0.

Post-harvest Processing

Tilapia has a very tough skin, so also the scales. Head of tilapia is bony and has less of meat in it. Tilapia head and skin taste bitter. Eor skinning the fish is held flat and the entire dorsal ridge including the fins are cut deep up to half an inch up to the tail. With the help of a blunt knife, the skin at the junction of head and anterior dorsal ridge is loosened on both the sides. The skin is pulled thin and strong when the entire skin of each side comes out as a whole long strip.

Degutting

Tilapia has very small pluck and the belly cavity is restricted up to the anus, which is at the center of the body. Belly cavity and belly muscles are very thin. So they are removed completely along with the belly bones. The peritoneal cavity is located deep in the chest cavity. The peritoneal lining is very tough, which should be detached with a blunt knife and the thoracic cavity is washed properly.

Filleting

Skinless, degutted tilapias are washed properly with potable water and allowed to drain on drainage board. The headless whole round tilapia is placed flat on a cutting board. With the help of filleting knife broad fillets of both the sides are separated from the skeleton starting from the anterior mid-dorsal line and gradually coming up to the tail.

De-boning of Skeleton after Filleting

The tilapia carcass after removal of the two broad fillets is subjected to deboning. Either of the two methods as described below, can be followed to get minced meat.

1. Scrapping – With the help of a scrapping knife, the residual meat can be scrapped. This method yields scrapped meat about 3 per cent weight of the fillet less tilapia carcass.

2. The fillet less skeleton are put in a scalding tank for 5 minutes at 100°C. The boiled carcasses are taken out and allowed to drain properly. When cooled, the flesh is loosened with the help of a blunt knife and meat is separated. The meat so produced is made into mince with the help of a meat mincer. The minced meat is used for the preparation of ready to eat restructured products like meat ball, cutlets etc.

Tilapia steaks – Tilapia is a flat and fleshy fish. A kg per cent numbers) of headless and skin less tilapia give 30 numbers of fish steaks. Each tilapia is cut into 4 steaks. Tilapia steaks go very well for fish fry and curry preparation.

Minced meat – Tilapia meat scrapings and smaller chunks can be comminuted and made into minced meat. Binders, spices and condiments are used to make meat balls.

Yield (dressing) in per cent — Head on skin less dressed fish - 65

Head less dressed round — 50

Fish steaks — 50

Fillets — 20

Minced meat – 20

The fillets may be chilled to as low temperature as possible without freezing. The fillets may be placed in polythene bags, sealed and placed them in an expanded polystyrene bag and sealed.

Tilapia fillets can very well be converted into fish fingers, chunks, nuggets. Fillets can be breaded and battered. They are shallow fried and served. Chunks and nuggets are the smallest pieces which are breaded and deep fried.

Laminated Bombay Duck

Of total marine fish landings in India, about 10 per cent is constituted by Bombay duck (*Harpodon nehereus*). Large quantities of fish are landed in the states of Gujrat and Maharashtra and of the total catch of the fish about 90 per cent is constituted by the catch in these two states. The fish is also taken in appreciable quantities on the Andhra and Orissa coasts and estuaries of Bengal.

At present, the fish is mostly sun-dried by the traditional method. A portion of the sun dried products is exported to countries like, Mauritius, Sri Lanka, Mayanmar, Singapore and the rest is marketed internally. A small portion of the catch is now converted into dried laminated Bombay duck, which has got a good market in countries like, United Kingdom. There is every possibility of expanding the trade in dried laminated Bombay duck in view of the demand for the product abroad. As in the case of any other product quality will be the deciding factor in sustaining and expanding the market for the product. To get top quality product it is essential to adopt standard method of processing.

Process

1. Use only fresh Bombay duck. If there is any time lag before processing, keep the fish at ice temperature during the period.
2. Wash the fish thoroughly. Remove the guts and wash thoroughly again. Suspend the gutted fish from a scaffold for surface drying (for about 2 hours). At this stage remove the head, tail and fins using a sharp knife or scissors and split the fish longitudinally on the belly portion.
3. Dip the fish, thus laminated in sufficient quantity of 1 per cent brine (prepared from refined salt) for 20 minutes.
4. Drain the fish and spread on galvanized iron wire mesh trays (on a platform) or in any other suitable manner and dry to a moisture level of about 16-17 per cent.
5. Flatten the dried product by means of a roller press. Trim the sides to get pieces of uniform size.
6. Dry the product again for 1-2 hours, so that the final moisture content would be about 14 per cent.
7. Make the finished product into lots of 25 or 50 numbers and pack in polythene bags. Store properly under hygienic conditions.

In the commercial method followed at present, it is observed that the bone is removed from the fish just after splitting the semi-dried fish. Actually the bone which is very delicate and can not be made easily after pressing the dried fish need not be removed so that a very time consuming step in the method can be avoided.

Incorporation of BHT (Butylated Hydroxy Toluene) and NOGA (Nor dihydroguaiaretic acid) in dipping brine up to a concentration of 0.1 per cent effectively controls the development of discolouration (yellow brown) and consequent spoilage of the stored material.

It is always advantageous to use a tunnel dryer for drying the product in view of quality of the finished product. In sun drying, the fish is exposed to contamination from dirt, sand, flies and insects which results in sub-standard quality and early spoilage of the product. Also by sun-drying it is difficult to get the product of uniform quality with the desired moisture content. A tunnel dryer of 1 ton (raw material) capacity for production of laminated Bombay duck has been designed by the CIFT.

Processed Products from Sharks

The main processed products from sharks include;

1. Meat, whether fresh, frozen, salted or in brine and smoked
2. Fins to prepare shark fin soup.
3. Liver oil for cosmetics and pharmaceuticals.
4. Skin to prepare shark-skin soup, for leather and sand paper.
5. Cartilage, ground to powder and used to produce a supposed anti- cancer cure.
6. Teeth and jaws, in jewellery and sold as curios.

Main Shark Commodities

Shark fins are the most valuable of all shark products and therefore the main source of income in developing countries.

The belly flaps of piked dog fish is marketed in smoked form in Germany and shark meat was introduced as the "fish and chips' in the United Kingdom. Despite the nutritional content and appreciable taste, shark meat was considered a poor person's food and sharks were mainly caught, in the fifties, for their vitamin A rich liver oil. However, the waste up to 75-80 per cent of raw material led businesses and countries to improve fishing/processing technologies and marketing/distribution strategies, in order to generate a wider acceptance of shark meat. Since the late fifties, a wider acceptance has been achieved due to better handling, the use of ice and freezing, the awareness of wide spread malnutrition and thus the need to itilize the available protein for human nutrition, the contemporary shortage of bony fishes in some areas and the marketing efforts to promote shark meat.

Short fin mako shark is considered the world's best quality shark meat; it is marketed fresh in the United States and in Europe. Other largely appreciated species are thresher (*Alopias* spp) and porbeagle. The meat of smaller species like dog fish is also appreciated as it contains smaller amounts of urea and mercury than other species and is also easier to process. The backs of these sharks are marketed in Europe and Australia as fillets, steaks, portions and used in the fish and chips trade. The fresh whole carcasses are marketed in South America as *Cazon*. Other important sharks for the production of meat are requiem sharks.

Non-food uses of sharks includes shark liver oil products, cartilage, skin and teeth. The shark's liver is saturated with oil to maintain its buoyancy in water. Shark liver oil has been traditionally used as lubricant in tanning and textile industry, in cosmetics, skin healing and other health products, as preservative against marine fouling of wooden boats, as fuel for street lamps and to produce vitamin A, before synthetic vitamin A was discovered. Currently, demand is mainly for squaline, a highly unsaturated aliphatic hydrocarbon, present in certain shark liver oil (mainly of the family Squalidae). Squaline is used as bactericide, organic colouring matter, rubber, chemical, aromatics, in the textile industry, as an additive in pharmaceutical preparations, cosmetics and health foods. A related compound of squaline is squalane, a saturated hydrocarbon obtained by hydrogenation of squaline. Squalane is used in skin care products as it is a natural emollient.

Shark cartilage, processed into powder and tablets is used as a health supplement and alternative cure for several diseases, and beneficial in inhibiting the growth of tumours by impeding vascularization of malignant tissues (angiogenesis). Cartilage from the sharks is believed to be of the best quality as it is believed to be richer in chondroitin than other species. Chondroitin is an acid mucopolysaccharide used for various health problems.

Shark skin is used to produce leather. The market was buoyant until a few years ago, when leather from shark was used to produce hand bags, shoes, wallets, cigarette cases, watch straps, coin and key fobs. With the increase in the market for shark meat, shark skin lost its niche. In fact shark carcasses are sold with the skin intact inorder to protect the meat and present oxidation. Further more, sharks have to be bled, dressed and iced immediately after catch to prevent urea from contaminating the meat, but exposure to freshwater or ice damages the skin. Therefore now a days the market for shark leather is limited.

Shark Fin Rays

Dried shark fin is a valuable product of export from India. Though fins from several varieties of sharks are exported, those from Raja (*Rhynchobatus djiddensis*) fetch the maximum price. Fins from *Scoliodon walbheemi, Carcharinus malanopterus* and *Zygaena malleus*) are some of other varieties exported. The fins are exported mainly to Singapore, Hong kong and UK. At the importing countries the fins are processed for their fin rays, which are utilized in soup preparations. Of late, it is known that there is good internal demand also for shark fin rays, especially in major hotels.

Process for Shark Fins

1. Cut the fins (spinal and caudal fins) from the big sized sharks (of about 5 feet or more in length)
2. Remove the adhering flesh and wash thoroughly in freshwater.
3. Dust the fins with salt in the ratio of 1:10, the cut portions being liberally sprinkled with salt. Alternately, dust the cut portion with little lime and salt. Set aside for 24 hours.
4. After spreading on mats to prevent admixture with sand and other extraneous matter, dry the fins in sun to the desired degree (moisture content 7-8 per cent.

5. Grade the fins according to species, colour and size.

6. Pack in suitable lots in moisture-proof containers, such as, polythene lined gunny bags.

Extraction of Fin Rays (Both fresh and dried fins can be used)

1. In the case of dried fins, soak the material for 2-3 days in water acidified to pH 2.5 to 5 with acetic acid so as to hydrolyze the collagen in the fins to gelatin. If the muscle and skin do not get softened, even on soaking for 2-3 days in acidic water (as in the case of shark fins stored for more than a year when the re-hydration capacity is less), the period of soaking may be extended to 5-6 days.

2. Treat the fins (fresh or softened dry fins) in hot 10 per cent acetic acid at 60°C for one and half to two hours, depending on the size of the fins.

3. Scrape off the skin and rinse with water.

4. Heat again in fresh 10 per cent acetic acid for about one hour.

5. Separate the rays from the flesh while washing with cold water.

6. Wash off excess acid with water.

7. Collect the rays and dry under sun or preferably in an artificial dryer at 50-60°C (for about 4 hours) to a moisture content of 5-8 per cent.

8. Store the dried fin rays in polythene bags. Such fin rays will keep in good condition for more than a year without any significant change.

Shark Liver Oil

The oil is extracted from the fatty livers of the sharks, skates, rays and dog fishes.There are about 69 species of cartilaginous fishes and 57 of them are available in Indian waters.

The livers vary in size. A single liver may be anywhere from a few ounces to 400 pounds in weight as in the case of saw fish measuring 25 feet. The oil content also varies from season to season and from shark to shark. Wide variations are found in the vitamins.

The simplest method to extract oil from shark livers is to chop the liver and boil it on open fire and skim off the oil that oozes out. Steaming the liver shortens the time. In the oil factory, the livers are finely disintegrated by the chemical means and are digested by a special process, This mass is then separated by high speed centrifuges to yield a fine polished oil, free from last traces of moisture and liver debris.

Flow Sheet

Shark liver → graded according to potency → Liver disintegrator → Lurd liver (pulped liver of 1mm size) → Steel tank with perforated steam coil, boiled for 2 hours → Oil separator for separating oil, water and debris → Crude oil containing stearine → filtered in hand press for separation of stearine → Refined to remove free acid by neutralizing with alkali → Sharple's super centrifuge 15000 rpm, oil entered from the bottom, refined oil out from the top → added with refined ground nut oil to standardize vitamins potency → Centrifuge again in super centrifuge(Dilution with refined ground nut oil depending on the potency of vitamin A) → Filling in bottles and sealing in bottling machine → Packing and storing.

The standard laid down according to IP

1. Iodine value–90-125 mg/l
2. Acid value–Not more than 2
3. Vitamin A — 10000 IU/g
4. Refractive index — 1.459-1.466
5. Saffonication value — 150-200
6. Un-saffoniable matter – up to 10 per cent

Some of the products from factory

☆ Sea gold–1500 IU Vitamin A, 100 IU Vitamin D –For human consumption

☆ Stay fit 1000 IU Vitamin A, 100 IU Vitamin D–For human consumption

☆ Univax – 1500 IU Vitamin A, 150 IU Vitamin D—For live stock

☆ Trivax – 6000 IU Vitamin A, 600 IU Vitamin D – For live stock

Extraction of Sardine Oil

Sardine oil extracted by the existing commercial practice is generally of inferior quality with low storage life.un-desirable colour and objectionable odour. During the fishing season, many oil extraction units spring all along the coast, but most of these would not be equipped with to carry out the extraction on a scientific basis. The limited use of sardine oil must be responsible for this lack of interest on the part of the industry in adopting scientific methods.

Oil sardines constitute nearly 20 per cent of the total marine fish landings in India, the average annual catch being of the order of 2,33,000 tons.

Assuming that about 50 per cent of the total catch would be available for extraction of oil, the quantity that could be produced worked out to be about 11700 tons per year on the basis of an average oil content 10 per cent in the fish. But the quantity produced at present is much less.

Since sardine oil after separation of the stearine, has been found to have characteristics similar to some important vegetable drying oils, like linseed oil and thus have potential commercial application as a cheap substitute to them. The method of extraction of high quality oil from sardine at little or no extra cost is;

1. As far as possible use only fresh sardine for extraction of oil. Wash the material well.
2. Take water in the vessel used for extraction in the ratio of 1 part of water to 1 part of fish (by weight). Boil the water (Aluminum or tinned copper vessel are suitable for extraction of the oil).

3. Add fish to the boiling water and continue the boiling till the oil get separated at the top. Stir occasionally while boiling.

4. Collect the separated oil by means of spoons or trays.

5. Cook again for about 30 minutes with occasional stirring. Allow to settle. Collect the supernatant liquid and keep separately.

6. Take the cooked material in canvas bag, press under screw press and collect the press liquor. Mix the press liquor with the supernatant liquid collected before.

7. Add sufficient quantity of common salt to the oil-water mixture to break the emulsion. Collect the separated oil and mix with the sample of oil collected earlier.

8. Heat the oil on a water bath to remove last traces of water.

9. Store suitable lots in containers.

The characteristics of the oil extracted by the improved method are given below. For comparison, the characteristics of samples of commercially available oil are also given.

	Oil Prepared by Improved Method	Commercial Oil
Physical Characteristics		
Colour	Lemon yellow to yellow	Yellow to black
Clarity at room temp.	Clear	Turbid to clear
Odour	Characteristic of the oil	Rancid/Putrid
Chemical characteristics		
Saponification value	192-195	192-193
Iodine value	152-175	139-161
Peroxide value	0.32-2.7	0. 4382-7.2
Un-saponifiable matter	0.82-1.15	0.84-1.55

The oil prepared by the method besides having marked improvement in quality in many ways, keeps well for one year (or even more) without significant changes in the analytical characteristics. This is important since any industry that may utilize the oil will require a constant supply. Since the extraction is seasonal, a constant supply can be ensured, only if the oil can be kept in good condition at least till the ensuing fishing season.

Minced Fish

Meat Picking

Meat picking is one of the ways where is potential for using many of the low cost fishes. The resulting product "minced meat" is a value added product, which can again be used as a raw material for manufacturing products like fish cakes, fish pastes, fish fingers etc.

Minced meat can be referred to as a comminuted product. Minced meat production is the means of removing the edible flesh of fish from rest of the body by mechanical application. Comminuting the fish is one of the best ways of utilizing the landed fish, since almost all the flesh of the fish can be practically removed from the bones, skin etc with the help of a meat separator. In actual fishery practices, there is a significant waste of potentially edible animal protein. Most of the bigger vessels look for a particular species and the other fishes that come up with the catch are thrown over board. This is because these fishes do not get better market price and the vessel operators feel that it is more economical to get rid of the fish rather than to preserve them on board. It is in this field that the comminution has a vital role to play in bringing up the so called "trash fish" and the under-utilized fish to the level of proper utilization in the form of human food material.

For comminuting, fishes which are smaller in size, too many bones and other types of fishes which do not have a better market value may be selected as raw material. Generally fish varieties having white and light meat and with less fat content are preferred for minced meat production. If fatty fishes are used, anti-oxidant treatment is advised to obtain a better shelf life under frozen condition.

Fishes, like, croakers, catfishes, mackerels, sardines, Bombay ducks, horse mackerels, jew fishes, anchovies, barracudas and flying fishes and other un-conventional fishes, not fully utilized as human food and many of the currently under utilized fishes offer a vast scope for production of fish mince for the development of diversified surimi-based seafood products.

White-fleshed lean fish like the Alaska Pollock (*Theragra chaleogramnia*) is best suited for minced meat preparation. In India, threadfin bream (*Nemipterus japonicus*) and sciaenids are good source of minced meat. Grading of different species of fish based on quality of minced obtained from them are given below.

1. Teleosts – (a) Croakers, Lizard fishes, Perches, Shads, Ocean perches, Threadfin breams, Horse mackerels and flying fishes are considered as very good; (b) Barracuda, Eels, Soles, Ribbon fishes, Sciaenids, Carangids, Seer, Yellowfin tuna, are considered good; (c) Oil sardine, Mackerel, Silver bellies, Anchovies, Bigeye tuna are considered poor.

2. Elasmobranchs (i) Hammerhead shark, Dog fish are considered very good; (ii) Man-eater shark, Rays are considered good; (iii) Spiny dog fish is considered poor

3. Freshwater fishes- (i) Carps, Tilapia, Mullets, Milk fish are considered good; (ii) Murrels, Freshwater shark are considered poor

Among invertebrates, Cuttle fish, Squid are considered good, while Prawns and Krills are poor.

Preparation of Mince

In earlier days, mincing was done with knives. Later rocking knives were mounted on top of a rotating bowl. Now a days, a variety of machines are available for mechanically separating fish flesh from the bone or shell by forcing the fish against screened surface, where flesh passes through the opening as a finely ground paste.

Meat separators of Maader and Bibun make operate on a belt-drum principle, where the fish is forced against a perforated drum (hole sizes can range from 1 to 10 mm in diameter). The flesh passes through the holes, while the skin and bones are ejected through a discharge chute. Paoli machines break up the bones and shells and separate the flesh by the micro-groove principle, while Beehive machine use a feed screw or auger and a perforated drum to separate pre-chopped material. Using high belt tension and large-hole drum generally maximizes product yield, as well as the amount and size of impurities. Following deboning, mince may be refined by passing through it through strainer which will remove bone fragments and small pieces of belly linings. The pore size in strainers typically rangees from less than 1 to 2 mm in diameter. Straining coarse minces will reduce the meat particle size and produce a more homogeneous pasty product.

Another machine used is the stamp type in which fish are compressed against perforated plate in a continuous operation. For better mince quality, it is advisable to cut open the fish along the back bone, thus allowing removal of exposed viscera as well as the head.

Mincing mechanically disrupts these protein structures and enhances the sensory tenderness. The disruption of cellular membranes allows a rapid distribution of subsequently added salt or other additives throughout the meat particles. Swelling of myofibrils, therefore, takes place much more rapidly. This is a pre-requisite for comminuted meat products, and mincing is the first step of processing to achieve this. This minced meat has excellent water holding capacity, because mincing with subsequent salting (1.5-4.5 per cent NaCl) leads to disintegration of the myofibrillar protein. Mincing of meat permits mixing with other foods allowing the extension of meats. Starch and other poly-carbohydrates, non-meat proteins, water etc are possible extenders. Grinding or mincing also permits the formation of uniformly shaped restructured meat in moulds and enhance tenderness and a uniformly distributed flavour are easily achieved. Finally mincing permits the addition of substances that may enhance the microbial and sensory shelf life although mincing is counter productive in terms of shelf life.

During mincing two things to be taken care of. Firstly the cellular membrane is essentially a lipid bi-layer through which oxygen can not easily penetrate into intracellular space. In the case of mincing, the grinding disrupts the cellular structures. As a result, oxidation leads to rancidity. For this reason the storage potential of frozen minced meat is very limited. Secondly, on mincing the surface area is greatly increased and the micro-organisms present on the meat surface are distributed throughout the mince. Immediate chilling to zero to plus 1°C retards or inhibits their further growth, but a common means of preventing microbial growth is to heat the mince immediately or within 24 hours of production.

The presence of dark spots in minced fish is due to the presence of bits of skin, belly lining or organ meat in the flesh is esthetically objectionable and limits the application of the product. The blood content of minced flesh can vary widely among different species of fish. As a result the colour of minced fish can range from dark to white. Washing of minced fish will leach out a large proportion of the pigment, if performed immediately after de-bonning. The addition of commercial whitener, composed largely of starch and vegetable fat, may also be used to partially whiten minced fish

Minced fish may be used alone or in combination with filleted and flaked fish to produce excellent breaded fish sticks. It can be used to replace filleted fish in recipes for soups and in a wide variety of seafood dishes. Minced fish is an ideal extender for crab meat in patty formulations.

Seasonal fluctuation in supply of fish requires a certain quantity of minced fish to be stored for future use. The actual freezing method selected will vary with each type of minced fish produced and for each intended end use.

The greatest potential for marketing minced fish lies in the area of gel type products. These products resemble the higher fat emulsion-type products manufactured from red meats or chicken, such as hot dogs. Luncheon meats and jerky gel products are prepared by finely chopping the deboned meat with spices, flavourings and possibly by adding binders to achieve a homogenous mass. Salt is added during chopping to extract the proteins responsible for firm binding and good texture in the final product. The raw gel is then stuffed into casings or pans and cooked, possibly with smoking to produce the finished products. Gel products produced in this manner from minced fish possess a firm texture and bite. The colour will vary from off white to golden brown depending on the blood content of the tissue and any colouring agents added.

While gel products prepared from minced fish do compete favourably in price with their meat and chicken analogues, the biggest advantage is in nutrition. Gel products prepared from minced croaker are far lower in fat and calories and higher in protein than the traditional processed red meat items. Most of the calories in the minced gel are obtained from protein and not from fat or carbohydrate, as in the case of processed products prepared from red meat. The high protein, low fat, nature of minced fish, in addition to the high poly un-saturation of the fat present, offers nutritional advantages for utilizing this material in gel type products.

Shelf stability of gel-products will depend on the species of fish used, the use of additives or preservatives and on the packaging and method of storage. Fatty fish generally yield products with a shorter storage life due to their fat oxidation and rancidity development. Addition of various antioxidants and proper packaging to prevent freezer burn will improve shelf life in cold storage.

Storage Life

It may be generally said that a quick frozen minced meat block will keep at 0°F for about six months and when treated with antioxidants such as, BHA (Butyl hydroxyl anisole) and BHT (Butyl hydroxyl toluene) for fatty fish, the storage life can be extended to more than six months. The shelf life of frozen minced meat very much depends on the fat content and whether the flesh was washed after separation.

The quality of minced meat of freshwater silver carp changes during frozen storage. The moisture content of silver carp (*H. molitrix*) decreased and the protein content also decreased from 16.79 per cent to 15.79 per cent of *Labeo rohita* mince meat in 12 weeks. This may be due to denaturation of protein during the frozen storage. The Total Volatile Base Nitrogen (TVBN), consists of a mixture of ammonia, trimethylamine and dimethylamine in quantities, that depend on the species and degree of spoilage showed a steady increase (12.6 mg per cent to 16.8 mg per cent) during the different period of storage. The free amino acid increased during storage of minced meat (24.3 mg per cent to 90 mg per cent). The lipid content did not show much variation during 12 weeks storage. The free fatty acid showed the lowest value in 9 weeks of storage and highest value in 12 weeks of storage. The peroxide value registered level when compared to the other studies. The minced meat of silver carp showed some variation in biochemical composition during frozen storage. But these are reported to be within permissible limits.

Advantage

Mainly small varieties of fishes will be used for meat picking for better yield. A mechanical meat separator with provision for adjusting the operational pressure can almost completely remove every particle of flesh from the fish that had been passed through it. Some varieties of fishes which are commonly used for meat picking along with their meat picking yield and filleting yield are furnished below. These fishes are considered normally as too small for filleting and even if they are filleted, the filleting yield.

Sl.No.	Fish Varieties	Meat Picking Yield	Filleting Yield
1.	Croakers (Sciaenids)	40 per cent	31 per cent
2.	Pink perch (*Nemipterus* sp.)	43 per cent	28 per cent
3.	Lizzard fish (*Saurida* sp.)	50 per cent	33 per cent
4.	*Platycephalus* sp.	47 per cent	31 per cent
5.	Ribbon fish (*Trichiurus* sp.)	42 per cent	34 per cent

Minced Meat from Fatty Fish

Normally lean fishes are preferred to fatty pelagic species for production of minced fish. But due to non-abundance of lean fishes, other abundant small fatty pelagic fishes, such as mackerel and sardine have been used for production of washed mince. Horse mackerel (*Trachurus trachurus*) has been shown to be very good alternative to lean fish species for preparation of fish mince. Fatty fish contains high concentration of polysaturated fatty acids (PUFA), Eicosapentaenoic acid (EPA) and Doeosahexaenoic acid (DHA). Therefore, fish products are susceptible to loss of quality through lipid oxidation. The onset of rancidity is fast, particularly in fatty and semi-fatty fish like horse mackerel. The use of synthetic antioxidants raised the question regarding food safety and toxicity. The use of natural antioxidants is emerging as an effective methodology for controlling rancidity and limiting its deleterious consequences. Natural phenolic compounds with antioxidant properties, such as, rosemary extract, tea, catechin, tannins etc have been gaining increasing attention due to their food safety.

A decrease in white meat fish species has opened a new avenue for utilizing new species in surimi industry. New technologies, like decanter technology and new washing techniques have allowed the processing of surimi from white non-fatty fish and also from fatty fish, such as, mackerel. Also several advances have been made to increase the yield of surimi production, thereby improving economics of surimi process. The global decrease in white fish supply has strengthened the demand for other product forms, namely, fillets, blocks, while surimi industry has learned to use surimi of other quality, that is, lower gel functionality and darker colour to process surimi products, thus favouring the production of surimi from alternative resources.

Sturgeons and Caviar

Caviar is prepared by removing the egg masses from the freshly caught fish (modern harvesting methods do not generally involve killing of the animal) and passing them through a fine mesh to separate the eggs and remove lumps of tissue and fat. Then 4 to 6 per cent salt is added to preserve the eggs and bring out the flavour. The denomination of *Malossol* ("little salt") in caviar packages indicates the low content of salt in high quality caviar. The caviar is then packed in cans, glass or porcelain containers. In some cases it is pasteurized to obtain long term storage.

There are three types of caviar; Beluga (from *Huso huso*), Osetra (mainly from *Acipenser gueldenstaedtil* and *Acipenser persicus*) and Sevruga (from *Acipenser stellatus*). They are all graded according to the size of the eggs and processing methods.

Grade I – Caviar firm, large-grained, delicate, intact of fine colour and flavour.

Grade II – Fresh caviar with normal grain size of very good colour and flavour.

Grade III – Pressed caviar (Payusnaya).

In the pressed caviar grade, external factors have caused the fracture of more than 85 per cent of roe skins before being removed from the fish. The product consists of blend of roes from Osetra and Sevruga, which is heated to 38°C in a saline solution and stirred until it has absorbed salt and regained its natural colour. Then it is put into "talecs", fabric pouches in which it is pressed to remove excess salt and oil. The resulting pressed caviar appears as a dry, spreadable black paste. It contains four times more roe than fresh caviar of the same weight, as it takes four pounds of fresh caviar to prepare one pound of *Payusnaya*. Because of its strong taste, it is favoured to Grade I and II by some connoisseurs.

A cheaper caviar produced is the *Jastichnaja*, which is unripe caviar, which may be found closer to the ovaries of the fish. *Jastichnaja* is the caviar obtained from roe that has not been properly separated from the connecting tissues. It is more salty in flavour and irregular in egg size than other caviar.

Caviar Substitutes and other Fish Roes

Roes coming from a fish other than *Acipenseriformes* is not caviar, and is often classified as "Caviar substitute". Appreciated fish roes include those of salmon, trout, carp, pike, tuna, mullet, cod and other white fish, lump fish and flying fish. "Caviart' is a caviar product made of sea weed.

Cod roes are marketed fresh or smoked. Smoked cod roes are used to prepare *tarama*, which is a mixture of roes, oil and other ingredients (bread, garlic, lemon juice, pepper etc) and Kaviar, namely cod roes in tubes.Lump fish roe makes a cheap caviar substitute. Salmon and trout roes may also be considered as relatively upmarket caviar substitute. Mullet and tuna roes are processed into a dry salted paste called *bottarga* in Italy, *pontargue* in France, a gourmet delicacy. Herring roes are considered as delicacy in Japan. Finally fresh and frozen sea-urchin gonads are used to prepare a sauce or may be added to speciality recipes.

FAO Fish stat + data show a huge increase in revenues from exports and re-exports of caviar substitute and other fish roes.

One of the consequences of the break-up of the Soviet Union in late 1991 has been the weakening of the long-established sturgeon and caviar management system. Soviet Union used to produce and export 90 per cent of caviar entering international trade. Following the break-up, the value of world export of caviar fell and continued to follow a constantly declining path. The depletion of the resource in its main production basin, the Caspian was largely responsible for this.

At the same time, the trade of caviar substitutes and other fish roes expanded to reach the record value in 1996 (160.8 million US $). The scarceness of caviar and its rocketing prices led consumers to explore caviar substitute and other roes which are sometimes considered as delicacy in themselves.

Prawn Wafers

Prawn wafer is a product that can be prepared from minced prawn meat mixed with starch and other ingredients in suitable proportions, steamed and then dried after cutting into desired size and shapes. When fried in edible oil the product swells up and become very crisp. It has the characteristic odour of prawns.

CIFT has worked out a method for preparation of prawn flake. Cooked prawn meat (canned prawn meat rejected during pre-shipment inspection for reasons other than spoilage and non-edibility can well be processed into prawn wafers. If any metallic taste is detected in such material, wash it twice in hot water before homogenizing with water. Fish wafers can be prepared by taking in place of prawn similar quantity of cooked fish meat). The sample prepared by the method was found to contain 15-20 per cent protein and was satisfactory as regards, texture, flavour and keeping quality. An outline of the method is given below.

Composition

Cooked prawn meat (cooked in 7 per cent boiling brine for 4-6 minutes)–2 kg, Corn flour –1 kg, Tapioca starch –2 kg, Refined common salt –50 g and water –3.5 litres.

1. Homogenize the cooked prawn meat with 1 litre of water for 10 minutes in a mechanical grinding machine.

2. Add the corn flour, tapioca starch, salt and rest of the water and blend the whole mass for one hour.

3. Spread the homogenized mass uniformly in aluminum tray in a thin layer of 3-4 mm thickness and cook in steam for 10-15 minutes.

4. Cool in room temperature.

5. Cut the layer into desired shapes and dry under sun or preferably in artificial dryer (at 45 to 50°C) to moisture content below 10 per cent.

6. Pack suitable lots of the dried product in sealed polythene bags or glass bottles and store in cool and dry place till marketing.

Permitted food colours can be incorporated, if needed at the time of mixing the other ingredients with the blended prawn to get the desired colour Generally this type of product is used as side dish.

Soup Powder from Trash Fish

The utilization of trash fish (sole, silver bellies, anchoviella, sciaenids etc) for preparation of protein rich food products initiated to develop one such product is good quality of soup powder out of trash fish. Soup powder prepared from different food materials like vegetables, meat and egg are popular in different parts of the world. These dry products are rich in dietary constituents like protein, vitamins, fat and minerals. The soup powder prepared out of trash fish is also a rich source of animal protein and other nutritive factors. The method tried by the CIFT for preparation of the product is mentioned below.

Process

1. Preparation of the material–Wash the fish in good water to remove blood, slime etc. Remove the head, viscera and other waste parts. Wash thoroughly again.

2. Cooking–Cooking the fish in equal quantity of water for half an hour after adding 2 drops of ortho-phosphoric acid for every litre of water, so as to make the pH of water nearly 5.5.

3. Pressing of the cooked meat – Press the fish taken in a canvas bag under the screw press. Remove the big pieces of bones, if any, after pressing.

4. Re-cooking – Disperse the press cake in equal quantity of water after adding ortho-phosphoric acid to the water as above and cook again for 15-20 minutes and press. Repeat the process of cooking and pressing once again. The press cake after third operation can be used for preparation of soup powder. Repeated cooking and pressing remove a large amount of the volatile substances and fat which are responsible for fishy odour.

5. Blending of the press cake – Disperse the press cake in about one and half times its weight of water and blend the material in a blender.

6. Incorporation of fried ingredients and re-blending – Fry the onion and coriander in vegetable fat in quantities as shown in the composition below.

Pour the blend of cooked fish into it and again boil about 10 minutes. Cool to room temperature and again blend in the waring blender.

7. Drying – Pour the whole mass in thin layer in aluminum trays and dry in an artificial dryer at about 70°C.

8. Powdering and incorporation of other ingredients – Powder the dried mass and the other ingredients (starch, salt, skim milk powder, glucose, pepper, ascorbic acid, sodium carboxy methyl cellulose, sodium bisulphate and mono-sodium glutamate) and powder well to get a homogenous product.

9. Packing–Pack the soup powder in air tight polythene lined aluminum foil bags or in cans.

Composition of Ingredients

Fish press cake–100g; Salt–26.6g; Fat–17.7g; Coriander powder–2.2g; Starch (maida) –44.0g; Skim milk powder –17.7g (as dietary supplement); Glucose –8.8g (sweetener); Sodium Carboxy methyl cellulose –0.44g (emulsifying agent); Sodium bisulphite –0.22g (as preservative); Monosodium glutamate –1.32g (flavouring agent).

The product prepared from threadfin bream (*Nemipterus japonicus*) contained 35 per cent protein on wet weight basis.

Preparation for Consumer

The product can be prepared by boiling one part of it in 20 parts of water for 5 minutes.

Beche De Mer

Beche De Mer or *Trepang* is the trade name given to processed sea cucumber or holothurians of the phylum Echinodermata, a phylum also includes the more interesting, better known sea lilies, sea urchins, star fishes and sand dollars.

Trepang is an exotic delicacy used to flavour soups, noodles and other dishes. It is prepared from the body walls of certain species of large holothurians. It is rich in protein (soluble in pepsin and, therefore, highly digestible) and has a low fat content. Beche De Mer is also credited with having curative powers for such ailments as high blood pressure and muscular disorder. The cuvrierian tubules of certain species of beche-de-mer have been traditionally used by fishermen of Cuba and Maldives as a plaster for minor wounds and sprained wrists. Its reputation as an aphrodisiac has also enhanced its popularity.

At the beginning of the fishery, sea cucumbers were picked by hand during low tide from inter-tidal region and from shallow water lagoons of less than 1m depth. As resources in these waters became less abundant, snorkeling and other methods began to be used to exploit the resources in waters up to 20-25 m. A pointed metal spear mounted on a long wooden pole and a fishing hook fixed to a block of lead and attached to a fishing line are popular amongst Maldive fishermen.

When processing is done by the fishermen's own place, women too participate in all activities after the sea cucumbers are de-gutted and cleaned. Smoking of the cooked beche-de-mer is often done in the kitchens. Techniques for cleaning and processing received by the local exporters from the foreign buyers make differences in the processing techniques adopted by fishermen.

Processing of sea cucumbers need only simple equipment and the processing is also very simple, but needs to be carefully carried out if a good quality of product is desired. Proper cleaning and sun drying is required to improve the quality.

Beche-de-mer, a traditional delicacy in countries where ethnic Chinese communities exists, has many nutritional values. The following is the nutritional composition of the dried product.

Protein–43 per cent, Fat–2 per cent, Moisture–27 per cent, Minerals–21 per cent and insoluble ash–7 per cent

Fish Ensilage

The fish waste about 20 per cent of total post harvest lost in Indian fish production can be recycled as fish silage, the liquid fish meal. The fish silage can replace fish meal in the formation of balanced diet for fishes under culture. It can also be incorporated in cattle food.

Fish silage is a liquid product made by the action of enzymes and bacteria in the fish in the presence of an added acid. The enzymes break down fish proteins into smaller soluble units and the acid helps to speed up their activity, while preventing bacterial spoilage.

Fish silage is prepared from low nutritional or low commercial value fish, offal, mainly from head, intestine and bones collected from fish processing or fish dressing factories. Silage can be made from both white fish and fatty fish.

There are two basic methods for silage production. These are (i) acid fermentation and (ii) bacterial fermentation.

Several acids can be used for fermentation of silage, either alone or in combination. Hydrochloric or Sulphuric acid can be used. They are reasonably cheap, but a lower pH is obtained with these mineral acids than from using some organic ones. This means greater corrosion problems and therefore, silage has to be neutralized before use.

Formic acid, an organic acid, a good choice for the fermentation because preservation is achieved at a slightly higher pH. It has some bacteriostatic action. The silage need not be neutralized before adding it to the feed. Formic acid is more expensive than mineral acids.

The principle behind the aforesaid method is fermentation by micro-organisms, that produce lactic acid. Among lactic acid producing bacteria, *Lactobacillus plantarum* is the best suited for production. They ferment the sugars present in the medium into organic acid, predominantly lactic acid, thus lowering the pH. At low pH, growth of putrefying organisms is inhibited by competitive forces and also by the action of certain antibiotics produced in the system. Fish contains only a small quantity of starch or carbohydrate. For this reason they are added in silage production to facilitate microbial fermentation. This is one of the reason why a starter culture containing *Lactobacillus* is invariably used in microbial fermentation, and molasses, tapioca meal and ragi etc are added to increase the fermentation period.

The raw material is first minced suitably by using a hammer mill grinder fitted with a screen containing 10 mm diameter holes, thereby obtaining small particles; immediately after mincing 3.5 per cent of 85 per cent formic acid by weight is added. It is important to mix thoroughly, so that all the fish come into contact with acid, to avoid putrefaction pockets of untreated material. The acidity of the mixture must have pH 4 or lower to prevent bacterial action. After the initial mixing, the silage process starts naturally, but occasional stirring helps to ensure uniformity.

The production tank can be of any size or shape, provided it is acid resistant. Steel containers used for making or carrying the silage may need a polyethylene liner to prevent corrosion. Concrete tanks treated with bitumen are suitable for holding large quantities of silage. The sizes and number of tanks depend on the quantity and type of raw materials available. The rate of liquefaction depends on the type of raw material, its freshness and the temperature of the process. Most species can be used, sharks and rays are rather difficult to be liquefied and they need to be mixed with other species. Fatty fish liquefy more quickly than white fish offal and fresh fish liquefy much more quickly than stale fish. The warmer the mixture, the faster the process. Silsage made from fresh white fish offal takes about 2 days to liquefy at 20°C, but takes 5-10 days at 10°C and also much longer at lower temperature. Once the silage is prepared, it can be handled like any other liquid, and transported in bulk or in containers. It can also be blended with cereals to make semi-dried feed.

Silage made from fatty fish is more homogeneous and there is little seperation even after prolonged storage. However, oil in it deteriorates more rapidly. If the oil has to be removed and used for other purposes, it can be separated by heating and centrifuging. Fish silage can be concentrated to reduce its bulk.

The production process involves, (1) raw material should be as fresh as possible. This may include whole fish, filleting wastes, offal or other suitable protein material. The fish are comminuted by mincing, cutting or chopping. This operation can either be manual or mechanical. The particle size should be reduced to 3-4 mm. The minced meat is treated with acid and thoroughly mixed. The mixture is constantly stirred until the desired pH is reached. The container is left preferably covered for the mixture to liquefy. This can take 3 to 4 days. In the case of fatty fishes, removal of oil can be achieved by raising the temperature of the silage to 65-75°C. The coarse suspended solids are removed by centrifugation to remove the oil.

The average composition of fish silage are moisture (70-81 per cent), crude protein (15-17 per cent), fat (0.5-13 per cent) and ash (2-4 per cent). Whole fatty fish like sprats and sand eels have a higher protein and fat content and correspondingly lower water and ash content.

The silage concentrate is a highly digested protein hydrolyzed, which is convenient as a protein supply for weaning calves and pigs as well as poultry.

Fish silage of correct acidity keeps at room temperature for at least two years without putrefaction.

Fish silage is used in the same way as fish meal in animal feed.

The main advantage of the fish silage process is that, in areas, where there is no fish meal factory, fish offal and waste fish can be utilized for producing fish silage instead of throwing them. The capital cost of fish meal plant is fairly high and requiring engineers and technical staff, where as the cost of silage equipment is fairly low and can be made by unskilled workers. Smell is a problem when making fish meal, but there is no smell while making silage. Transport of fish meal is cheaper, but silage is more expensive to carry because it is liquid and four to five times as bulky as meal. Marketing of fish meal is long established and the product is well known.

Fermented Fish Products

Hydrolyzed protein prepared by acid hydrolysis of cheaper protein from vegetable sources, such as, wheat proteins, which are rich in glutamic acid, are used as flavouring ingredients in soup mixes, meat and other food products in the western countries, chiefly in the United States. Food and pharmaceutical grades of fish albumin are obtained by alkali hydrolysis of fish meat for use as egg albumin substitute. In the pharmaceutical industry, proteolysates are prepared for oral administration for therapy in peptic ulcer and other disorders, in the nutritional therapy for convalescence, and also for intravenous and parenteral administration. Proteolysed liver preparations are also made. Peptones obtained by controlled hydrolysis of meat, casein or other proteins are common constituents of bacteriological culture media in laboratory and could be for industrial bacterial culture as well. Protein hydrolysate has been used or proposed for use as adhesives and binders and in foaming formulations.

Choice of Proteins for Protein Hydrolysates

The choice depends on the use for which the hydrolysates are intended, but the followings are the general consideration.

(i) Cheapness and the availability of the raw material. Casein, beef protein, whale meat, blood fibrin are being widely used. Livers from abbatoirs are used for obtaining specialized pharmaceutical preparations rich in Vitamin B 12 and iron.

(ii) Availability of proteins in the raw material in sufficiently high concentrations. Proteins form the major component in meat and fish. It generally does not need processing to separate proteins from non-nitrogenous components preparatory to hydrolysis with such raw materials. Vegetable protein sources have to be treated for obtaining protein concentrates or isolates before they are taken for hydrolysis.\

(iii) Balanced proportion of amino acids. In this respect, the animal and fish proteins are far better balanced than the vegetable proteins, especially in essential amino acids. Fish protein is rich in lysine.

Autolysis

Proteolytic enzymes inherent in fish are distributed in its various parts, potency being variable with the part. The proteases of the pyloric caeca, the intestinal mucosa and the stomach are significantly potent in autolysis of whole fish, while in the absence of these enzymes in the eviscerated fish, cathepsins assume importance. The optimum temperature for autolysis is in the region of 30-40°C, the pH optima being in the acidic range for flesh and chiefly on the alkaline side for viscera.

Processes which utilize autolytic degradation of fish generally depend upon some means of preventing undesirable microbiological (putrefactive) spoilage. In ensilage, sulphuric, formic or organic acids produced by fermentation of carbohydrate substrate serve to lower the pH to the acidic side, while in traditional fish sauce production in South East Asia, common salt serves as a preservative.

Fermented Fish Products

Fermented of various types are indigenous to most of the countries of South East Asia forming important adjuncts to predominantly rice diet.

(a) Fermented fish - Whole fish parts, chunks or organs of fish are salted and allowed to ferment for periods varying from a few days to more than a year with or with out adjuncts, such as roasted rice powder, yeast, pineapple, papaya etc.

(b) Fermented fish paste – Fish (including crustaceans, mollusks etc) are salted and repeatedly dried and ground into fine paste and allowed to ferment for periods up to several months until the desired flavour has developed. Various adjuncts may be added.

(c) Fermented fish sauce –Fish salted and allowed to undergo autolysis for several months until all tissue is broken and a liquid sauce consisting of protein hydrolysate is formed which is clarified and also aged in some cases.

The products are many and varied, individual methods of preparation differing with the region. In the main, fish utilization is effected by control of putrefactive spoilage by salt or other adjuncts and the products appear to be safe for consumption, being free from pathogens. The products are valued in the areas where they are produced for their characteristic flavours. They would, however, be new and foreign elsewhere and may not necessarily be acceptable.

The most typical of the fish sauces of the region are the *Nuoc mam* of Vietnam and *Nam Pla* of Thailand. Their production is characterized by extensive hydrolysis of fish proteins.

Nuoc mam is prepared mainly from small marine fishes of the clupeid and carangid species. The fresh uncleaned fish are mixed with salt and packed in vats. The amount salt used varies with the kind of fish, 4 to 5 parts of salt being usually used for every 6 parts of fish. After 3 days, the self brine is drained off through the tap at the bottom of the vat. The fish which settle down are packed thoroughly and submerged in part of the self brine which is returned, the contents are covered and heavily weighted down. Aging is now allowed to take place which may be from 6 to 12 months depending on the fish. After maturing the pickle is run off as the finished sauce and the un-dissolved residue is sold as manure. *Nuoc mam*, is a clear brown liquid rich in salt and soluble nitrogen compounds. The fish sauce is used as a condiments and flavouring agents and as substitutes for solid salt

Nuocmam prepared on a semi-commercial scale under aseptic conditions was not acceptable to its regular users, even though extensive hydrolysis had taken place. Although the enzymes in the fish may be sufficient to autolyse the fish, the agency of the micro-organisms was equally necessary for producing the typical sauce flavour.

Bacto-peptones

Fish protein hydrolysates have been prepared, tested and found suitable for use in media for bacterial culture.

Tarr and Deas prepared bacto-peptones from some marine fish with (i) tryptic and peptic enzymes derived from fish digestive tracts, (ii) hydrochloric acid digestion and (iii) caustic alkali digestion. Only the enzymatic hydrolysate consistently supported the growth of six types of *Streptococcus haemolyticus* and one *Colstridium botulinum* type E culture.

Proteolysed Liver Preparation

Guttman prepared aqueous extracts and enzymatic hydrolysates from fresh livers of salmon, cod, haddock and some mammals. The extracts and hydrolysates were concentrated to syrupy consistency under vacuum and finally dried to a powder at low temperature under vacuum. All the preparations were good sources of B Vitamins including B12.

Rajagopalan and Sarma prepared papain and pancreatin digests of shark liver oil residues. These were adjudged suitable alternative to mammalian Vitamin and mineral composition.

Fish Albumin

Minced fish muscle is heated with stirring at 160 to 175°F for one hour in an aqueous solution containing 0.5 per cent acetic acid. The proteins get partially hydrolyzed and connective tissues are extracted. This is washed in cold water to remove digested proteins and some of the acid. The residue is pressed to not more than 40 per cent water content, the press cake is broken up and is extracted free of fat by ethyl alcohol or trichloro ethylene, and dried under vacuum with the material temperature about 122°F. The product is insoluble technical grade fish albumin. This has industrial uses in paints, varnishes, textiles, papers and other manufactures.

Food and pharmaceutical grades of fish albumin are produced from the technical grade albumin by mild alkali digestion. 100 kg dry protein is taken with 500 litres of water to which 6 to 8 kg of caustic soda is added. Digestion is for one hour at 86°F and at 176 to 194°F next hour. Viscosity tests show the completion of digestion. The digest is neutralized with acetic or lactic acid and the fluid is spray dried under controlled conditions. The edible grade fish albumin consists mostly of polypeptides with virtually no free amino acids. Its use is as egg white substitutes on account of its foaming properties.

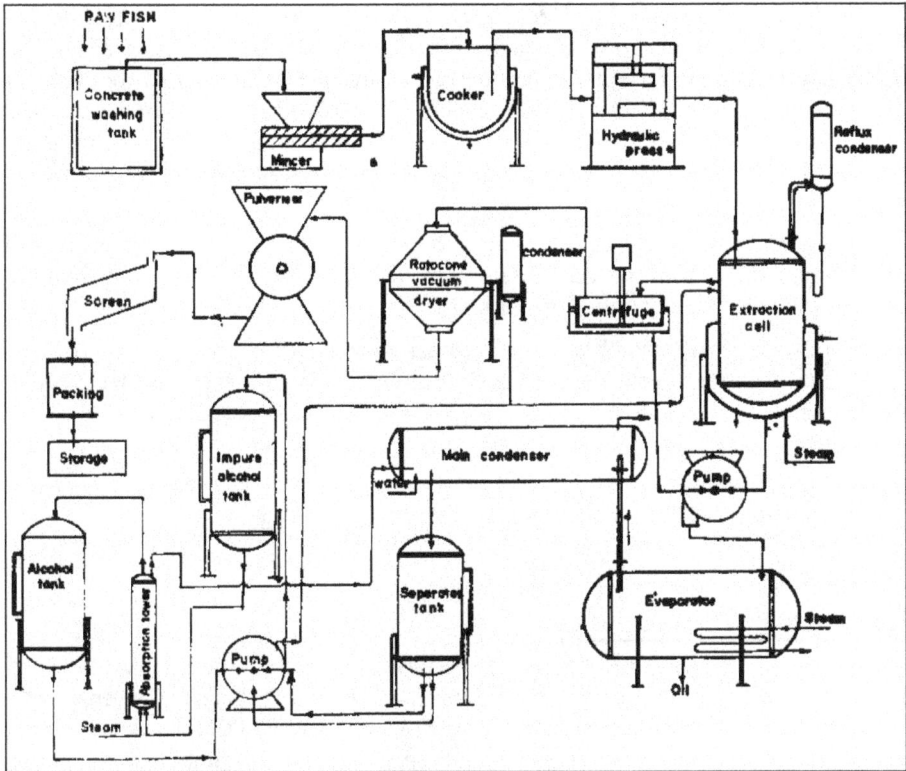

Flow Sheet for FPC Plant

Alkali Solubilisation of FPC

A procedure similar to conversion of technical grade fish albumin to food grade albumin has been tried for solubilizing solvent extracted fish protein concentrate. Ten parts of FPC were solubilized in 100 parts of 0.2 N NaOH at 95°C for 20 minutes. When the solubilised FPC was used in protein beverage products, the beverage had an off flavour. This could be eliminated by acidification of the solubilized FPC, centrifuging filtration of supernatant through charcoal and finally recombination of the centrifuged residue and the filtered fraction.

Enzymatic Hydrolysis

Whole fish or fish fillets are minced, mixed with the minimum quantity of water adjust to pH 5 and brought to 60-65°C. Papain is added as a dispersion in water. The quantity of enzyme will depend on its activity. The slurry is maintained at the same temperature for the duration of the hydrolysis (which depend on the degree of hydrolysis desired) after which it is quickly brought to boiling and boiled for 15 minutes. There after it is decanted followed by centrifugation in a basket type centrifuge to remove bones and larger pieces of undigested residue. It is then clarified in a Westphalia clarifier to remove the major portion of suspended material and pressure filtered with hyflosupercel (0.5 per cent) to get a sparkling clar hydrolysate. The hydrolysate is vacuum concentrated in a forced circulation evaporator at 25 Hg and 3 psig steam pressure from a solid content of about 8 per cent to one of about 30 per cent with Dow Corning Silicone Antifoam C added to control excessive frothing. The concentrate is taken inpolythene lined metal trays and vacuum dried in a shelf drier at 27.5 in Hg and 45-60°C. The dried material, a cream yellow powder is stored in air tight bolltles.

The trials on spray drying of a fish protein hydrolysate from cooked, processed oil sardine have been reported. The final product was free flowing, light cream in colour and had the appearance of skim milk powder.

The hydrolysates, so obtained have been proposed for use in formulating (i) industrial and laboratory bacteriological culture media, (ii) pre-digested protein preparations for convalescence and (iii) food beverages for nutritional use.

The nutritive value of fish protein hydrolysates have been reported to be not significantly different from the protein of skim milk powder. It was observed that tryptophan tended to be the limiting amino acid.

Fish Protein Concentrates

By Biological Procedures

The most extensive and concerted studies yet on enzymatic hydrolysis of fish protein have been carried out by U.S. National Marine Fisheries Service. The enzymatic processes have been designated as "biological procedures for FPC". The development of such processes have been favoured because, (i) the end product properties, such as, solubility make it more suitable for certain applications than solvent extracted fish protein concentrate, (ii) the process may be cheaper than solvent extraction and more suited for small scale operation.

Among the more active enzymes as determined by a 24-hour hydrolysis, pancreatin, pepsin and papain were found to be relatively economical.

The studies with the whole hake, it was found that, native proteolytic enzymes play a major role in hydrolysis and also that solubilisation of the pre-cooked fish could be achieved only with excessive amount of commercial enzymes. Raw whole fish was therefore, the chosen substrate for enzymatic hydrolysis. Yields of dry solids from red hake solubilized by different proteolytic enzymes ranged from 10 per cent for autolysis alone to about 14 per cent with alkaline protesses or pancreatin

From amino acid assays for hydrolysates obtained from the wide range of enzymes used, tryptophan under acid conditions, histidine at slightly alkaline conditions and isoleucine with alkaline proteases at higher pH tended to become the limiting amino acids.

Protein efficiency ratio determination revealed that the products from red hake were inferior to casein when used as the sole source of protein but equivalent to casein when used as supplement to wheat flour.

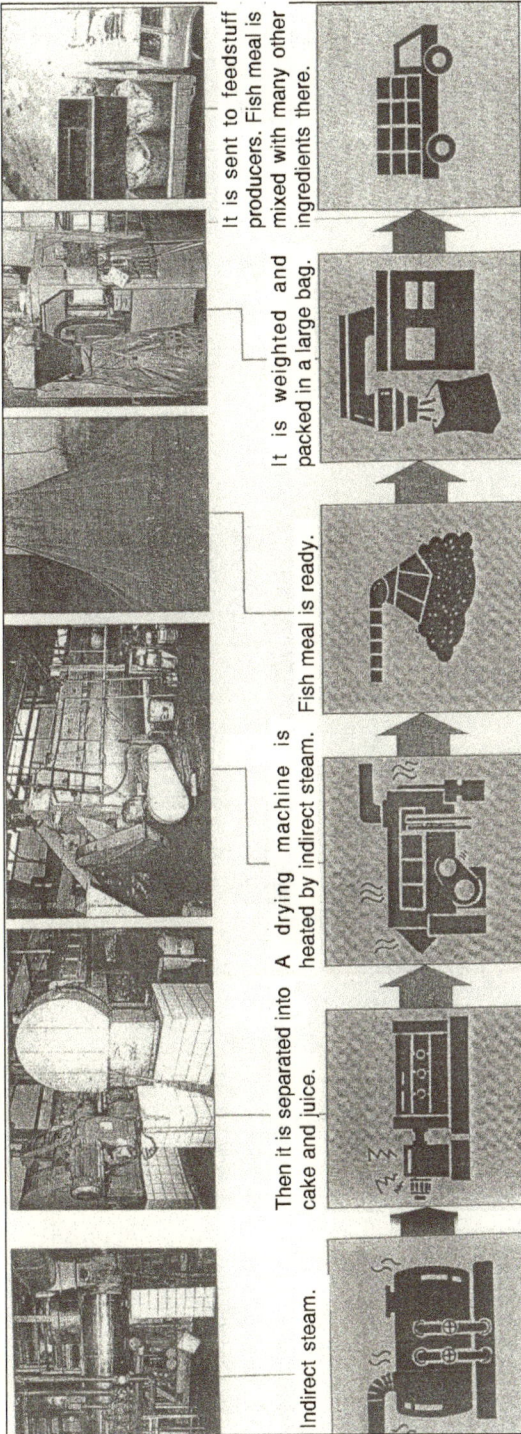

Fish Oil and Fish Meal Plant

Juice is transferred to a desludger where sludge is removed to make clear juice for further evaporation.

Juice is centrifuged to separate the oil, juice and precipitates. They are transferred to further processing systems respectively.

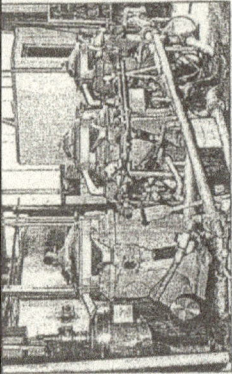

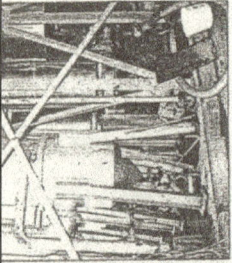

Juice is evaporated in vacuum. This makes fish soluble to be used as feedstuff.

Fish oil is refined and sold to an oil factory where it is processed into margerine.

Processing capacity: 100 tons per day

No. of employes: 14

Site are: 1,620 m^2

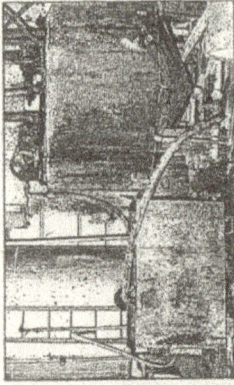

Fish Meal

In the suggested process outline for the raw fatty fish, raw whole fish are converted to a slurry with the addition of equal weight of water during communition. The chosen enzyme is an alkaline protease of *Bacillus subtiis* added at 0.1 per cent by weight of fish. The slurry is adjusted to pH 8.5 by adding calcium hydroxide. The pH is maintained at 8.5 during hydrolysis by adding 5 N NaOH solution. The temperature is maintained at 55°C for 5 hours. The hydrolysed slurry is neutralized to a pH of 6.5 with sulphuric acid.

The undigested residue is dried as bone meal and the oil is also recovered. The dissolved solids are concentrated to a solids content of about 50 per cent and spray dried.

It is considered uneconomical to obtain a totally soluble produce by using press cake instead of raw whole fish as the native proteolytic enzymes get inactivated during cooking and very high levels of commercial need to get the desirable results. A partially soluble end product has therefore been suggested if press cake is to be utilized.

Fish Protein Concentrate from Oil Sardine

Sardine press cake can profitably be utilized for human consumption after converting it to fish protein concentrate. The solvent of odour bearing compounds and residual fat from the sardine press cake prepared under sanitary conditions can result in a more concentrated light coloured flour containing more than80 per cent protein of almost blunt taste. The product in addition to being nutritious can be incorporated into a variety of foods, like breads, chapathis, soups, flakes etc without imparting any fishy flavour.

Fish Meal

During the glut seasons, the whole fish, as such, were used as manure. The most common form of fertilizer is as fish guano, the residue left after oil extraction by the crude method. This is generally beach dried and is found to contain 8-10 per cent nitrogen and a good amount of phosphates. Being well cooked this disintegrates easily and mixes with soil quickly and is believed to be 15-20 times richer than ordinary cattle manure. The trimmings and wastes from the sardine canning factories, when this is taken up on a large scale, will constitute a good source of manure.

The press cake left after extraction of oil from oil sardine forms one of the chief sources of raw material for the manufacture of fish meal. Lean fish and wastes from fish canneries also are important raw materials. The valuable protein source with all the essential amino acids and minerals like, phosphorus and calcium can be made use of as the main ingredient in the feeding stuff of domestic animals and poultry.

Chapter 16

Plants and Machineries for Fish Processing Industry

Fish Protein Concentrate Industry

Fish protein concentrate, FPC in short is a highly purified protein, generally bland and odourless, derived from fish by solvent extraction or other means. The product may eventually take number of forms, all of which will be comparatively inexpensive and will show long term stability under all climatic conditions. It will be highly nutritious and suitable for diverse applications as ingredient in staple diets and manufactured foods. To the fishing industry, the advent of FPC promises to open up a steady and profitable outlet for vast unutilized resources.

An FPC plant differs from conventional fish processing facilities in that it represents a continuous sophisticated operation, more akin to a biochemical plant with a high standard of sanitation. It calls for specialized knowledge in food technology and for new approaches in respect of fish harvesting. The investment is relatively high, but so are the potential rewards in terms of diversification of fishing activities, of product values and markets.

FPC production generally lend itself to a high degree of mechanization and automation, with little manual attention. The operation of plant will be governed largely by quality control to maintain the stringent demands of Food and Drug regulations and of trade specifications. For this purpose it will be necessary to develop and apply suitable techniques of recording and controlling analytical data, such as, bacteria counts, amino acid compositions, residual lipids and solvent content, trace elements etc. to permit rapid response to changes in raw material characteristics and rapid detection of any malfunctions.

More and more developing countries are taking an interest in producing and distributing FPC to combat malnutrition. The protein requirements of food manufacturers indicate promising applications of FPC as long as there is no compromise in respect of product quality.

Food and Drug approval to permit utilization of all suitable major resources and species(covering three entire families of species, namely cod, herring and smelt) that are under-utilized or not exploited at all are necessary. One approach to large scale fish farming for FPC production has been piloted in the Vergin Island, involves the artificial upwelling of nutrient rich water. In this manner, a high yield of plankton may be induced, which may well sustain anchoveta and other useful species.

Isopropanol extraction process in some form or other seems to have come to stay, in which case, the means of storing the fish in the solvent aboard vessel have to be thought of. The raw material can be preserved the moment it leaves the sea, transported in tanks and pumped without detriment. Even pre-treatment at sea can be made. The partial removal of bone, moisture and oil could be carried out on board, leading to an intermediate product that can be stored without spoilage for long periods and that can be shipped long distances to centrally located extraction plants.

Such pre-treatment facilities might be installed in a large fishing vessel, or on a factory ship served by a number of catchers. There is, in fact, no technical reason why FPC should not be produced in its final form at sea.

FPC production will evolve gradually, but surely, along with special products for animal feed. Such products have properties, which fish meal does not offer, such as, solubility and low lipid content. The reason is that the growth of FPC production will depend on progress in developing suitable fishing techniques, on the formulation of products and their acceptance by consumers, and on the continuing evolution of suitable manufacturing processes. These developments will take time and depend on one another.

The most advanced process is based on extraction with isopropyl alcohol, can readily be applied to commercial production in several alternative forms. A number of other exciting processes have also developed. Of particular interest are those which give at least some degree of heat coagulation ability and solubility, and which promise substantial reductions in production cost. Procedures based on enzymatic and chemical hydrolysis to produce soluble fish protein concentrates for food and feed applications have been developed. Such products can be expected to find their way into various kinds of beverages.

Heat treatment as well as alcoholic solvents tend to denature proteins, with the result that they lose their coagulating and water binding properties. Instead of extracting the unstable fats, there fore, they may be stabilized by powerful anti-oxidants or even utilized for conversion by lipolytic micro-organisms into additional protein.

To achieve full utilization of protein contained in the raw fish, more than one protein concentrate may be produced in the same plant. For example, the isopropanol extraction process leads to an insoluble FPC, which can be solubilized by subsequent hydrolysis. From the by-products of the same process at least feed grade of soluble protein may be recovered.

It is of utmost importance that an FPC plant be supported by an effective fishing and fish handling operation, the task of which will be to ensure that the plant receives acceptable species of fresh fish on a realistic schedule in accord with the capacity and process facilities of the plant, since the FPC plants will not lend themselves very well to intermittent operation as is often the practice in the fish processing industry. The magnitude of the investment involved, the nature of the unit operations, the operating and personnel requirements, product control and sanitation measures all demand fully continuous operation, preferably on a year round basis, interrupted only by regular cleaning routines.

Fish Protein Concentrate (FPC) Plant

Earlier attempts made on the production of FPC were based on the extraction of fish meal by solvents. Moroccan process involved the extraction of fish meal with a solvent mixture of hexane, ethyl acetate and iso-propanol. Chilean process involved the extraction of meal with 95 per cent ethanol, whereas in the Peruvian process fish meal was extracted with hexane. All these processes could produce FPC of Type B quality, which was not completely free from fishy odour.

Next attempts were made on the development of suitable process for the extraction of fat and odour using wet material to produce Type A quality FPC. In Astra process iso-propanlo was used, whereas in Viobin process raw fish was first extracted with ethylene dichloride and finally extracted with iso-propanol. Halifax process of Fisheries Research Board of Canada used iso-propanol at elevated temperature for the extraction of fat from raw fish. Of all these methods the process developed by U.S. Bureau of Commercial Fisheries (1966) gained considerable popularity. This method, a modification of Halifax process consisted of the extraction of wet whole fish using isopropanol to produce Type A FPC. This process ultimately gained commercial acceptance due to the support of FAO and FDA of USA.

In India, Ismail Madhavan and Pillai (1968) developed a process for the production of Type A FPC from locally available fish by using an azeotropic mixture of hexane and ethyl alcohol. The process is much simpler and efficient compared to other processes described above for economic production of FPC due to the following reasons.

1. The process uses a non-toxic solvent mixture which is easy to recover by simple distillation and does not require any costly equipment like rectification column as in BCF process.

2. Hexane alone is not efficient in deodourizing fish muscle, but is very efficient in removing fat, and alcohol alone is inefficient in removing fat, but it is the most efficient solvent in removing odouriferous compounds and water from press cake. So, the azeotropic mixture is very effective in performing both these functions.

3. Both these solvents are easily available and comparatively cheaper in India than iso-propanol. Moreover, both these solvents are inert to the material of construction.

Considering all these advantages, their process has been adopted for designing a pilot plant to study the economics and other technical aspects of production of FPC on commercial scale.

Equipments and their Function

FPC plant comprises of the following equipments

(a) Concrete Wash Tank (300 litres capacity)

Fish from the boat isreceived in this tank to remove sand, dirt and other extraneous matter using potable water. Preliminary hand sorting to remove fatty and poisonous fish is also done here. The tank is made up of concrete (1:2:4) and is provided with water inlet, false bottom and 50 mm diameter outlet vaive.

(b) Meat Mincer (100 kg/hour capacity)

The purpose of mincing is to reduce the size of fish muscle and thereby help in cooking process, mixing different sizes and types of fishes, head and tail portions, and thereby rendering the mass homogeneous. The mincer which can reduce fish to 3 to 6 mm size particles, is fitted with helical screw, reduction gear motor, reversing switch, cutter, perforated disc etc.

(c) Cooker (100 litres capacity)

Minced meat is taken in a hemispherical steam jacketed open type stainless steel Kettle with tilting arrangement for cooking the mass at proper pH. The purpose of cooking is to render the mass soft and release water and fat. The vessel is fitted with steam fittings, like steam joints, safety valve, pressure gauge and steam connections

(d) Hydraulic Press (15 kg/sq.cm capacity and 0.5 sq m base area)

Cooked mass is taken in a canvas bag and pressed under hydraulic press to remove substantial portion of water and fat. This amounts to a reduction in weight by 50 per cent. The removal of water is essential as more water in cake, the less efficient is the extraction. The press has the arrangement to adjust pressure on the cake, collection of press liquor, hydraulic pump motor set, press proper, ram, pressure gauge and basket capacity of 0.5 sq m×0.5m.

(e) Extraction Cell (450 litres capacity)

It is a stainless steel steam jacketed reaction vessel fitted with temperature controller, solenoid valve, stirring and refluxing arrangement. The vessel is fitted with anchor type agitator assembly, removable false bottom, wide charging and discharging doors, standard steam fittings on the steam jacket and dial thermometer on the vessel proper.

The function of extraction cell is to extract oil and soluble odour bearing compounds from the press cake, by a solvent mixture of hexane and ethyl alcohol at its azeotropic boiling point and the evolved solvent at the time of extraction is condensed back by reflux condenser fitted to the cell. At the end of extraction, solvent is drained out from the bottom and any adhering solvent is distilled off by blowing open steam from the bottom of the cell and the evolved vapor is condensed in the reflux condenser. So a provision is made for steam connection at the bottom of the cell. The vessel is fitted with all standard accessories, flame and explosion proof drive system and designed for 5kg/sq.cm test pressure. In the pilot design one extraction cell is used with a provision for two more in future.

(f) Evaporator (500 litres capacity, heating surface 1.5 sq.m)

Mild steel horizontal type evaporator with removable hair pin type steam coil fitted with solvent inlet and outlet connections, vapor column, pressure gauge, dial thermometer, steam fittings etc has been designed for recovering solvent obtained from extraction cell. Solvent mixtures from extraction cell is charged into the evaporator and steam is passed into the coil when solvent vapor at azeotropic composition is evolved which goes to condenser and residual oil and solid matter is removed at the bottom. As the solvent is non-corrosive, the material of construction is mild steel and the vessel has been designed for a test pressure of 5 kg/sq.cm.

(g) Main Condenser (Cooling surface 7.0 sq.m)

This is a shell and tube type horizontal condenser with copper tube and mild steel shell designed for condensing solvent vapor evolved from evaporator by passing cooling water inside the tube. The condenser is connected to the evaporator by a 125 mm diameter vapor column. The tube plate is fixed and tubes are fitted on 25 mm triangular pitch. There are cooling water connections, vent cock and vapor seal U-tube for condensate outlet.

Another condenser, called reflux condenser, is fitted to the extraction cell. This ia vertical shell and tube type condenser with copper tube and mild steel shell and cooling medium inside the tube. It has got 1.4 sq.m cooling surface, cooling water connections, vapor pipe and vapor seal U-tube. The design pressure for shell and tube side for both the condensers is 3.5 kg/sq.cm.

(h) Separater Tank (350 litres capacity)

This is an ordinary mild steel tank fitted with level gauge to collect condensate containing two solvents and allowed to separate in two layers. The gauge glass helps in viewing the separation mark so that the bottom layer is drained out (containing alcohol) into impure alcohol tank and hexane remains in the vessel. Exact quantity of hexane can also be drained to the extractor from this vessel. This cylindrical steel vessel is designed for a test pressure of 3 kg/sq.cm.

(i) Impure and Pure Alcohol Tanks (280 litres capacity each)

These are identical vessels as described above except the size and height at which they are fitted. The alcohol obtained from the separator is impure in the sense that it contains odour bearing compounds and as such, can not be used for extraction purpose without further purification. It has to be stored separately. The solvent after passing through absorption tower gets de-odourized and this pure alcohol can be reused again for extraction. Pure recovered alcohol is stored in the pure alcohol tank from where it is drained to the extraction tank. Both these tanks are fitted with gauge glass, other standard fittings and designed for the pressure of 3 kg/sq.cm.

(j) Absorption Tower (100 litres volume)

It is a tall cylindrical tower, filled with active carbon having a size of 1520 mm tall and 307 mm in diameter. The tower is built up three packed beds of carbon which are fixed on flanges and the tower height can be altered by changing the number of beds. Carbon on each bed is supported on fixed support rings, removable perforated plates screen and glass wool filter. There are two pipe connections at the bottom, three pipe connections at the top and the shell is designed for 3 kg/sq.cm pressure.

The impure alcohol passed from the bottom of the tower gets purified (free from odour) by the process which is collected from the top in the pure alcohol tank. After few days operation the carbon gets deactivated and its power of removing odour is lost. At this stage the tower requires revivification, which can be done by passing steam through the tower from the bottom. By this process of steam stripping, all solvent and odour is removed from the tower packing and the tower gets reactivated again. But after few months the tower become dead. In that case the old packing is removed and new packing is added. Depending on the presence of the amount of odour bearing compounds, the height of the tower can be changed to get the desired purification.

(k) Solvent Pump (centrifugal type of 25-30 litres capacity per minute)

Two CI pumps fitted with flame proof motor are used for pumping impure alcohol from separator tank to pure alcohol tank via absorption tower and measured quantities of two solvents from separator and pure alcohol tank extractor. Other pump is used for pumping spent solvent-oil mixture to evaporator.

(l) Basket Type Centrifuge (basket size 570 mm diameter and 280 mm height)

Heavy duty top suspended and bottom discharge centrifuge with perforated stainless steel basket lined with metallic filter cloth and fitted with flame and explosion proof three phase motor having 6000-9000 rpm has been used to remove adhearing liquid from the extracted mass. The solvent is taken to evaporator and solid to rotocone vacuum drier. It has got arrangement for slurry feeding and washing and outer stationary basket for the collection of liquid.

(m) Rotocone Vacuum Drier (25-30 kg/hour capacity)

Double cone rotary vacuum drier fitted with variable flame and explosion proof motor is used for removing last traces of solvent from the extracted material from centrifuge. It has been observed that the traces of solvent from the FPC can be removed only under vacuum and exposing new surfaces constantly by agitation. This drier is also used for mixing FPC with other starchy material. The main body is of stainless steel with outer mild steel steam jacket heating arrangement. The drier is provided with charging and discharging doors, rotary pressure and vacuum joints, steam fittings, vacuum gauge, dial thermometer, condenser unit and structure, vacuum pump and other pipe connections for water and steam. The vessel is designed for 5 kg/sq.cm test pressure.

(n) Pulverizer (50 kg/hour capacity)

A beaten type pulverizer is used for powdering dry FPC to 100 mesh particle size at temperature less than 60°C. The machine is complete with three phase motor, body and collector of powder.

(o) Shaking Sieve (50 kg/hour capacity)

Two deck stainless steel continuous gyratory type shaking screen, fitted with charging arrangement, motor and sieve cleaning brush is used for sieving powder obtained from pulverizer to screen to 80 and 100 mesh sizes and separating the over size for regrinding.

The dry FPC so produced is packed in suitable container manually.

The design has been prepared after thorough study of technical aspect of different processes developed so far for the production of FPC of acceptable quality at minimum cost under Indian condition. The first consideration is the choice of solvent which is cheap, non-toxic and available in India and easy to recover. The azeotropic mixture of hexane and alcohol from extractor can be recovered by simple distillation, which on cooling separate into two layers, thus avoiding costly rectification as required in BCF process (1966). The process adopted for pilot plant is different from BCF type types of process (extraction of raw meat) and Chilean types of process (extraction of fish meal). The first extraction stage is inefficient in BCF process, because of non-release of fat from raw meat, whereas there is no special advantage of using fish meal. Moreover, cooked pressed cake extraction gives a better quality product by avoiding oxidation of fat. Extraction is more efficient than BCF process and avoidance of one important step of drying make it a cheaper process.

The most difficult technological problem in the preparation of FPC is the complete removal of residual solvent apart from the removal of fat and odour. In this process hexane component efficiently removes fat up to 0.5 per cent and water and odour is efficiently removed by alcohol. At the end, a wash with alcohol of the extracted mass can remove the residual hexane considerably and the rest is almost completely removed by steam stripping. Finally the mass is dried under vacuum in a constantly agitated rotocone vacuum drier at steam temperature where practically no hexane can remain. So the presence of residual solvent if any may be the alcohol (non-toxic) and practically zero per cent hexane, the harmful effect of which has not been established so far. For purification of ethyl alcohol an active carbon tower has been suggested.

Fish Meal Manufacture in India

The seventy per cent of trash fish like, caranx, silverbellies, sole, jew fish, ribbon fish come along with trawl catch have only very limited demand from the processors of fishery products. If properly processed, this valuable protein material can be used as a source of animal protein not only for man, but also for farm animals and poultry.

During the peak season of oil sardine fishery, the catches become so large that this fish is sold at cheap rate and used as manure directly without recovering the valuable body oil. The extraction of oil in the pure form by improved methods for utilizing the same for preparing useful industrial products and conversion of guano into valuable fish meal are the major needs of this vital industry. The composition of trash fish available varies largely from season to season and from place to place. These lean fishes and also the waste, such as, head, offals, trimmings etc from fish cannaries are the main source of raw material for the production of fish meal.

The special properties possessed by the fish meal is summarized as follows;

1. It contains high quality protein with a favourable distribution of all essential amino acids with high nutritive availability
2. It contains B complex vitamin
3. It contains liberal amount of phosphorus, calcium and other essential trace elements.
4. Fish meal in correctly formulated ratios helps keep hens laying and is believed to assist in maintaining the fertility to hatchery eggs.

Traditional Method of Fish Meal Production

Fish meal production is not a recent introduction in our country. India had been exporting sizeable quantity (about 6000 tons) annually of fish meal. But in recent years there has been a considerable fall in the export of this commodity. This is mainly because of fish meal exported was being manufactured in a crude manner. Fish landed all along the sea coast by the individual fisherman from the country craft as well as from the mechanized fishing boats. But due to the difficulties in transport and preservation, much of the catch is sun dried at or near the landing centers and ground to get the meal. With its several disadvantages, the method of production generally gives product of inferior quality, low biological value containing sand and other extraneous matter to a considerable extent, sometimes contaminated with pathogenic bacteria and comparatively high moisture content. The export can be re-established only if a fish meal free from *Salmonella* and other pathogenic bacteria as required by the importing countries and having the required protein content is manufactured.

The specification of fish meal laid down by Government of India for export is;

Protein – 45 per cent (minimum), Moisture – 12 per cent (maximum), Insolubles – 7 per cent, Fat – 7 per cent, Salt – 7 per cent and Size – 0.25 mm (particle size).

Modern Methods

Fish can be reduced in two ways (i) by wet reduction process and (ii) dry reduction process. In the later process, the fish is dried below the moisture level of 10 per cent and pulverized. The process, however, is suitable for lean fish, that is, with low oil contents. For oily fish, the fish has to be cooked and pressed so that some of the oil is extracted along with the press liquid. In the wet reduction process fresh miscellaneous fish or other fish wastes are cooked in an open pan, either steam jacketed or having other arrangements for heating with sufficient quantity of water for 30 minutes. The cooked fish is then pressed in a screw press to remove the stick water. The moisture content of the cooked and pressed material will be reduced to nearly 50 per cent from the original content of about 70 per cent. The pressed fish cake is then dried in the sun under hygienic conditions or in a rotary drier till the moisture content is reduced to about 10 per cent. The material is then pulverized in a hammer mill to required particle size and sieved to remove any scales, fibers etc and packed in polythene lined gunny bags.

It has been observed that the fish meal prepared from fish by the wet reduction process is superior in several respects to that prepared from sun dried fish.

Working of a Modern Fish Meal Plant

1. Hashing machine – For preliminary chopping of raw material for fish of more than 25 cm in length.

2. Continuous indirect cooker with strainer – Here fish is cooked indirectly by steam in order to denature the flesh protein of raw fish meat, that is, for making easier separation of fish into its main constituents- water, protein and bone and oil.

3. Continuous screw press – The cooked fish is subjected to pressure and thereby a separation of water, protein and bone and oil takes place. The separated press water with oil is pumped to the stick water tank and the press cake is led to the tearing machine.

4. Vibration strainer – The mixture of oil and fish stick water separated off in the screw press is passed through the vibration strainer and thereby a great part of the solid particles are removed and the same are added to the press cake in the conveyor to the drier.

5. Press water tank (1500 litres capacity) – The stick water and oil from the vibration strainer is pumped here for accumulation before it is sent to the centrifugal separator for separating the oil.

6. Centrifugal separator – The oil and the stick water present in the press water is separated by the centrifugal force generated by a rotating bowl at 6000 rpm. The nominal flow rate (capacity) of this centrifuge is 600 litres per hour.

7. Pressure concentrator – The fish stick water is concentrated for incorporating the same in the finished fish meal to increase the output of the fish meal by about 20 per cent and also to improve the quality of the meal.

8. Tearing machine – It is used for tearing the press cake coming from the screw press before the same is taken to the drier.

9. Continuous direct heated drier – The drier is a "Hot air type" in which the product passes through rotating cylinder so that it tumbles through the hot gas. The press cake particles on being dried in the current of hot gas, lose weight and are sucked by blower and conveyed to the milling plant.

10. Magnetic separator and milling plant – Here the scrap of iron in the fish meal is separated by a magnet before the scrap is ground into powder by hammer mill. There after the meal pneumatically conveyed to the bagging machine by the pneumatic conveyor attached to the meal.

11. Automatic weighing and bagging off machine – For bagging off fish meal.

Composition of fish meal

	Grade A	Grade B	Grade C	Grade D
Protein	60 per cent	50-59 per cent	40-49 per cent	Below 40 per cent
Fat (maximum)	10 per cent	10 per cent	10 per cent	10 per cent
Acid in-solubles (max)	10 per cent	10 per cent	10 per cent	10 per cent
Salt (maximum)	6 per cent	6 per cent	6 per cent	6 per cent
Free fatty acids (below)	1 per cent	1 per cent	1 per cent	1 per cent

The plant having a capacity of intake of 35 tons of raw material to produce 7.5-8 tons of fish meal and 10 per cent by weight of oil of raw material from oily fish was imported from Denmark. The plant is completely automatic, that is, cooking, pressing, drying, milling and bagging all automatic and within half an hour of feeding the raw material, the final product of finely powdered fish meal is obtained.

Small Scale Fish Meal Drier

When a regular supply of raw material is uncertain and the continuous operation of big capacity plants can not be economic, only batch type of driers, which can handle relatively small quantities of material and suitable for operation in the fishing villages can provide an answer to the proper and effective utilization of trash fish by converting them into fish meal.

Drier

The drier was a jacketed drum of 153.5 cm length, 76.7 cm internal diameter and 91.21 cm external diameter with 7.5 cm jacket made of 3 mm MS plate, 8 baffles of 7.5 cm height were fixed along the entire length at equal distance on the inside surface of the drier. The outer surface of the drier was insulated by asbestos rope to reduce heat loss. The drum was provided with a charging door of 30 cm diameter on the body which could be closed at will with a lid and two sampling doors at the two end sides of the drum with closing arrangements. The drum was supported on either ends on ball bearings of 15 cm diameter and rotated by a 5 hp motor through a reduction gear to obtain a final 15-20 rpm to the drum. Steam supply to the drum jacket was provided at the drive end of the drier through a rotary pressure joint. An exhaust fan of 12-17 cu.m/minute capacity provided at one axial end of the drier and connected through a regulator drew out moisture and air from the drier, atmospheric air being drawn through holes provided at the other end of the drier. The drum jacket was provided with usual steam fittings, like, pressure gauge, safety valves, drain ducts etc.

The unit was properly installed in proximity to other auxiliary units like, boiler, cooker, screw press, pulverizer etc, aligned properly spaced.

In the case of conversion into fish meal without pre-cooking, the fish were fed as such into the drier when press cake was used for drying, especially in the case of oily fishes, the fish was cooked in a steam jacketed kettle for about half an hour at a steam pressure of about 0.7 kg/sq. cm, pressed inside a canvas bag using screw press and then fed into the drier. Before feeding the drier was pre-heated by admitting steam into the jacket, keeping the drain duct open allowing air to be expelled and the condensate to drain out. During the course of drying, samples were taken at regular intervals and moisture determined to follow the progress of drying.

The final product as discharged from the drier was in a well pulverized form except for some bigger bones. The fish when fed into the drier and rotated come in contact with the hot walls of the drier and undergo rapid cooking and simultaneous disintegration of the fish. Moisture is released from the tissues at a very rapid rate which must be exhausted out at a very high rate as otherwise, condensation of the same takes place inside the drier thereby rendering the process too time consuming. This was achieved by operating the exhaust fan provided at full rate. However, when most of the water has been removed from the material, which is evidenced by the stoppage of visible exhaust of moisture vapor, the exhaust rate must be reduced as otherwise air drawn through the drier at this high rate cools down the material and increases the time taken for drying. Also there is a likelihood of the fish meal, which will be in a highly pulverized form by this time being blown out by exhaust.

The production of fish meal from whole fish and press cake has several advantage, namely, considerable saving of labor by avoiding elaborate cooking and pressing operations, prevention of loss nutrients with the stick water and overall better appearance of the end product. Hence this method appears to be promising for operation in fishing villages, particularly for lean fish.

But for a higher value of fat, the whole fish without cooking and pressing is superior to other products (cooked and pressed product) especially as regards the content of protein, minerals and available lysine, since much of the nitrogenous matter including available lysine might have been lost during cooking and pressing operations.

	Cooked, Pressed and Dried	Dried and Pressed
Moisture content (per cent)	10.54	8.34
Protein per cent (NX6.2)	70.26	71.16
Ash (per cent)	17.3	19.83
Sand (per cent)	-	11.69
Fat (DWB) (per cent)	5.775	8.14
Pepsin digestibility (per cent)	91.58	93
Available lysine g N/100 g protein	4.87	5.86

Determination of Moisture Content of Fish Meal

The "Cenco" moisture balance is an instrument for measuring the moisture content of material that do not change their chemical structure by loosing water on exposure to in free red radiation. A graduated scale gives a continuous percentage reading of loss in weight in the sample due to the loss of moisture. Since drying and weighing are simultaneous, this instrument is very useful in measuring the moisture content of substances that quickly reabsorb moisture.

Component of the Instrument

1. Auto transformer – Used to regulate the input power.
2. Heat lamp toggle switch – Used to switch the infra red lamp switch on and off.
3. Pointer adjusting knob – Used to position the pointer.
4. Knocked grip assembly – Used to hold the sample pan in place.
5. Sample pan – Used as a receptacle for the material.
6. Retainer pan – Acts as an arrest for the sample pan.
7. Handle – Used to raise or lower the lamp housing.
8. Lamp housing – Rotates the lamp and provides a drying chamber.
9. Infra red lamp – Provides the necessary heat for the removal of moisture.
10. Scale – Interacts with the pointer to indicate the percentage loss and weight (per cent of moisture of the sample).
11. Pointer – Interacts with the scale to indicate the percentage of moisture of the sample.
12. Scale adjusting knob with handle – Used to rotate the scale.
13. Scale lamp toggle switch – Used to illuminate the scale.

Operational Procedure

1. Turn the scale lamp on by means of toggle switch (13).
2. By turning the scale adjusting knob (12) rotate the scale until 100 per cent mark coincide with the index (fixed line).
3. Move the pointer to the index by turning the pointer adjusting knob (3) in a direction opposite to that in which the pointer must move to coincide with the index.
4. Rotate the scale until, the 0 per cent mark coincide with the index. The pointer is now above the index. This compensates for the weight of the sample.
5. Raise the lamp housing and carefully distribute the test material on the sample pan until the power return to the index. Approximately 5 grams of material will be required.
6. Lower the lamp housing and turn on the infra red lamp by means of the toggle switch no. 2.
7. Readjust the auto-transformer control to the proper setting for the specific material being tested. If the proper drying temperature is unknown and hence the auto-transformer in it is best advisable to set the transformer to the rated voltage of the lamp.

Discolouration or smoking of the sample for longer time caused by excessive heat indicate the release of volatile matter other than moisture and the sample may give wrong results. Repeat the trial at a reduced setting of the auto-transformer until the proper drying temperature and percentage loss in moisture is determined.

To determine the percentage reduction in weight during the drying operation, rotate the scale by turning the scale adjusting knob (13) until the pointer returns to the index. Read the percentage of weight reduction directly from the scale, which is the percentage of moisture loss based upon the initial weight of the sample.

To ensure complete drying, wait one or two minutes, after the weight stops changing. Record the final moisture content and switch off the infra-red heat lamp.

Fish Ensilage Plant

Fish ensilage is an alternative animal feed in place of conventional fish meal from miscellaneous fishes and fish waste. Unlike fish meal it does not require any dilution with other feeding stuff due to low concentration of protein compared to fish meal. Fish ensilage is prepared from edible varieties of miscellaneous fishes, like, jew fish, silver bellies, sprats and fish wastes.

Equipment and their Functions

Fish ensilage pilot plant comprises the following equipments having capacity to handle 250 kg raw material per batch.

(a) Concrete washing tank is required for washing the raw material to remove sand dirt and other extraneous matter. The tank is fitted with water inlet pipe and an outlet drain pipe with a gate valve for removing dirty water.

(b) Meat mincer – Helical type continuous meat mincer is used for mincing raw fish. The purpose of mincing is to reduce the size of the meat, thereby helping in cooking process and for getting homogenous minced mass from different types of fish. The mincer is fitted with reduction gear motor, electrical drive system with reversing switch, different sizes of perforated extruder plates, working platform etc.

(c) Stainless steel reaction vessel – This is a steam jacketed type reaction vessel with an air tight cover lid. The vessel is fitted with steam/water jacket, anchor type agitator drive system, charging and discharging doors, dial thermometer and other standard fittings

The vessel is used for two main purposes, namely, cooking and. fermentation. The raw minced mass is mixed with required amount of water (30 per cent approximately) and molasses (15 per cent) and cooked (20 minutes at 90°C) using steam while stirring. The cooked mass which is sterile is cooled by passing cold water in the jacket, fermenting agent added and is allowed to ferment for 1-2 days vigorously under constant agitation. The vessel is installed at higher level (1.5 m) than the concrete storage-cum-fermenter tank for transferring the material by gravity flow.

(d) Concrete storage-cum-fermenter tank is a closed vessel with wide charging and discharging doors. Discharge door is fitted with discharge pipe and a gate valve. The tank is fitted with handle. This serves as storage cum fermenter tank. Partially fermented mass from reaction vessel is drained by gravity using a hose pipe into the vessel and fermentation is allowed to continue for 6 to 7 days till completed (pH 4). From this vessel fish ensilage is filled directly in the final storage and distribution container.

(e) Rotocone blender-cum-drier – this equipment is used for the preparation of solid feed mix, Fish ensilage is mixed with deoiled rice bran in the proportion of 1:1 or 2:1 and dried at 70°C to 10 per cent moisture level simultaneously in this equipment. The equipment is fitted with steam jacket, charging and discharging door, dial thermometer, standard steam fittings and geared motor electrical drive system for rotating the drier. Product from drier can be directly fed into polyethylene lined gunny bags for storage and distribution.

Requirement of Space and Equipment

A. Total Space Requirement (minimum)

(i) Total land (15mX10m) = 150 sq.m

(ii) Built up area (9.5mX6.5m) =62 sq.m

B. Machinery and Equipment

(i) Wood/coal/oil fired low capacity boiler (100 kg/hour, 5.5-7 kg/sq.cm)-1 no complete with fittings and accessories

(ii) Cemented washing tank, 400 litres capacity complete with fittings – 1 no.

(iii) Heavy duty screw type meat mincer, 100 kg/hour capacity complete with accessories and fittings – 1 no

(iv) Stainless steel reaction vessel, 412.5 litres capacity, complete with agitator drive system, accessories and fittings —1 no

(v) Concrete storage-cum fermenter tank, 375 litre capacity, complete with fittings – 3 nos

(vi) Rotocone blender-cum-drier, 375 litres capacity, complete with accessories and standard fittings – 1 no

(vii) Water pump set and overhead tank (200 l) – 1 set

(viii) Platform balance, 100 kg capacity – 1 no

(ix) Ice box, 1 cubic meter capacity – 2 nos

(x) Plastic buckets of 15 litres capacity – 6 nos

(xi) G.I. tubs of 50 litres capacity – 6 nos.

(xii) Plumbing, steam and water line fittings

(xiii) Electrical wiring and fittings

(xiv) Installation of machineries

In this integrated plant provision has been made to prepare both liquid fish ensilage and dry feed mix to bring down production cost. In order to reduce plant cost, three concrete storage-cum-fermenter tanks have been used in place of costly stainless steel fermenter and only one stainless steel reaction vessel has been designed to perform several functions, like, cooking, mixing, sterilization and initial fermentation. This vessel has been installed at higher level than that of concrete tanks for transferring partially fermented product from stainless steel fermenter by gravity alone For the production of solid feed mix, only one additional equipment, namely rotocone blender drier has been added up to the main plant. In the design provision has been made to increase the plant capacity considerably by incorporating few more concrete fermenter tanks without much extra cost.

Mechanized Peeling Table for Prawn Processing Factories

The conventional method of hand peeling of prawns in processing plants is often time consuming and laborious and the raw material and wastes are to be handled manually. Depending on the length of the peeling shed, the distance of waste disposal bin and amount of work load, considerable time is wasted on the handling of the material. Delay caused due to handling also affects the quality of raw material and the finished product. To save time and improve the peeling operation, a prototype mechanized peeling table was designed and operated successfully.

The table consists essentially of two belt conveyors for carrying whole prawns and waste material to and from the individual peeler continuously. The main structural frame work of the table is made out of 50 mm standard angle iron with table top of wooden plank covered with aluminum-magnesium sheet. The two belt conveyors (upper and lower) are supported in four rollers, two drive rollers and two idlers. The drive rollers are driven by a geared motor through sprocket wheel and chain drive system so as to run the two conveyors at different speeds, the lower being faster than upper one. All the four rollers are supported on self lubricating ball bearings. There are several support rollers for the conveyors, which operate on ball point bearings for free rotation. Raw material charging hopper at the inlet end of the upper belt is fitted with a self proportionating impeller driven by the same drive mechanism which discharges a required quantity of material on the upper conveyor, so that no unpeeled material is left out on the conveyor, when it reaches the discharge end. Suitable discharge hoppers are fitted at the other end of the conveyors to collect waste prawn and waste material separately. On the working platform, three discharge channels are fitted terminating on the lower conveyor for the discharge of waste material.

The specification of the table are;

> Dimension – 180cm L X 75cm B X 90cm H
>
> Working platform per head – 45cm X 60cm
>
> Width of conveyor belt – 30 cm each
>
> Speed of upper belt – 30cm/minute
>
> Speed of lower belt – 60cm/minute
>
> Capacity of charging hopper – 30-35kg/hour
>
> HP of geared motor – 2

The operation of the table is very simple. It can be started by switching on the geared motor when both conveyors and feeder start moving. Material from the hopper is delivered to the individual worker by the upper conveyor. Workers standing on the side of the table can pick up prawns according to their likings from the moving belt, peeled and discharge the waste to the waste channel and meat into the bucket provided. The waste is automatically removed to the waste disposal bin by the lower conveyor continuously and under size prawn left on the upper conveyor or may be collected or allowed to go to the waste bin. At the end of a day's operation, the table top and conveyors can be washed well with spray of water.

The time saved by using mechanized table compared to the conventional method of peeling 5 kg of prawns are 20 minutes, which worked out to a saving of about 33 man hours per ton of material.

Even after accounting for electrical charges, there was a net gain of about 60 paisa per worker per 8 hours for the plant owner even though the peeling process as such is not automatic. The other advantages are; (i) raw material flow is regular and uniform and the discharge of waste material is automatic. (ii) Bacterial contamination is much less, because of easy cleaning of metallic top and rubber coated conveyors. (iii)Hand peeling and size grading can be done simultaneously and (iv) due to stream-lining of the entire peeling operation, effective quality control measures can be successfully adopted.

Mechanized Fish Processing

In a fish processing line, cutting operations like, gutting, deheading, filleting, skinning and boning are the most time and labor consuming operations. The mechanization of cutting operations is necessary as in savings in labor and a more economical utilization of the raw material will have to compensate for the rising costs of labor and raw material.

The ever increasing demand for food fish in the developed countries will depend on the growth of per capita income and on the prices of fish products. An increase in price will lower the level of consumption. Present processing methods will have to be improved to maintain the predicted growth of the market by increasing prices.

The rapid expansion of ground fish filleting industry in the mid-1960s in various countries led to an over-production of frozen fillets. It will be the main problem for the fish processors in the future years to find new ways to produce more economically to meet the present product prices and to compensate for increasing costs of labor and material.

If the overall demand for food fish do not increase due to rising prices of fish products, there will be an increase in the consumption of some fish products compensated by a decline in consumption for other fish products. This could be caused by a decline in landings of certain species or by a change in consumer habits. New processing methods could open up possibilities for new products and the creation of new products might call for the development of special equipment.

There is a noticeable trend towards more ready prepared convenient fish products. The consumption of fish sticks and fish portions in the USA rose in the five years from 1963 to 1968 from 78000 tons to 119000 tons per year. The per capita consumption of these products increased by 43 per cent over the 1963 consumption.

More economical processing methods must help in the future to keep the costs of production from rising and thus further expand the market for these products.

Processing costs mainly depend on the amount of labor involved and on the yield of finished product in comparison to the raw material used. Future fish processing methods will therefore concentrate on a reduction of labor and on an increase in yield. In manual operations there is very little room left for improvement. Most plants are already equipped with conveyorized cutting tables to reduce manual handling of the fish to the cutting, trimming, boning and packing operations

Time and motion studies could ledto some, but not a significant, increase of the production rate per man hour.

In some cases incentive systems allowing the worker to participate in the production and yield gains could be a step towards reduced production costs. However, the more sophisticated the system, the less versatile it gets in handling a variety of different fish species and end products.

For mechanization of fish cutting operations, it is necessary to specialize and concentrate on certain species of fish. A fish filleting machine has to perform the filleting cuts close to the bones of the fish in order to achieve a high yield. This can only be done by presenting the fish in proper position to the cutting tools, so that the knife enters the fish at the correct spot. Bone structure and outer shape of the fish are the characteristics for the design of the guides which hold the fish in its place.

Variability in size have to be compensated for by a flexibility of the guides and by automatic controls and self-adjustment of the cutters. Moreover, the same components have to be suitable or adjustable for differences in texture. A machine should process both firm and soft fish. The sequence of cuts for the removal of fillets from the bone depends on the bone structure.

Therefore, the design of an universal fish filleting machine, which will handle any species of fish with equally good results is impossible.

After the fillets have been removed from the bone, the differentiating characteristics of shape and bone structure disappear and the removal of skin is not affected by these problems. Skinning machines are therefore much more versatile.

For mechanized processing, the commercial fish species can be divided into four groups having similar characteristics.

1. White fish – Cod, haddock, Pollock, hake, whiting
2. Red fish – Ocean perch, sea bream, rock fish
3. Flat fish – Flounder, plaice, sole
4. Herring – Sardines, pilchards, sprats

To make full use of the labour-saving advantages of mechanization, equipment to handle all the species that have to be processed in one plant has to be installed.

To achieve more economical mechanized processing, a change in fishing operations would be necessary to provide a steadier supply of both white fish and flat fish to the processing plants simultaneously. Then separate specialized processing lines, with a lower capacity each, for white fish and flat fish, could be installed and used more economically, as if plant operations had to change the plant's full operational capacity from white fish one day to flat fish the next, half of the equipment would be idle half the time. Either changes in fishing operations or a change to specialized plants, handling one groups of species only, will be necessary if flat fish filleting machines is available. There will be no filleting equipment which could individually handle both white fish and flat fish.

Automation and mechanization of fish processing operations in the future will be possible to a greater extent with processing lines specialized on certain groups of fish species.

Processing of White Fish

The unpleasant task of gutting fish onboard the fishing vessel has already been mechanized. The wide range of sizes on cod from 30 cm (12 inch) to 120 cm (47 inch) can not be covered by a single machine. There are two types of machines with sufficient overlappig range. One for the smaller fish from 30 cm(12 inch) up to 70 cm (28 inch) and the other for larger fish from 50 cm (20 inch) up to 120 cm (47 inch). The speed of the gutting machines with 40 fish per minute, for the smaller fish and 25 fish per minute for the larger fish is sufficient to handle as much fish as a crew of 5 to 6 men could gut by hand.

The gutting machines will perform the slitting of the belly and removal of entrails as done manually. The head may be left on the fish, but a head cutter could be attached to remove the head leaving the lug bones on the fish. This way the gills can be removed together with the head. The removal of the gills is an advantage for the retention of quality in storage, and quality will be of growing importance. It is not likely that a mechanized operation will remove the gills from a fish with head on, because of the complex problems for such mechanized cut.

For mechanized filleting there will be a selection of different types of filleting machines for the same type and size range of fish. For the shore plant, yield will be the most important consideration, whereas on board factory trawlers the throughput per operator will be more significant. New cutting methods will be used in filleting machines. These machines are less complex and will produce clean full nape fillets with a higher yield. It will not be possible to cover the entire size range of white fish in one machine. There will be machines with overlapping ranges, from 30 cm to 55 cm and from 45 cm to 80 cm. These machines will take fish which has been gutted and deheaded onboard. These processing machines for shore plants will concentrate on product yield, which is the most important factor for the production costs ashore.

Other processing machines for factory trawlers will combine more operations in one machine to reduce the number of operators for the production line. In some cases, such combination can be accomplished only by employing cutting methods which do not give the maximum product yield. A reduction in crew will bring more savings for a factory trawler than a slight difference in yield. Therefore mechanized fish processing lines will be different for shore plants and factory trawlers.

For the production of completely boneless fillets there will be attachments to the filleting machines for the automatic removal of the pinbone section. It will not be possible to remove the pinbones without some meat from the fillets. For a high yield in boneless fillets, the cut should be close to the pinbones, but for a mechanized operation, such close cut would lead to a higher percentage of pinbones left in the fillet than the requirements for boneless fillets could allow. The pinbones can be removed with sufficient meat around them for a safe cut, and the meat can be reclaimed by a meat separator.

Meat separators allow the separation of bones and meat. There are types of meat separators, which allow a recovery of 95 per cent boneless meat from normal pinbone V-cuts and which produce a coarse fibrous minced meat.

The minced meat can be added to the fillet blocks for the production of fish sticks and breaded portions. If pinbone V-cuts from cod are utilized this way, the amount of minced meat in the final product will be about 12 per cent of the fillet weight. The overall yield will be increased by about 5 per cent.

If fillets and minced meat are mixed properly and the fillets are partially shredded to increase the surface for a more even distribution of minced meat throughout the block, the amount of minced meat could be increased to 30 per cent of the total without changing the character or quality of the fish sticks or portions. This will open new ways in fillet production. More boneless products could be produced. Parts of the fillets like the boneless tail ends or back loins of larger fillets could be used for IQF products, while the remaining flaps and smaller fillets could be mixed with the minced meat from the pinbone V-cuts. Other products like, fish cakes, fish balls, fish sausage etc can be produced from minced meat alone.

There will be a variety of different products made from fish. Instead of the usual fillets, sections of the fillet will be used for the type of product they are most suited for, and where they will give the best returns.

The packing of fillets in blocks and consumer packs is still a time-consuming manual operation. As long as both individual fillets and a certain weight are required for a pack, such an operation could probably not be mechanized. The packing of individual fillets with varying weights or fillet blocks with constant weight could be mechanized

For fish blocks as raw material for portions and sticks, there is a new method to produce an absolutely straight, smooth and voidless block with accurate measurements. First a rough block is frozen, which does not have to be carefully packed. Then this block is inserted into a chamber of the correct block measurements and is pressed under high pressure into the chamber, into this chamber, where the frozen fillets within the block slowly arrange themselves to completely fill the chamber. Just one dimension will vary depending on the weight of the rough block. This dimension is trimmed by a saw cut. This method allows the production of voidless blocks onboard, where, though the difficulties of getting correct weights, the production of voidless blocks has been impossible.

The production of portions and sticks from blocks calls for more savings in labor and cutting waste. The development will eventually lead to an automatic production of portions from fresh fillets with a portion forming machine and subsequent freezing and breading eliminating the intermediate stage of the block entirely.

For the blocks as export commodity further rationalization could be achieved by producing slabs instead of blocks. Slabs could be formed and portioned from fresh fillets and filled into cartons automatically.

For producing salt cod, mechanization would be possible only by establishing processing centers, where splitting and drying operations could be carried out on a larger scale, using splitting machines and drying plants. By this individual fisherman could be relieved of the time-consuming task of splitting and salting the catch, and more first grade and higher priced light dry salted cod could be produced.

Filleting Machines

There are machines available, which already bring a high degree of mechanization. One type of machine is available for smaller red fish (7 to14 inch), while the other type of machine is used for the larger fish of 12 to 22 inches in length. These machines automatically perform, beheading, gutting, filleting and skinning operations. The only manual labor left is feeding raw fish to the machine. Automatic feeding of these machines would not bring any further rationalization, since it would just change the job of the operator from feeding to observing.

Flat Fish Filleting Machines

Flat fish filleting machines have been in use, already in Europe for filleting plaice. Flat fish species, like, flounder, sole and yellowtail are different, and the machine designer has faced the problem of handling different types of fish in different size ranges and redesigning the machine for easy adjustments for a change from one species to the another.

For the entire range of sizes from 12 inch up to 24 inch, there are two different types of machine, one for the smaller fish of 12 inch to 16 inch and the other for larger ones from 13 inch to 24 inch. These machines will de-head and fillet the fish, 32 to 40 fish per minute with one operator. The fish can be gutted or un-gutted and with or without bleeding cut in the belly. The fillets produced by the machine will be the same as hand cut fillets with skin on. The flat fish fillets will have to be placed on to the skinning machine by hand.

The problem of splitting tail ends of flat fish fillets, when improved, will improve the appearance of the fillets and will make them more suitable for IQF products.

Mechanized Herring Processing

Herring, a very popular food fish, is processed for cured, marinated and canned products. The use of herring for food will be an important new branch of the industry in Europe.

Mechanization of herring processing has been brought to a high degree. Well proven processing methods and equipment for the different products are available.

There are filleting machines for all sizes of herring, from 5 to 15 inches, which will handle 200 to 300 herrings per minute with just two or three operators. These machines are very versatile and can be adjusted to produce single fillets, butterfly fillets or dressed herrings without heads.

Nobbing machines operating at the same speed can dehead and eviscerate the herring for canning or salt curing. Pre-cooking and can filling can be performed by one mechanized unit.

The processing line for canned herring fillets, for instance requires very little labor. There are two persons for the filleting machine. Washing and brining of the fillets is done automatically and continuously. Four operators are needed for the pre-cooking and can filling unit.

Automatic feeding of the herring filleting machine will bring savings only if the herring is of top quality all the time. With poorer quality, including broken herrings or herrings distorted in storage, any automatic feeding system will make mistakes leading to faulty cuts in the filleting operation. At a speed of 300 herrings per minute, a person watching the operation can not intervene and correct a mistake made by the automatic feeder because the herring is moving too fast. So an inspection of the fillets will become necessary, which requires one or two persons, which is just the same as an automatic feeder would save. The high speed herring filleting machines are designed to make manual feeding as simple as possible to increase the efficiency of the operators. A new beheading device in the herring filleting machine allows a direct head cut independent on variations in sizes at the speed of 300 per minute. This is important for a high yield in finished product.

For canned herring products, light weight aluminum foil cans, made of plastic laminated aluminum foil with a welded cover of the same material are now used. They have the advantage of easy opening without a can opener, easy disposal, being completely neutral in taste and resistant to aggressive contents. They are equally suitable for semi-preserved and fully sterilized products. The cans can be produced by a deep drawing press placed at the side of the canning line, so the cans can be formed, filled and sealed without handling and storing of empty cans. The special sales appeal of these new cans for the consumer will help to open the market for herring products.

The market for fish products and the development of new fish processing machines depend upon each other. New machines facilitate the economical production of new products and the development of new fish products often requires new processing equipments. Research and development works determines whether an idea could be developed into reliable and economically working processing equipment. Correct adjustments and preventive maintenance to keep up the good performance of all processing equipment will reduce time and will increase the throughput and the product yield.

Complete automation of all operations from receiving the raw material to packing the finished product is not within the reach in the near future. Automation will be limited to several steps of operation like, gutting, beheading, filleting and skinning, which could be performed by one automatic unit, but manual handling before and

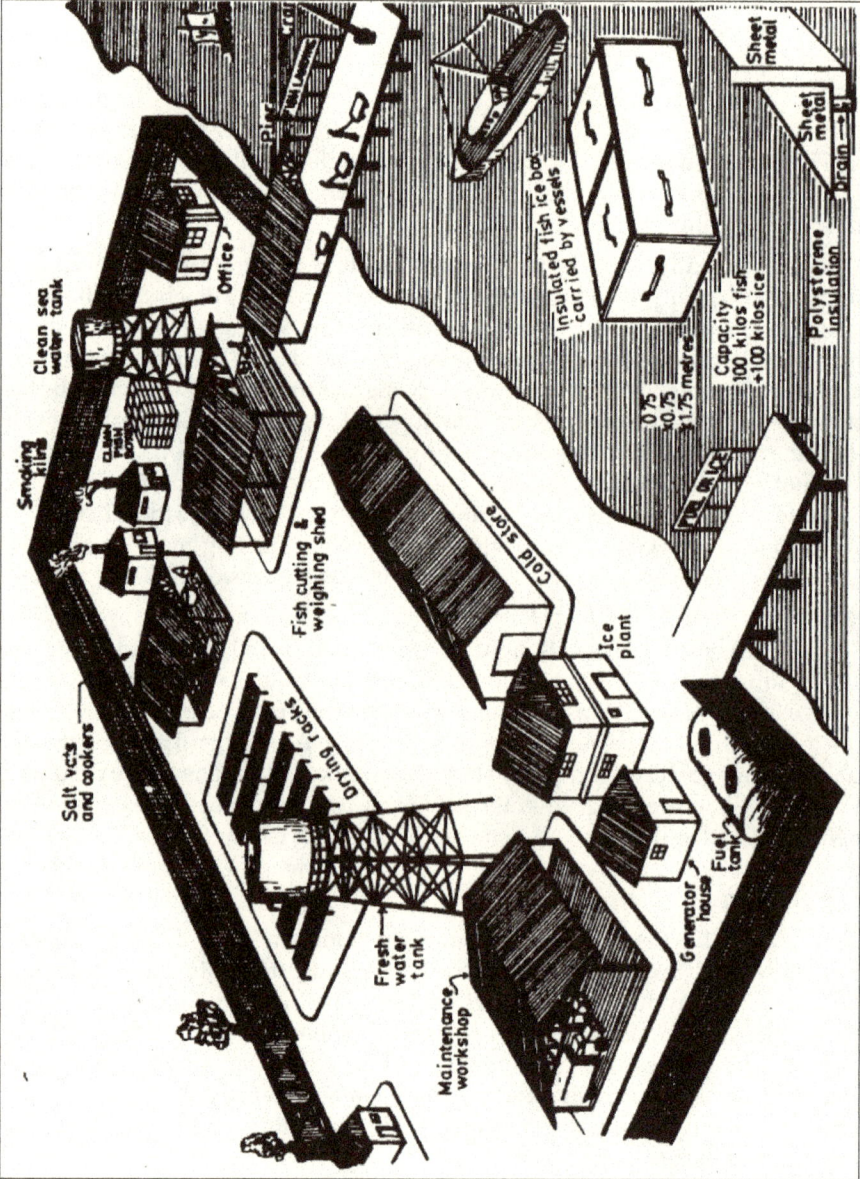

Lay Out of Composite Fish Processing Units along Fishing Harbour

after such a unit could not be avoided because of the difficult and the always changing properties of the raw material, fish.

Processing of Herring for Food

The European markets, such as, Germany, Holland, East Europe and Scandinavia are far ahead in the world's production and consumption of herring products. Many years ago these countries found that herring not only can be used for ordinary food purposes, but it is one of the better fish, which can be used for a very diversified range of delicious products. Germany alone produces more than 210000 tons of herring and has developed, in addition to a big home market, many other markets around the world.

Modern Fish Processing Plant

To have a maximum of automation, the fish is required to be handled when fresh as quickly as possible, with a minimum of manual labor involved. The plant should be build in such a way that the fish goes into the feeding section, runs through the different stations and leaves the plant to go directly into the rail car, truck or storage without any delays or unnecessary handling.

In order to operate a food herring plant, it is necessary to have a first class raw material, which has to be handled with much care. For higher ambient temperatures, even on short distance hauls, ice-boxing or refrigeration in sea water containers is needed.

The low fat content in the herring form the biggest problem for continually economic production. As the fat content differs from catch to catch and from day to day between 8 per cent and 13 per cent and sometimes up to 15 per cent resulting in production times and speeds having to be changed very often.

The production line contains the following sections;

1. Automatic washing and scaling
2. Automatic feeding into the filleting machine
3. Automatic filleting
4. Automatic washing of the fillets
5. Automatic brining of the fillets
6. Hand feeding of the fillets
7. Automatic pre-cooking
8. Automatic sauce filling (i)
9. Hand packing the fillets
10. Automatic sauce filling (ii)–weight checked
11. Automatic clinching
12. Automatic coding
13. Automatic can closing
14. Automatic can washing

15. Automatic sterilizing
16. Automatic can washing
17. Automatic can drying
18. Automatic can packing
19. Automatic master carton coding
20. Automatic transportation of the finished product

Seperated from the general processing area is the sauce division, the production of which is the key for the whole operation. It contains the following equipments, like, kettles with a capacity of 2700 litres of sauce per hour, ranges, scale, mixing equipment, automatic pipes, tanks for vinegar and oil, refrigerator. Up to 100 different ingredients are used and a good storage for spices is needed.

The pre-cooking tunnel, the autoclaves, the washing machines and the sauce kitchen are operated with steam. A high capacity of boiler is needed.

The most important facts are that the herring has to be chilled to 2°C. The fat content will determine the pre-cooking time and should be as high as possible. The average processing time from the raw product to the finished pack is two to three hours.

Herring can be processed in many different ways, including whole herring frozen or salted, herring fillets frozen, marinated, herring fillets in different cream in glass containers, herring fillets in different creams or sauces in sterilized cans. More than 1000 different products are produced all over the world, including, herring fillets in tomato cream, paprika cream, sweet pepper sauce, French onion sauce, mushroom cream, curry cream, cheese cream, bacon cream etc.

Chapter 17

Quality Control of Fish Products

Fish Food Quality

Importance of fish and fish food products lies primarily on the protein content of fish. Fish proteins are well balanced with regard to most of the essential amino acids needed by human body. They are also rich in lysine, which is deficient in India's cereal based diets. Fish protein is particularly suitable for enhancing the biological value of cereal proteins. Thus fish forms an excellent supplementary protein food to the cereal rich diet. Highly unsaturated oil (lipids) of seafood has beneficial effects on cardio-vascular system of human physiology. Poly-unsaturated fatty acids, present in oils of sea fish have special role in human physiology, as they are almost absent in vegetable oils. Seafood supplies different Vitamins, minerals and micro-nutrients.

One of the present day challenges facing the seafood industry are traceability for biological, chemical and physical contaminants. This is more to do with contaminants in the form of antibiotics and chemicals resulting in serious problems to the processors. One way to overcome this problem is to depend on related advanced technologies.

In general, quality is considered as a degree of excellence. It is often related to the price at which the commodity is purchased or the purpose for which it is used. With respect to seafood and seafood products, the quality is the sum of total of its composition, nutritive value, degree of freshness, physical damage, deterioration while handling, processing, storage, distribution and marketing, hazards to health, satisfaction upon consumption and yield and profitability to the producer and entrepreneur. In essence, seafood quality means all those attributes which a consumer considers to be present individually or collectively in the product. Inspection includes monitoring which is essential to measure the effectiveness of the quality control procedure and also those official devices that are employed to protect the consumer and facilitate the trade.

Quality control of seafood for exports has become mandatory in India from 1965 onwards through Compulsory Quality Control and Pre-shipment Act in 1963. Until 1977, it was end product sampling of seafood. However, it alone was not sufficient enough to ensure safety. Hence, the In-process Quality Control System was introduced from 31st December, 1977, which specified requirements for pre-processing and processing facilities, and end product standards. Subsequently, the importing countries made the quality assurance of food products very strict and that resulted in evolution of quality assurance programs based on Hazard Analysis Critical Control Points (HACCP) concept. At present, in the enhancement of food safety this is considered to be the best system.

Quality control of seafood differs a lot from that of any other foods. The food items, such as, fruits, vegetables etc. are harvested in optimum conditions, that is, the right type of food is harvested at the right time and right place. This indicates that it is possible even to select the right species to rear them to the optimum level of growth and harvest them in a pre-determined place as per pre-determined schedule. But in the case seafood, the harvesting is from the sea and other water areas from the stocks of unknown identity as to age, sex etc. under the most difficult conditions. Hence, the intrinsic quality varies in all possible combinations. Another important aspect, which complicates the quality control of seafood, is the unpredictability of catch. At times, the glut catches may exceed the handling capacity, to such an extent that the ice for preservation may not be available in sufficient quantities. The poor catches will make the processing units idle for weeks together. Another peculiarity of tropical seafood is the presence of large number of species in limited quantities making quality control more difficult.

Seafood harvested from unpolluted environments may not harbor any pathogens. However, the bacteria inherent to skin, gills and intestinal flora will act upon the seafood muscle when it is dead and convert the seafood protein to many simpler products resulting in offensive odour and the colloquial expression for its spoilage. The determination of mesophilic aerobic bacterial counts will indicate the extent of spoilage. At times, the seafood may not show any indication of spoilage, but can harbor pathogenic bacteria due to unhygienic handling. The scientific study indicates significant variation in consumer perception of hazard and the scientific understanding of the same based on epidemiological data. There are a number of bacteria of seafood origin that pose problems to human health. The pathogenicity varies from organism to organism, but it mainly results in infection, at times toxin in food and in intestines. The incubation period varies as low as one hour (*Bacillus cereus*) to as long as 5 days (*Compylobactor jejuni*) and so is the duration of disease from one to eight days.

Seafood Borne Human Bacterial Infections

Positive correlation with waters contaminated with human waste; positive correlation with warm temperature and positive correlation with the consumption of raw seafood products exists.

There are a number of challenges to overcome in the upkeep of the quality of seafood. In this context RFID (Radio frequency identification) technology is helpful to face the challenges, needs thorough examination and before that, good understanding of how RFID works, its applications in other areas and intrinsic challenges associated in implementing this technology and case studies related to implementation of this technology in seafood is need of the hour.

Radio frequency identification or RFID is a generic term for technologies that use radio waves to automatically identify people or object. There are several methods of identification, but the most common is to store a serial number that identifies a person or object, and other information on a microchip that is attached to an antenna (the chip and the antenna together are called an RFID transponder or an RFID tag). The antenna enables the chip to transmit the identification information to a reader. The reader converts the radio waves reflected back from the RFID tag into digital information that can be passed on to computers that can make use of it.

There are other systems to address quality related issues in seafood industry. At present only RFID technology is dealt for exposing the seafood industry to latest advances in seafood quality assurance. The RFID has enormous scope in seafood industry to overcome challenges related to quality as evidenced by case studies in Vietnamese Seafood Industry and Sweden's automation of seafood outlet at Laxbutiken dining self service, where each LAXOMAT is fully automated with radio frequency identification technology, so there are no greeters, sales clerks or stock crew to get between customer and the food.

General Principles to Produce Good Quality Fish Products

Quality Control in Fish Food Service

When the definition of quality control is applied to fish food service, it becomes the standard to which all steps of the operation must, of necessity, conform in order to ensure that changes in characteristics of food item do not take place. This role promotes quality control to a broad, encompassing and highly significant activity in the development work. Many factors are responsible for poor quality food. Most of them can be traced to poor sanitation, faulty handling, malfunctioning equipment, incorrect preparation, and carelessness. The following are the prime factors responsible for significant quality changes.

1. Spoilage due to microbiological, biochemical, physical, or chemical factors.
2. Adverse or incompatible water condition.
3. Poor sanitation and ineffective ware washing.
4. Improper and incorrect pre-cooking, cooking and post-cooking methods.
5. Incorrect temperature.
6. Incorrect timing.
7. Wrong formulation, stemming from incorrect weight of the food or its components.
8. Poor machine maintenance programme.
9. Presence of vermin and pesticide.
10. Poor packaging.

All these factors, either singly or in combination, will contribute to poor quality and effect changes that will be evident in the food's flavour, texture, appearance and consistency.

Raw Materials – Controlling Microorganisms in Raw Materials

Major losses in raw agricultural and fishery products throughout the world result from microbial, insect or rodent attacks. The increasing limitation being placed by regulatory agencies on the use of pesticides, antimicrobial agents and rodenticides, make it more necessary than ever that physical methods be used for protecting raw materials from contamination prior to their receipt at food processing plants.

Corlett (1974) proposed a scheme for the use of microbiological testing in acceptance of raw materials and ingredients. While buying raw products or ingredients, the general list of specifications, such as, organoleptic, physical and chemical characteristics including a microbiological specification should be considered. Raw materials and ingredients must be tested on a routine basis to ensure that they meet the suppliers specification.

Sanitation

Contamination may take place through the following channels;

1. During production on farm or at sea.
2. During harvesting, slaughtering, or catching.
3. In handling prior to processing
4. During processing.
5. During distribution and selling, also in retail.
6. In cooking and feeding in home.
7. Storage.

A sanitary program for a crab-meat plant must include the following points to which special attention should be given

1. Surrounding of packaging plant.
2. Water supply.
3. Construction of buildings and equipment from the sanitation point of view.
4. Personnel hygiene.
5. Plant cleanup at the end of processing.
6. Sanitation during processing.

The performance of a plant sanitary program is reflected in the end-of-line bacteria counts. Counts exceeding the bacteriological standards indicate inadequate plant sanitation. Since crab meat is considered to be a cooked product, ready to be served as food, without further preparation, public health authorities are very particular about the number and types of bacteria present in the product. Absence of *Salmonella aureus*, a food poisoning bacteria is required.

Crab meat processed under the following sanitary programme should comply with all requirements for wholesomeness.

1. Surrounding Area of the Packaging Plant

(a) The surrounding area of crab meat processing plants is to be free of debris, offal and other food material which serves as nourishment for insects and rodents. Hiding and breeding places of vermin must be identified and destroyed.

(b) Continuous effort is required to eradicate undesirable insects and rodents appearing on the premises.

(c) Offal bins and surrounding areas should be cleaned daily.

(d) Windows, doors and other openings in the plant are to be screened to present a physical barrier to the entrance of these animals into the plant area.

2. Water Supply

(a) Water for the processing of crab meat must be from an approved source and must meet the bacteriological requirement of less than 2 coliforms in 100 ml of water.

(b) Chlorination of water supply is recommended to a residual of 5 ppm available chlorine even if the source is unpolluted. This chlorination is an added insurance against the possibility of infrequent pollution, especially in the water lines inside the packaging area.

(c) Water used for the manufacturing of ice must meet the bacteriological standard indicated above.

3. Construction of Buildings and Equipment Requirement

(a) Buildings and equipment should be so designed to facilitate cleaning.

(b) Surface of floors shall be sloped for draining purposes and constructed of concrete or other approved material.

(c) Drains shall be of adequate size, suitable type, and where connected directly to a sewer, equipped with traps.

(d) Inside surfaces of walls shall be constructed of smooth, waterproof, light coloured material that can be thoroughly washed up to a height not less than 4 feet.

(e) Ceiling in working areas shall be free from cracks, crevices and open joints and constructed of smooth, washable, light coloured material.

(f) There shall be no exposed pipes over working surfaces on which crab meat is prepared or packed.

(g) Adequate ventilation shall be provided in working rooms to prevent vapor condensation and to lower the air temperature in the packaging area as required.

(h) A minimum illumination of twenty-foot candles shall be provided on all working surfaces in the processing areas.

(i) Equipment is to be manufactured from approved material designed to enable quick dismantling and sanitizing.

3. Personnel Hygiene

The success of the plant's sanitary programme necessarily depends to a large extent on the personal hygiene of employees. Food handlers are the primary source of Stapphylococci and coliforms, thus the importance of personnel as a source of those undesirable bacteria can not be over emphasized.

(a) A person should not enter into the working area of a cannery if he/she (i) is known to be suffering from any communicable disease; (ii) is a known "carrier" of any disease, or (iii) has an infected wound or open lesion on any portion of his/her body.

(b) Every person engaged in processing of crab meat shall wash hands thoroughly with warm water and liquid or powdered soap followed by dipping in an approved disinfectant solution at the beginning of the day's operation and after each absence from duty. Hands should be dried with disposable paper towels.

(c) Rubber or plastic gloves shall be worn by employees in the processing area and shall be disinfected at each break during the work shift.

(d) All employees engaged in processing operation shall wear clean overalls, smocks, or coats, and head gear of an approved type. Protective outer garments such as waterproof aprons, coats and pants shall be properly cleaned after each work shift. Employees should ensure that sleeves or other parts of their clothing shall not come in contact with the meat.

(e) No person shall smoke or spit in a working area.

(f) Personnel should be conscious of own personal habits; hands must be kept away from nose, mouth and hair.

(g) Care should be exercised to avoid overcrowding of working areas. It is impossible to maintain satisfactory plant sanitation in areas that are overcrowded by employees.

(h) Adequate toilet and hand washing as well as disinfecting facilities shall be provided. Hand washing and disinfecting practices should be closely supervised after lunch, coffee breaks and visits to washrooms.

4. Plant Cleanup at the End of Processing

Plant surfaces directly or indirectly in contact with meat become contaminated with meat or tissue fluids as well as with bacteria during the daily operation. The purpose of the general plant cleanup is to remove the dirt from the working surfaces and kill the remaining bacteria with disinfecting solution. Elimination of microorganisms from the working surfaces is crucial in keeping the end-of-line bacteria count low the following day. Post packing cleanup consists of two parts; in Part A the dirt is removed from the surfaces while in Part B disinfection is carried out.

Hygienic Conditions with Respect to Bacteriological Characteristics of Cooked Frozen Prawn

Rigid schedules of testing precooked frozen foods for bacterial quality are followed in almost all the importing countries. A total bacterial count of 1.0 x 10000 to 2.0 x 100000 per gram has been suggested in general, as the standard for acceptance quality of cooked frozen prawn in many countries. In addition, limits for *E. coli* count ranged between 10-20/g and the faecal *Streptococci* of 100/g with the complete absence of pathogenic organisms have been found in many standards. Certain regions of United States have laid down stricter bacteriological standard for cooked frozen foods, namely, 50000 organisms/g for total plate counts, less than 10 coliforms and no coagulase positive *Staphylococci* or *Salmonella shigella* organisms. Australian standards put the upper limits at 250000 for total counts, 20 for *E. coli* and 100 for *Staphylococci* with absence of pathogenic organisms. The standard prescribed by the U.S. Armed Forces for precooked frozen foods allow only 100000 total viable organisms/g, 10 coliforms/g and the complete absence of pathogenic organisms. Indian standards for cooked frozen shrimps recommended maximum total plate count of 2.0 x 100000/g and 100 *Enterococci*/g.

In the prawn processing factory, where the water-ice mixture used for cooling the cooked prawn was traced out to be the main source of contamination. It is clear that the cooling operation is the most vital step in the processing of cooked prawns. If the cooling medium, that is, ice cold water is highly contaminated with bacteria, not only the partial sterility which the material has attained during cooking is lost, but also chances of faecal contamination are quite common. The cooked material should be processed within the minimum possible time. Holding the cooked prawn in the ice over night or for longer periods although useful in maintaining organoleptic quality during peak season may result in recontamination, if the bacterial quality of ice is not good. During times of heavy catch, it is difficult to handle all the material on the same day, the uncooked material may be stored for processing on the succeeding days. The personal hygiene of the workers is also important in producing and maintaining the quality of the processed product. The hands of the workers should be washed thoroughly from elbow down with a detergent and adequately disinfected using chlorine solution containing a minimum of 200 ppm available chlorine at intervals or after each absence from the processing hall. There should be separate units for the processing of raw and cooked materials preferably with separate utensils.

All the utensils and equipments used should be cleaned and disinfected thoroughly. There is every chance of contamination if the microbial quality of the re-glazing water is poor. It should be chlorinated to a residual level of 50 ppm and should be changed intermittently. It is always preferable to re-glaze them by pouring water over the frozen block taken in the carton rather than by dipping in water.

Experiments conducted on a commercial scale, processed as per the scheme developed below to find out the bacterial quality of the material

Grade raw material → Washing in potable water → Cooking → Immediate cooling in water ice mixture, chlorinated to 20 ppm → Immediate peeling. All the utensils used for processing may be washed as per CIFT cleaning schedule → Washed in water chlorinated to a level of 10 ppm → Dipping for 15 minutes in water containing 20 ppm available chlorine → Draining arranging in trays and adding glazing water chlorinated to 10 ppm → Quick freezing → Re-glazing by pouring water chlorinated to 10 ppm available chlorine → Packing and storage at minus 20°C.

Chemical Quality of Water Used in Fish Processing Industry

An abundant supply of good quality water is the life line of fish processing industry. This fact has been well recognized, reflected in the large number of reports on the bacterial quality of water. But the chemical quality of water used in fish processing, though not attracted the attention for its importance, is known to influence the finished product.

For example, high amount of copper and iron cause blackening of canned prawns. The primary fish processing centers, where collection and dressing of the material is done, are scattered all along the coastal area and due to the absence or inadequacy of public water supply, these processing centers depend on wells, bore wells for water. Even in the canning or freezing factories, which are usually located in urban areas depend on private source of water, at least occasionally, is unavoidable.

Examining nearly one hundred samples of water used in the fish processing industry, the extent of dependence on private sources of water and stresses on the necessity of improving the water quality.

Surveying in and around Cochin to find out the various sources of water used in various fish processing establishments and analyzing for pH, total dissolved solids (Ts), chlorides (Cl), total hardness (TH), temporary hardness (Te H), permanent hardness (PH), hardness due to calcium and magnesium (Ca-H and Mg-H), alkalinity, sulphates SO4), copper (Cu) and iron (Fe); it was revealed that of the total samples, only 40 per cent were within the limits of acceptability, when all the quality criteria prescribed by the Indian Standard Specifications (IS :3257-1966) for ice manufacture were considered together, the limiting factor being alkalinity. Sulphate content was less than 100 ppm in 89 per cent of the samples and copper less than 0.2 ppm in 93 per cent.

Criteria	Per cent of Samples within Limit	Limit of Acceptability
pH	99	6.5 to 9.2
Chlorides (Cl)	71	1000 ppm
Alkalinity	69	250 ppm
Total hardness	42	100 ppm as $CaCO_3$
Sulphate (SO_4)	90	200 ppm
Copper	100	1 ppm
Iron	63	0.3 ppm

It is seen that only 32 per cent of the sample is soft, while really hard water is 25 per cent. Though the exact influence of hardness of water on fish quality is not known, it is known to reduce the effectiveness of quarternary ammonium bactericides and to influence toughening of peas during blanching for canning or freezing.

The total dissolved solids is mainly sodium chloride. This is especially so in the case of bore-well waters. The proximity of wells to the sea shore and the direction of tide contribute to the amount of sodium chloride usually encountered.

However, the amount of sodium chloride may not influence the quality of processed fish as storage in refrigerated sea water or sea water ice is an accepted method of fish preservation. At least in 25 per cent of seafood processing units only, water from private sources is used. No treatment other than chlorination in a few cases was imparted to the private supply. Water softening or demineralization was not at all used. The survey revealed the poor quality of water used in majority of the processing units.

Seafood Borne Bacterial Pathogens

Seafood may be a vehicle for many bacterial pathpgens (Feldhusen, 2000). Crabs, shrimp and oysters are the most often implicated sources of pathogens that naturally occur in marine or freshwater environments and most illness are associated with eating under-cooked or raw shell fish, particularly raw molluscan shell fish (Huss *et al.*, 2003). At least ten genera of bacterial pathogens have been implicated in seafood borne diseases. Shell fish, especially filter feeding bivalve mollusks (oysters, scallops, mussels, clams, cockles) can accumulate pathogenic bacteria in the alimentary tract. Since the alimentary tract of these bivalves forms the major edible portion for humans, these mollusks can serve as extremely effective vehicles for a wide range of organisms pathogenic to humans.

Bacterial Pathogens Associated with Seafoods

Pathogenic Bacteria	Seafood Transvector	Clinical Presentation
Vibrio parahaemolyticus	Crustaceans, fish	Diarrhoea, nausea, vomiting
Vibrio cholerae	Prawn, squid, shellfish	Diarrhoea, vomiting
Vibrio vulnificus	Fish shrimp, mussels	Wound infection, septicaemia
Clostridium botulinum	Fish, fish products	Weakness, paralysis
Aeromonas spp.	Shellfish, seafood	Diarrhoea, vomiting
Plesiomonas shigelloides	Fish, shellfish	Watery diarrhoea
Listeria monocytogenes	Raw seafood, salted fish	Diarrhoea, vomiting
Bacillus cereus	Seafood, squid, prawn	Diarrhoea, nausea, vomiting
Clostridium perfringens	Fish/shellfish	Diarrhoea, seldom lethal
Salmonella spp.	Prawn, fish, mollusks	Diarrhoea, vomiting, fever
Shigella	Fish, mollusks	Severe watery/bloody diarrhea, cramping
Yersinia enterocolitica	Fish/shellfish	Diarrhoea, vomiting fever
Campylobacter jejuni	Molluscs	Diarrhoea
Escherichia coli	Fish/shellfish	Diarrhoea, fever depends on strain
Staphylococcus aureus	Seafood	Vomiting, diarrhoea, abdominal cramps

In par with rapid strides in food preservation techniques, the safety of the consumer is gradually diminishing. Although seafood was regarded as comparatively safe among other foods, recent surveys indicate that seafood-poisoning outbreaks are also steadily increasing all over the world. Apparent absence of illness or pathogen does not guarantee safety of the consumer. Food is a dynamic system and microbial growth in food is a complex process governed by environmental, physiological and genetic factors. Hence relative risks for each type of food or process vary considerably and now it is possible to assess the risk in quantitative data providing critical insight into food-microbe interaction for eliciting better strategies for human health and consumer safety. Reducing the number of seafood –related outbreak requires continued and coordinated efforts by several agencies involving water quality, disease surveillance, consumer education, seafood harvesting, processing and marketing.

Method of Examination for Shrimp Spoilage

Fresh shrimp – Examine a random representative portion of the lot to be evaluated.

Frozen shrimp – A representative number of packages units or sub-divisions should be collected on a code by code basis for organoleptic evaluation.

1. Method of Thawing: In general frozen shrimps should be thawed in a spray of cold or cool (50-70°F) water. In-shell shrimp will thaw quite quickly in a spray of cold tap water. Some species of peeled shrimp containing soft tissue are sometimes damaged by this spray stream during the thawing cycle. These type of shrimps are best thawed in a pan of water with a stream of running water constantly rinsing the shrimp. Unless it is absolutely necessary, do not allow shrimp product to thaw in the open air at room temperature.

2. When thawing commercially prepared packages of frozen shrimp, the analyst should always examine a minimum number of three units for net contents, using the procedure of net contents of frozen seafood described in the Association of Official Analytical Chemists (AOAC) 12 ed. 18.001. If the analysis of three units indicate a weight shortage, additional units should be examined.

3. Method of organoleptic examination

   (a) Count or determine the number of individual units per container or package and record this number. In the case of very small units, such as tiny shrimp (100 to 500 units per pound) it is permissible to count the number of shrimp per unit weight and calculate the number of shrimp per pound or package.

   (b) After the products thawed, rinsed and brought close to room temperature, it is ready for organoleptic analysis.

   (c) For large shrimp (0-100/lb) the flesh of shrimp to be examined is broken into with the thumb and fore finger, and the freshly exposed muscle tissue is brought closely to the nose, smelled for odours of decomposition and an organoleptic classification is made.

   (d) For small shrimp (100-500/lb), small portions, usually 2-3 ounces are taken and rubbed between the hands for a short period of time, then brought close to the nose, smelled for odours of decomposition and the portion classified. Two or more portion per pound found decomposed cause the entire sub-division to be classed as decomposed.

Classifications Used to Judge each Shrimp (or portion) Examined

Class-I (Passable) – This category includes fishery products that range from very fresh to those that contain fishy odours or other odours characteristic of the commercial product; but these odours are not definitely identifiable as those of decompositions.

Class–II (Decomposed) – Slight but definite. This is the first stage of definitely identifiable decomposition. An odour is present that is not really intense, but is persistent and readily perceptible to the experienced examiner as that of decomposition. Shrimp in this category are not acceptable for human consumption.

Class–III (Advanced decomposition) – The product possesses a strong odour of decomposition, which is persistent, distinct and unmistakable. Shrimp in this category are not acceptable for human consumption.

Each sub-division of the sample (package, carton or container) should be examined separately. Segregate the examined portion into various classes on the basis of odour. If 5 per cent or more of the shrimp are class 3, or if 20 per cent or more of the shrimp are class 2 or if the percentage of class 2 shrimp plus 4 times the percentage of class 3 shrimp equals or exceeds 20 per cent, the sub-division should be classified as decomposed.

Percentages to be reported on the basis of either counts or weight, when the shrimp are uniform in size, and on weight basis when the shrimp are non-uniform in size.

Examination for *Salmonella* in Shrimp

Sample Preparation

Start the examination by aseptically consisting 25 grams of shrimp from each of the sub-samples into a sterile plastic bag and cover with sterile lactose broth.

Place the bag in a large plastic beaker or other suitable container and shake 15 minutes on a mechanical shaker set for 100 excursions/minute with a stroke of 4 cm.

Pour off lactose broth from the bag into another sterile plastic bag and add more lactose broth up to a total of 3500ml.

Place the plastic bag containing the lactose broth into a plastic beaker or other suitable container. Incubate 24 plus/minus 2 hours at 35°C.

Isolation of *Salmonella*

Gently shake the incubated sample mixture; transfer 1 ml of the mixture to 10 ml selenite cystine broth and another 1 ml of mixture to 10 ml tetrathionate broth.

Incubate 24 plus/minus 2 hours at 35°C.

Streak a 3-mm loopful of incubated selenite cystine broth on brilliant green (BG) agar, Salmonella-Shigella agar and bismuth sulphite (BS) agar.

Repeat with a 3-mm loopful of tetrathionate broth

Incubate the plates 24 plus/minus 2 hours at 35°C.

Examine the plates for the presence of *Salmonella*-suspicious colonies.

(a) Brilliant green agar – Most *Salmonella* colonies appear as pink, opaque or translucent colonies, with the surrounding medium pink to red. Typically some *Salmonella* colonies will appear colourless. Excluding the lactose-positive *S. arizonae* organisms, less than 1 per cent of *Salmonella* bacteria ferment lactose and will appear as yellow green or green colonies.

(b) Bismuth sulphite agar – Typically *Salmonella* colonies appear as dark brown to jet black colonies, sometimes with metallic sheen. The surrounding medium is usually brown at first, but may turn black in time with increased incubation, producing the so-called halo-effect. Some strains may produce green colonies with little or no darkening of the surrounding medium.

Examination of Frozen Peeled and Deveined Shrimp for Insect Filth

Procedure

A block of shrimp can either be partially thawed in its own container over night at refrigerator temperature or be analysed directly from frozen storage without the pre-thawing. Place about one half of a block of shrimp (approximately 2 pounds) on a 12 inch diameter standard No 8 mesh sieve nested on top of a standard No 140 mesh sieve. Wash shrimp thoroughly with a force stream of hot water. Transfer the material retained on the No 140 sieve to a 1 litre Wildman trap flask using water.

Wildman Trap Flask – Consists of 1 litre Erlenmeyer into which inserted (a) close-fitting rubber stopper supported on stiff metal rod (5 mm) diameter and approximately 10 cm longer than height of flask Rod (of greater diameter is not desirable because of its greater displacement of liquid) is threaded (10-32) at lower end and furnished with nuts and washer in the rubber to prevent striking flask.

Trap off as per method described in AOAC chapter 44.004 using water and 30 ml heptane.

Operation of Wildman Trap Flask

Unless otherwise directed in specific method, cool mixture in flask in room temperature. Bring the volume of liquid to about 600 ml in 1 litre flask. Add required volume of floatation liquid (Heptane 30 ml) by pouring down stirring rod. Tilt flask approximately 45°From vertical and mix 1 minute at the rate of 200-250 strokes/minute with brisk rotary motion so that liquid is brought to a roll. Avoid splashing through surface of liquid with rubber stopper. Add enough liquid to bring floatation liquid well into neck of flask.

Unless otherwise stated, let mixture stand 30 minutes, intermittently stirring the bottom layer every 3-6 minutes, during first 20 minutes of standing. Spin stopper to remove sediment and trap off raising stopper as far as possible into neck of flask, being sure that oil layer and 1 cm of liquid below interface are above stopper. Hold stopper in place and pour off liquid into the beaker. Rinse out material on rod and in neck of flask with liquid extraction medium in which floating was performed and added to beaker.

Do not wash out neck of flask with 95 per cent alcohol or otherliquid which may interfere with surface relationships of the 2 phases, this will cause loss in recovery in subsequent trappings.

Filter trapped material and rinsing with suction through rapid paper in Hirsch funnel. Add floatation liquid as specified to trap flask and stir vigorously. Add enough liquid extraction medium to bring floatation liquid into neck of flask. Trap off again, rinse and filter as above.

Filtration Technique

Clearing of plant materials with sedimentation or floatation procedures – Some food material may be trapped off with filth particles. By proper clearing, filth may be made to stand out in contrast with white background of filter paper by one of the following techniques.

1. For heavy filth, moisten filter paper with water or 50 per cent alcohol (This method does not clear material completely, but it leaves rodent pellets and other filth soft and pliable)

2. For light filth examination, wet paper with glycerol-alcohol (1:1), immediately after filtering. Place enough liquid on paper to fill fibers but not enough to cause flowing of extracted material. This clearing agent does not harden filth material on paper, as do many oils which might be used as clearing agents (Whatman filter paper No. 8 with lining is most favourable).

Microscopic Examination of Filter Papers

Make examination at 30 X (unless otherwise specified) using Widefield Stereoscopic microscope, on properly cleared paper. Continually tease and probe particles while observing through microscopy. Turn over all large pieces of material, such as, bran which might obscure filth elements.

Filth

A. Filth flies (whole or equivalent)

B. Filth fly fragments

C. Cockroaches-whole, equivalent or excreta

D. Hairs-Rat or mouse.

Testing of Fish Ham and Sausage

Fish sausages are finely ground fish flesh, either of single species or mixed homogenized with starch, sugar, fat spices and preservatives, generally filled in cylindrical synthetic or natural casings and pasteurized. Similar products containing small pieces of quality fish and lard are termed "fish ham". The products are highly relished in Japan.

But being a pasteurized product, which is often consumed as such without any further cooking, strict quality control measures have to be enforced so as to avoid food poisoning hazards. Besides physical characteristics like absence of damages, pin holes, curliness and air pockets, as well as jelly strength, texture and flavour, chemical characteristics like pH and acid values, moisture, carbohydrate and fat contents and volatile bases have to be assessed. A very important test that has been carried out before marketing the lot is the bacteriological examinations.

In accepting or rejecting any lot of finished product presented for inspection, usually a single sampling procedure is followed where a random sample of prescribed size from the lot is taken and analyzed. The lot is accepted if the random sample conform to the quality specification laid down by the plan.

The usual procedure of lot acceptance sampling by physical attributes alone for fish ham and sausage is as follows; The finished product is accepted if it possess the following physical attributes. Colouring is moderate and granules of pigment in the connective meat are not conspicuous, agreeable smell is present, taste is good, spices are well matched, elasticity is moderate and no free oil and juice and air spaces are present in the product. The failure of the product to comply with one or more of the above classifies the sample as "defective".

Drained Weight of Fishery Products for Quality Control Inspection

Canned and frozen prawns are subjected to compulsory pre-shipment inspection, during which compliance of the products with their declared drained weights is verified. The material used in the processing of these products being biological in origin, the drained weights are susceptible to variation due to number of factors, such as, the raw material is iced and dispatched from scattered centers along the coast to a distance of 160 km or more; the prawns are cooked in the brine of known salt concentration for a fixed time, cooled, size graded and weighed into the cans followed by hot brine and the cans exhausted, sealed and sterilized. In case of freezing, the prawns are size graded weighed in the trays and kept in freezer after filling the trays with glaze water. Since the determination of drained weight involves destructive and time consuming procedures, application of control chart for drained weights on the processing lines is not possible. Studies have shown that gross weight of the products are significantly correlated with their drained weights and since the determination of the former does not have the disadvantages of the determination of the latter,the drained weights can be controlled through the application of control chart for the gross weigh

Quality Control Laboratory for Seafood Processing Plants

In the present set up of the industry in India, the major items that have to be tested in the quality control laboratory are raw materials, finished products (frozen and canned), miscellaneous items like, bleach liquor, detergents and fish meat. The characteristics to be tested can broadly be divided into four categories, namely, physical, chemical, bacteriological and sensory.

Physical characteristics relate to size, weight etc. Chemical characteristics require the estimation of sodium chloride and acidity in the brine of canned products; moisture, protein, sodium chloride, fat and acid insoluble ash in fish meal and chlorine content in bleach liquors. Microbiological requirements necessitate the determination of total plate count, counts of *E. coli*, coagulase positive staphylococci, faecal streptococci and tests for salmonella and commercial sterility. Sensory evaluation relate to factors like spoilage, discolouration, material appearance, colour, odour, texture and flavour. A quality control laboratory attached to seafood processing factory shall have the facilities and personnel to carry out the aforesaid tests.

A quality controller and a laboratory assistant are the minimum staff required. The Quality controller shall be a graduate or post graduate in Industrial Fisheries/ Fisheries/Food Technology and shall have some experience in fish processing and quality control. The Laboratory Assistant shall be a matriculate with some experience in seafood analysis.

The laboratory shall be kept separated from the processing area so that fumes odour etc from the laboratory do not contaminate the product. As far as possible the area should be free from foul odours, dust and too much noise etc.

An area of 20 square meter is reasonably sufficient for the laboratory. It is preferable to have it divided into 2 to 3 separate rooms to accommodate various categories of tests. Water and electricity shall be available.

Working tables with attached washing sinks, almirah to store chemicals, drying rack, office table, chairs etc are the usual pieces of furniture required.

Air ovens, incubators, sterilizer, balances, water bath, arrangement for burners (LPG gas cylinder and Bunsen burners) are the major equipments required.

Glass wares like Petri-dishes, test tubes, graduated pipettes, burettes, conical flasks, beakers, measuring cylinders, funnels, crucibles, microkjeldal distillation apparatus, digestion flasks, soxhlet extraction apparatus are usually required.

Chemicals required to carry out the routine chemical estimations are silver nitrate, potassium chromate, sodium hydroxide, phenolphathalein, sodium thiosulphate, acetic acid, potassium iodide, soluble starch etc. For micro biological work, items like, agar agar, peptone, tryptone, different sugars, bile salts, sodium thiglycollate, some dyes and indicators are essential.

References

Agriculture Marketing Adviser –Preliminary guide to Indian fish, fisheries, methods of Fishing and curing, Govt. of India, Delhi, 1941.

Balachandran, K.K. and K.P Madhavan –Canning of lactarius, Fishery Technology, Vol. 13, No. 2, 1976.

Balachandran, K.K. and P.K Vijayan Processing aspects of Indian mackerel-a review, Fishery Technology Vol. 13, No. 2, 1976.

Bhattacharyya, S.K D.R Choudhury and A. N Bose–Preservation of Bombay duck (*Harpodon nehereus*) and Rohu (*Labeo rohita*) by gamma irradiation, Fish Technol, Vol. 15, No. 1 1978.

Bijoy, V. M – Minced meat of silver carp; Quality changes during frozen storage, Fishing Chimes, Vol. 28, No. 6, 2008.

Chakraborty, P.K. and M. A James – Pilot plant for producing fish ensilage and the economics of production, Fishery Technology, Vol. 13, No. 2, 1976.

Chakraborty, P.K., S.A Pillai and K.K. Balachandran–A small scale fish meal dryer, Fishery Technology, Vol. 7, No. 2 1970.

Chakraborty, P.K. and H.K. Iyer –A mechanized peeling table for prawn processing factories, Fishery Technology, Vol. 8, No. 1, 1971.

Chakraborty, P.K –A pilot plant for fish protein concentrate, Fishery Technology, Vol.13, No. 1, 1976.

Catarci, Camillo – Sharks, world markets and industry of selected commercially Exploited aquatic species with an international conservation profile, FAO Fisheries Circular, No. 990, 2004.

Clucas, I.J — Design and trial of ice-boxes for use in fishing boats in Kakinada, India, BOBP/WP/67, 1991.

D. Costa, A, T.Tomiyama and J,A Stern — The applicability of the vacuum distillation technique for the determination volatile acid and bases in fish flesh, Chilling of fish, Fish Processing Technologists Meeting, 25-29 June, 1956.

Dora, K.C. and R. Kundu— Minced fish, its production and use, Fishing Chimes, Vol. 30, No. 1, 2010.

Farber, Lionel and P.Y. Lerke – The objective assessment of raw fish quality, Chilling of fish, Fish Processing Technology Meeting, 25-29 June, 1956.

Ghosh, Swagat *et al.* – Masmeen, a traditional tuna product of Lakshadweep Islands, Fishing Chimes, Vol. 30, No. 3, 2010.

Ganguli Subha and Arun Prasad- Bacterial flora in fishes, Fishing Chimes, Vol. 30, No. 9, 2010.

Ghadi, S.V., V.N. Madhavan and Kumta–Diversification in utilization of trash fish by gamma irradiation, U.S Fishery Technology, Vol. 11, No. 2, 1974.

Huss, H.H - Quality and quality changes in fresh fish, FAO Bulletein No. 348, 1995.

Hisham, J *et al.* – Tuna fishery in Lakshadweep, Fishing Chimes, Vol. 30, No. 10, 2013.

Hennings, C. — On the use of electricity in testing the quality of fresh fish, Fish Processing Technologists Meet, 25-29 June, 1956.

Iyer, T.S.G, D.B. Choudhuri and V. K. Pillai– Studies on technological problems associated with the processing of cooked frozen prawns, Fishery Technology, Vol. 6, No. 1, 1969.

Joseph, Anita – Quick freezing for quality, Seafood Export Journal, Vol. 19, No. 8, 1987.

Jadhav, M.G. and N.G. Magati – Preservation of fish by freezing and glazing, Keeping qualities of fish in prolonged storage, Fishery Technology, Vol. 7, No. 2, 1970.

Joseph, L. and Hassan Shakeel-The Beche-de-mer fishing in the Maldives, BOBP Issue No. 41, 1991.

Khuntia and B.K. Choudhury – Autolysis in fish-A spoilage enhancer, Fishing Chimes, Vol. 32, No. 1 2012.

Karthiayani, T.C. and K.M.Iyer - Seasonal variations of bacterial flora of fresh oil sardines (*Sardinella longiceps*), Fishery Technology, Vol. 8, N0 1, 1971.

Kandoran, M.K. *et al.* – Canning of smoked eel, Fishery Technology, Vol. 8, No. 1, 1971.

Kamat, S.V., S.G. Gaonkar and U.S.Kumta - Studies on radiation pasteurization of medium fatty fish, storage properties of white pomfret (*Stromateus cinereus*) fillets, Fishery Technology, Vol. 9, No. 1, 1972.

Kamat, S.V. and U.S.Kumta - Control of radiation induced oxidative changes in white pomfret vacuum packing, Fishery Technology, Vol. 9, No. 1, 1072.

Krishnamurti, B. and G.A.Samuel – Meat picking, one way of utilization of low priced fish, Proc.Trg. Prog. In utilization of low priced fish, CIFE, Bombay, 1997.

Mishra, R.N.- Shrimps, its post-harvest technology and quality control, Proc. Nat. Seminar, shrimp seed production and farming, CIFA, 1991.

Madhavan, P. *et al.*– A review on oil sardine, Fishery Technology, Vol. 11, No. 2, 1974.

Muraleedharan, V. and A.P.Valsan – Preparation of smoke cured fillets from oil sardine, Fishery Technology, Vol. 13, No. 2, 1976.

Mathen, Cyriac -Quality and shelf life of dried sharks produced in India, Fishery Technology, Vol.7, No. 2, 1970.

Mathen, Cyriac, —A survey of chemical quality of water used in the fish processing Industry, Fishery Technology, Vol. 8, No. 1, 1971.

Madhavan, P. and K.K.Balachandran- Canning of tuna in oil, Fishery Technology, Vol.8, No. 1, 1971.

Marriott, Douglas – Process refrigeration in the fishing industry, APV Parafreeze Ltd BMEC Forum, 1979.

Montgomery, W.A – A rapid method for the estimation of volatile bases in fish muscle Chilling of fish, Fish Processing Technologists Meet, 24-29 June, 1956.

Norman, D. Jarvis - Principles and methods in the canning of fishery products, Research Report No. 7, U.S. Deptt of Interior, Fish and Wildlife Ser., 1943.

Niar, T.S, P. Madhavan, R.Balachandran and Prabhu, P.V. - Canning of oil sardine (*Sardinella longiceps*) in natural pack, Fishery Technology, Vol. 11, No.2, 1974.

Nandakumaran, M, D.R. Chaudhuri and V.K.Pillai – Blackening of canned prawn and its prevention, Fishery Technology, Vol. 7, No. 2, 1970.

Nambiar, V.N. and K. Mahadeva Iyer – Common microflora involved in spoilage of canned prawns, Fishery Technology, Vol. No. 2, 1970.

Nambiar, V.N. and K.M. Iyer – Bacteriological investigations of prawn canneries, Fishery Technology, Vol. 8, No. 2, 1971.

Priyadarshini, M.B. – Post-mortem changes in fish after death, Fishing Chimes, Vol. 31, No. 7, 2011.

Prasad, M.M — Seafood quality, RFID as panacea for traceability problems, Fishing Chimes, Vol. 31, No. 6, 2011.

Prudhiarnand, Rengrudee – Fish Products, south East Asian Fisheries Development Center, 1982.

Prabhu, P.V.–On sun drying of Bombayduck, Fishery Technology, Vol.9, No. 1,1972.

Prabhu, P.V, A.G. Radhakrishnan and T.S.G. Iyer– Chitosan as a water clarifying agent, Fishery Technology, Vol.13, No. 1, 1976.

Pottinger, S.R and David.T.Miyauchi — Protective coverings for a frozen fish, Fishery Leaflet 429, U.S. Deptt. of the Interior, Fish Wildlife Service, 1981.

Reddy, A.D. *et al.* – Importance of fish in human diet, Fishing Chimes, Vol.30, No. 1, 2010.

Reddy, V.K.S. and D.V. Reddy - Chitosan as an anti-microbial agent, Fishing Chimes,

Vol. 30, No. 12, 2011.

Rattakool, Pongpen – Fish Preservation, SEFDEC, 1982.

Rao, C.V.N — Physics in fish processing technology, Fishery Technology, Vol.8, No. 1, 1971.

Rao, K.K – On the control of drained weights in some fishery products, Fishery Technology, Vol. 9, No.1, 1972.

Rohit Prathiva and K. Rammohan - Visual quality testing method used in field for grading yellow fin tuna, Mar.Fish.Inf.Serv., T and E Ser, No.201, 2009.

Report — Sashimi tuna exports, Suggestions of visiting importer's team, Fishing Chimes, Vol.8, No.8, 2008.

Rao, D.R, - Nutritive value in the utilization of fish, Proc. Trg. Program, Utilization of low priced fish, CIFE, Bombay, 1997.

Rao, D.R, - Smoke curing of fish, Proc. Trg. Program, Utilization of low priced fish CIFE, Bombay, 1997.

Rao, D. R, - A simple and effective method of utilization of low priced fish, CIFE, Bombay, 1997.

Sundaram, Sujit – Various uses of cephalopods, Fishing Chimes, Vol.29, No. 8, 2009.

Singh, R.R, *et al* – Fish silage, Technology for recycling of fish wastes, Fishing Chimes, Vol. 31, No. 12, 2012.

Sofi, F.R, *et al* — Fatty fishes as potential raw material for minced meat production and Minced based products, Fishing Chimes, Vol. 31, No. 12, 2012.

Srinivasa, G.T.K. *et al.*– Potential and future prospects for the processing and export of major carps, Fishing Chimes, Vol. 32, No. 1, 2012.

Sahu, B.B, *et al* - Artisanal processing for value addition and employment generation, Fishing Chimes, Vol. 31, No. 10, 2012.

Saisithi, Bung-Orn – Fish handling, preservation and processing, SEFDEC, Thailand, 1982.

Surendran, P.K and K.M. Iyer - Bacterial flora of fresh and iced Indian mackerel (*Rastrelliger kanagurta*) and its response to chlorotetracycline (CTC), Fishery Technology, Vol. 13, No. 2, 1976.

Solanki, K.K, N.K. Kandoram and Venkatraman, R. – Studies on smoking eel fillets, Fishery Technology, Vol. 7, No. 2, 1970.

Surendran, P.K. and K.P.Mahadeva Iyer - Use of antibiotics in the preservation of prawns, Fishery Technology, Vol. 8, No. 1, 1971.

Shenoy, A.V. and Arul James - Freezing characteristics of tropical fishes, Tilapia, Fishery Technology, Vol. 9, No. 1, 1972.

Savagoan, K.A, and H.E.Power. Effect of box stowage on quality of fish, Fishery Technology, Vol. 13, No. 1, 1976.

Seddon, Bernard - Refrigeration systems for fishing vessels, BMEC Forum, 1979.

Stansby, E.M — Changes taking place during freezing of fish, Fishery Leaflet No. 429, U.S. Deptt of the Interior, Eish and Wildlife Service, 1981.

Stansby, E.M. - Changes taking place during cold storage of fish, BMFC Forum, Fishery Leaflet, No.429, Refrigeration of fish, Part-3, U.S. Deptt of the Interior, Fish and wildlife Service, 1981.

Sripathy, N.V., - Application of proteolysis for fish utilization, Seafood Export Journal, Vol.7, No. 4, 1975.

Slavin, J.W – Some observations concerning the brine cooling of fish aboard a fishing Vessel, Chilling of fish, Fish Processing Technologists meeting, 1956.

Soudan, F. *et al.* – Spoilage index by means of systematic organoleptic examination, Fish Processing Technologists Meeting, 25-29 June, 1956.

Tarr, H.L.A – Fresh fish preservation, Chilling of fish, Fish Processing Technologists Meet, 25-29 June, 1956.

Unnikrishnan Nair, T.S. and A.P. Valsan - Time lag between catching and curing of fish and its influence on the finished product, Mackerel, Fishery Technology, Vol.8, No. 1, 1971.

Udupa, K.G and G.K. Kulkarni – Sequential analysis for testing quality standard of fish ham and sausage, Fishery Technology, Vol.9, No. 1, 1972.

Venugopalan, V *et al.* – Protein from blanch liquor, Fishery Technology, Vol.7,No.2, 1970.

Venkataraman, R. *et al.*– A preliminary study on insulated containers and their efficacy in long distance transportation, Fishery Technology, Vol.13, No. 1, 1976.

Verma, P.K.G. and R.Venkatraman. Canning of smoked dhoma (Sciaenid sp), Fish Technol, Vol. 15, No. 1, 1978.

Valsan, A.P, - The role and importance of fish curing in the utilization of low priced Fish, Proc. Tgr. Program in utilization of low priced fish, CIFE, 1997.

Index

A

Air circulation 90
Albumin 241
Aluminum foil 182
Antibiotics 175
Ash 4
Autolysis 12, 13, 14, 15, 17, 18, 239

B

Bacterial flora 20, 21, 22, 41
Bacteriological pathogens 280
Bacteriological quality 151, 152, 153
Bacto-peptone 241
Beche-de-mer 236
Biophysics 197
Blast freezer 67
Brine cooling 84

C

Canning 122, 126, 131, 135, 139.205
Caviar 233
Cell puncture theory 73
Cellophane 182
Cephalopod 215
Changes after death 8
Changes during freezing 72, 101, 102
Chilling 26, 219
Chitosan 210
Cleaning 155, 156
Cold storage 87, 88, 99, 100
Colombo cure 166
Compressor 71
Condition of fish 40
Contamination 147, 148, 154
Cryogenic freezing 79, 81
Curing 157, 161.164, 200

D

Desiccation 89
Dressing of fish 52
Drip 91, 92
Dry curing 165
Drying 109, 164, 199, 206
Dun 113

E

Elevator 133
Ensilage 237, 238, 260
Evaluation 40
Eye fluid 41

F

Fat 4
Fermented fish 239, 240
Fish drier 106
Fish in human diet 5
Fish meal 244, 246, 247, 254
Fish mince 103
Fish oil 244
Flavour 95, 96, 97
Fluidized freezing 59
Flume conveyor 134
FPC 242, 243, 246, 248
Freeze drying 193
Freezing 204, 219
Freezing machine 57, 77
Freezing plant 53, 107
Freezing tunnel 80
Freezing, airblast 58
Freezing, boxa 59
Freezing, immersion 66
Freezing, plate 60, 61, 63, 66
Freezing, spiral 58, 62
Frozen fish storage 67, 93, 94

G

Glazing 108, 177, 178, 180

H

Handling of fish 24, 35
Horse mackerel 103
Hydrolysis 12

I

Ice 50, 65
Ice box 187
Ice making machine 56
Ice, crushed 51
Ice, salt water 54, 56
Insulated container 185

K

Karmiboko 190, 191, 192 L

L

Laminated fish 223
Lipid oxidation 12

M

Machinery and equipment 261, 262
Major carp processing 218
MAP 184
Marinating 169, 220
Masmin 214
Mechanized fish processing 263, 264, 270
Minced fish 228, 229, 230
Mona cure 165

N

Nitrogen 3
Nutritive value 1

O

Oxidation 95

P

Pasteurization 173
Physical properties 197
Pink 113

Pit cure 165
Preservation 82, 83
Processing 201, 207, 221, 265
Protein 2

Q

Quality control 195, 272
Quality testing 31, 36, 37
Quick freezing 75, 76

R

Ready to eat product 203
Recent theory 74
Refrigerants 69
Refrigeration 49, 64, 70, 86
Renter's theory 73
Retarding spoilage 47, 48
Retort 119, 129, 140
Rigor mortis 9, 10

S

Salting 110, 112, 116
Sanitation 154, 196
Sardine oil 227
Sashimi 31, 35, 211, 212, 213
Sensory scores 29, 35
Shark commodities 224
Shark fin rays 225

Shark liver oil 226
Smoke cure 166
Smoked fish 104, 113, 115, 118
Soup powder 235, 236
Spoilage 28, 36, 149, 170
Sterilizing 120, 124
Storage 177, 188, 189
Storage temperature 66
Sun drying 105
Surimi 32, 33, 34

T

Testing for quality 284, 286
Thawing 77
Thermal death 130
Tilapia processing 216
Transport 186

V

Vacuum 130
Vacuum distillation 42, 43

W

Wafers 234
Wasps 181
Waxes 181
Wet cure 165